THE THEORY
AND PRACTICE
OF MODEM DESIGN

THE THEORY
AND PRACTICE
OF MODEM DESIGN

John A. C. Bingham

Palo Alto, California

WILEY

A WILEY-INTERSCIENCE PUBLICATION

JOHN WILEY & SONS

New York • Chichester • Brisbane • Toronto • Singapore

Library of Congress Cataloging in Publication Data:

Bingham, John A. C.
   The theory and practice of modem design.

   "A Wiley-Interscience publication."
   Includes index.
   1. Modems—Design and construction. I. Title.
TK5103.B56   1988   621.398'14   87-37262
ISBN 0-471-85108-6

Printed in the United States of America

10 9 8 7 6 5 4 3 2

This book is dedicated to Racal-Vadic Inc.
The leisure and peace of mind that come from its completion
are dedicated to my wife.

PREFACE

The subtitle of this book might be "Theory for Practitioners and Practicality for Theoreticians." The book is intended mainly for practicing and aspiring modem designers, but it should also prove useful as an ancillary book for a graduate course in data (digital) communications; it shows many ways in which all that theory can be put to use. A reader need have reached only an undergraduate level of mathematics; I have assumed familiarity with Fourier, Laplace, and z transforms, and matrix and probability theories, and I have tried to explain the more esoteric theories of convolutional coding, maximum-likelihood detection, and so on, in practical terms and at that same level.

Mark Twain said "reports of my death have been much exaggerated," and that also applies very well to modem designers. It has been predicted — for at least the past ten years — that the advent of the all-digital telephone network would soon make modems obsolete. This has not happened and is not likely to for at least another ten years. In the meantime the ever-increasing needs for data communications are continually eliciting exciting new applications of the existing analog network.

I am very pleased to have Ken Krechmer as a contributor to this book; he has specialized in modem marketing for ten years and has more interesting and potentially profitable ideas about where modems should go and what they can do than anyone else I know. His Chapter 2 is an excellent description of the practical commercial world in which we designers must operate.

It was also said a few years ago that modem theory had "gone about as far as it could go," but the continual expansion of signal processing capabilities has provided incentives for the development of new theories and the application of old ones. For example, convolutional coding — previously thought to be useless (or at least unusable) for the telephone network — was developed, and we achieved a 3–4 dB gain in signal-to-noise ratio (SNR). Now the big challenge is to devise a way of combining optimum detection techniques, such as maximum likelihood sequence estimation, with convolutional decoding.

Even if the need for modems on the telephone network should eventually dry up, data transmission in other media will become more and more important. Most of the sophisticated theories and algorithms for data transmission were developed for telephone modems — the medium is fairly precisely defined (at least statistically), the data rates are low enough to allow practical signal processing to keep up with the theoretical algorithm development, and the market was assiduously cultivated — but the last few years have seen an extension of the theories and algorithms into many other media. Modems are now used for coaxial cable, frequency-division multiplexing (FDM) group and supergroup bands, high-frequency (HF) radio, mobile radio, microwave radio (both line of sight and with tropospheric scattering), and satellite communications. Furthermore, the same basic principles are now being used in high-speed magnetic recording. In many cases the limitations in these media previously were power or processing capability; both of these are continually becoming cheaper, and eventually the only abiding limitation will be bandwidth — something that voice-band modems had to contend with right from the start.

All the modulation methods developed recently for the voice band have been "linear," and this book deals with linear techniques exclusively. Again the rationale is that voice-band modems are the bellwethers of the industry. Moreover, modulation techniques such as minimum-shift keying (MSK), continuous phase modulation (CPM), and tamed frequency modulation (TFM) can often be understood, and even implemented, more simply by linear methods; we will study a few examples of this.

I have worked for twenty years in data transmission, but I have been concerned mostly with the telephone network. I do not know very much about the detailed problems of the other media and will discuss them only when someone else has defined a problem well enough that I can extrapolate from my telephone experience, and suggest a solution. Occasionally in this book I will say that some subject is "beyond the scope of this book"; since that scope is defined partly by the range of my expertise, this will often be a face-saving way of saying "beyond my capability to understand or explain."

This is an incomplete book. I have asked some questions that I might have been able to answer if I had taken more time; even in the last few months when I was finishing — or rather, wrapping up — this book, several ideas have appeared in the literature that I would dearly love to have investigated and reported on. Also, I have sometimes suggested a solution to a problem — or at least a way of seeking a solution — that has not been adequately investigated or tested. The incompleteness, however, is an unavoidable consequence of the evolving nature of the subject. If I had succombed to the temptation to "finish" the book, I would have exhibited the second part of the engineering syndrome;

"No project will ever be finished because we will always think of an improvement."†

Parts of this book are also very timely, in that they discuss current controversies in the modem community. It might seem that they are therefore vulnerable to rapid obsolescence, but I hope that, regardless of how the issues are resolved, even a post-facto study of them will be useful for students of modem design.

The pronoun "we" used in most of this book is certainly not meant to be the "royal we"; it is an attempt to involve the reader in the fascinating learning process that I have gone through. If I am fairly certain that an idea or opinion is commonly held, I have used the passive voice ("it is known that . . ."), but where I am sure of only my own opinion, I have used the pronoun "I."

---

† The first part of the syndrome is "If it can be done, then it should be done." Beware of both!

ACKNOWLEDGMENTS

This book is dedicated to Racal-Vadic, and I am especially indebted to its president, Kim Maxwell, who originally suggested a 1200 bit/s full-duplex modem and oversaw its development; the VA3400 gave me immense satisfaction, earned me a lot of money, and financed the leisure (Leisure? Ha!) to write this book.

I am also indebted to Floyd Gardner for his advice and constructive criticism, and to my colleagues Jack Kurzweil, Steve Bradley, Chin-Chen Lee, Massimo Sorbara, and Max Lui, and many other experts around the world — Dick Brandt, John Cioffi, Adrian Clark, Payne Freret, Jerry Hays, Victor Lawrence, Noel Marshall, Dave Messerschmitt, Hemant Thapar, and Lee-Fang Wei — for enlightening discussions and valuable help. I also appreciate the support given by George Moschytz and James Massey while I was at ETH, Zürich, doing much of the work for Chapter 9.

I am very grateful to Hope Torley, who did all the drawings — skillfully and enthusiastically through many changes. Finally I thank Stanford University for the Terman Engineering Library — a wonderful and invaluable resource — and Apple Computer for their superb IIe computer, which did all my simulations and word processing.

CONTENTS

THE THEORY
AND PRACTICE
OF MODEM DESIGN

Chapter 1

Introduction

1.1 THE ARRANGEMENT OF THIS BOOK

The order in which the separate parts of modem design are treated in this book
is as logical as we could manage, but this book cannot be used strictly sequentially; much jumping forward and backward will be needed. For example,
marketing is dealt with first in Chapter 2, but what you should design is obviously very dependent on whether, and how well, you can design it, and, just as
obviously, on what people want (or can be persuaded to want) to buy; marketing factors must be considered at many stages of the design. Similarly, although
transmitters are dealt with first, many decisions on their design must anticipate
the problems of receivers: carrier and timing recovery, equalizer training, and,
perhaps, echo cancellation.

The rest of this introductory chapter contains definitions of the terms and
symbols that are used throughout this book, a brief discussion of modem
standards and of the organizations that develop them, a fairly detailed description of those features of the telephone network that are important for data
transmission, and a brief description of the principal ways in which the other
media differ from it.

Chapter 2, contributed by Ken Krechmer, a modem marketing specialist,
deals with the interaction of technology, applications, and standards in the
creation of new modem products.

Chapters 3 through 11 deal with most of the components of a traditional
modem as defined in Fig. 1.1. There is one omission; scramblers and de-
scramblers should be there, but there was no logical chapter to put them in; they
are discussed in Chapter 12, which is something of a catchall.

1

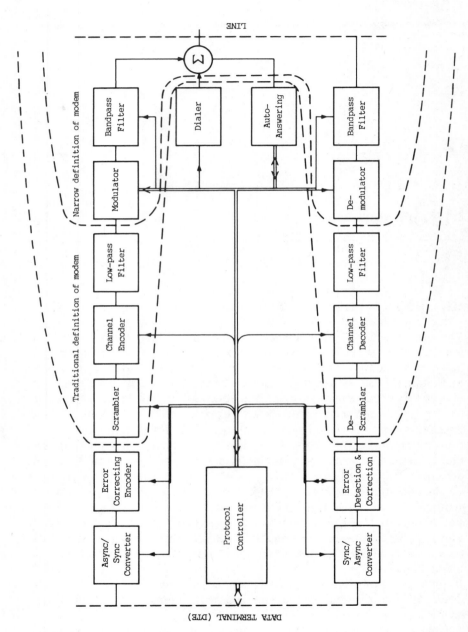

Figure 1.1 Components of a "full," a "traditional," and a narrowly defined modem.

2

Chapter 3 is entitled "Baseband Transmission," but the material in it is of fundamental importance for bandpass transmission as well. All the modulation methods considered in this book are linear, and it will be shown that, with the use of such methods, all passband channels have equivalent low-pass (or baseband) transfer functions and impulse responses.

Chapter 12 deals with the buffering, error correction, and scrambling that are needed to enable a synchronous modem to deal efficiently with asynchronous, character-formatted data. It then describes those functions that are needed to convert a private-line voice-band modem into one suitable for use on the Public Switched Telephone Network (PSTN).

Appendix I describes and lists some computer programs useful for both simulation and design; particularly important is a program for filter approximation that allows the design of all the filters in a modem to be treated as one comprehensive task. Appendix II contains some practical hints for diagnosing problems during the development stage.

The list of references at the end is dishearteningly long, but they serve three purposes—to acknowledge most (I hope) of the major contributions to the subject in the past twenty years, to trace and correlate previous developments of theories and methods, and to show where to find much of the detailed theory that there was no room for in the book. The references are cited by authors' initials rather than by the more usual anonymous numbers, in the hope that this will make them easier to recognize and then remember.

1.2   DEFINITIONS

1.2.1   Modem

The word "modem" is a concatenation of **mod**ulator and **dem**odulator, but there is a wide range of opinion as to what constitutes modulation and demodulation, and whether, indeed, a modem should comprise more than just a *mod* and a *dem*. The functions needed for transmission and reception of data are shown in block diagram form in Fig. 1.1, with narrow, traditional and broad interpretations of what constitutes a modem.

There are some conceptual advantages to be gained from requiring that the modulator be memoryless [Ma5]. This leads to a definition of a modem that is logically consistent but too narrow for practical use.

The traditional definition of a *mod* is "a device that accepts serial binary pulses from a data source and modulates some property (amplitude, frequency, or phase) of an analog signal in order to create a signal suitable for transmission in an analog medium." As shown in Fig. 1.1, this device comprises encoder, modulator, and filters. If the input data cannot be guaranteed to be sufficiently random for satisfactory operation of a sophisticated receiver, a scrambler will also be needed. It should be noted that this scrambler is for randomizing, and not for encryption or secrecy; these latter functions are the responsibility of the data source. A *dem* is similarly defined as the complementary receiving device. Most of this book is concerned with modems defined in this way.

In the late 1960s AT&T coined the term "data set" to mean the combination of a modem with all the control functions needed to operate on the switched telephone network. This was a useful distinction, but it did not catch on; "modem" is now often broadened to include all these functions.

In most standards, and discussions of interfaces a modem is called *Data-Communication Equipment* (DCE) and the data source and destination — whether they are video terminals, computers, multiplexers, or other devices — are called *Data-Terminal Equipment* (DTE).

1.2.2 Two- and Four-Wire Lines

A four-wire (4W) line is a pair of two-wire (2W) lines, one for transmitting and one for receiving, in which the signals in the two directions are to be kept totally separate. However, perfect separation can be maintained only if the four-wire configuration is sustained from transmitter to receiver. If the lines are combined in a 4W/2W combining network (often called a *hybrid* or a *hybrid transformer*) at any point in the signal path, impedance mismatches will cause reflections and interference between the two signals; this is discussed in more detail in Section 11.1

The terms "two-wire" and "four-wire" ("2W" and "4W") will be used in this book even if the medium contains no wire, and a "line" or channel will be considered four-wire, as far as the modem is concerned, if complete separation of transmit and receive signals is effected *outside* the modem.

1.2.3 Private and Switched Lines

Private, leased, or *dedicated* lines (usually four-wire) are for the exclusive use of "leased-line" modems — either a pair (in a simple point-to-point connection) or several (on a multidrop network for a polling or a contention system). If the medium is the telephone network, their transmission characteristics are usually guaranteed to meet certain specifications (see Section 1.4), but if the link includes any radio transmission, the quality of it may be as variable as that of a nondedicated (i.e., switched) line.

Dial-up modems can establish a point-to-point connection on the PSTN by any combination of manual or automatic dialing† or answering. The quality of the circuit is not guaranteed, but all phone companies establish objectives. The links established are almost always two-wire because four-wire dialing is tedious and expensive.

---

† In a few years' time, when touch-tone signaling is ubiquitous, the term "dialing" will be as puzzling as "clockwise," but nobody has yet suggested a satisfactory replacement.

1.2.4 Simplex, Half-Duplex, and Full-Duplex

These three terms are used to describe a modem, or a transmission channel, or a combination of the two, and care must be taken to describe any particular arrangement precisely. Full-duplex modems will not work on half-duplex channels, and half-duplex modems can sometimes work in a full-duplex mode!

"Simplex" means capable of passing signals in one direction only. A remote modem for a telemetering system might be simplex; a two-wire line with a conventional unidirectional amplifier is simplex.

"Half-duplex" means capable of passing signals in either direction, but not in both simultaneously. A telephone channel via a satellite often includes a device that suppresses echoes with very long delays by allowing transmission in only one direction at a time; this renders the channel half-duplex. Echo suppressors are slowly being replaced by echo cancelers, which are theoretically full-duplex devices; these are discussed more fully in Section 1.4.

Any modem that does not share circuitry between transmitter and receiver can transmit and receive simultaneously if a four-wire line is used. However, such operation is usually taken for granted, and to qualify as full-duplex, a modem must be able to operate thus on a two-wire line.

When a modem is connected to a two-wire line, its output impedance cannot be matched exactly to the input impedance of the line, and some part of its transmitted signal (usually badly distorted) will always be reflected back. For this reason half-duplex receivers are disabled (received data is clamped) when their local transmitter is operative.

Full-duplex operation on a two-wire line requires the ability to separate a receive signal from the reflection of the transmitted signal. This is accomplished by either FDM in which the signals in the two directions occupy different frequency bands and are separated by filtering, or by Echo Canceling (EC) in which a locally synthesized replica of the reflected transmitted signal is subtracted from the composite received signal. These two methods are described in more detail in Chapter 11.

The implication of the term "full-duplex" is usually that the modem can transmit and receive simultaneously at "full" speed. Modems that provide a low-speed reverse channel are sometimes called "split-speed" or "asymmetric" modems; since a 4800/4800 bit/s modem is full-duplex and a 4800/0 is half-duplex, a 4800 bit/s modem with a 300 bit/s reverse channel might be pedantically called a 17/32 duplex!

1.2.5 Synchronous and Asynchronous Data

"Synchronous" means that the data is always accompanied by a clock signal; data changes on one edge of the clock (the conventions are discussed further in Chapter 12) and should be sampled by the receiving device on the other edge. Synchronous data is almost always grouped in blocks, and it is the responsibility of the data source to assemble those blocks with framing codes and any extra bits needed for error detecting and/or correcting according to one of many different protocols (BISYNC, SDLC, HDLC, etc.). The data source and destination expect the modem to be transparent to this type of data; conversely, the modem can ignore the blocking of the data. Discussion of these protocols is beyond the scope of this book.

Asynchronous data is not accompanied by any clock, and the transmitting and receiving modem know only the nominal data rate. To prevent slipping of the data relative to the modems' clocks, this data is always grouped in very short blocks (characters) with framing bits (start and stop bits); the most common code used for this is the seven-bit *American Standard Code for Information Interchange* (ASCII) with even parity; this is a subset of International Alphabet No. 5 that is defined in the CCITT recommendation V.3.

All the encoding and modulation methods considered in this book are synchronous, so if a modem is to be able to accept asynchronous character-formatted data, its transmitter must include a means for synchronizing this data; ways of doing this are discussed in Section 12.1. The most common method generates *character-formatted synchronous* data; this sounds like an oxymoron, but merely means that the synchronous data stream retains a character format as far as possible by including start and stop bits of precise duration.

1.2.6   Symbol Rates and "Bauds"

In most modems the input bits are accepted in groups of from one to six bits, and each group is considered as one *symbol;* the symbol rate is therefore an integer dividend of the bit rate. The symbol rate can be measured in bauds,† but this is a much misunderstood and misused term. Since we need the "b" subscript for the bit rate, we will eschew bauds altogether and use f_s for the symbol rate in units of symbols per second.

1.2.7   Symbols and Terminology

The symbols shown in Table 1.1 will be used consistently throughout the book; others will be introduced when needed and may be consistent only within a chapter. As far as possible, initials of the quantities represented are used in order to serve as mnemonics. Some general rules should be borne in mind:

1. The symbols a and b are reserved for binary data; a usually for primary data and b, for the results of some coding operation.
2. The symbol c is used for the tap coefficients of an equalizer.

---

† A baud can be best defined by the analogy baud:hertz::symbol:cycle. The frequently used expression "baud rate" is tautological.

TABLE 1.1   Symbols Used Throughout the Book

$A(f)$	The amplitude part of $S(f)$
f_b	Bit rate [bit/s (bits per second)]
f_c	Carrier frequency (Hz)
f_N	Nyquist frequency, $= 1/2T$ (Hz)
f_s	Symbol rate [s/s (symbols per second)]
$h(t)$	The impulse response of an end-to-end channel, or any shortened version thereof
$H(z)$	The z transform of $h(t)$, $= h_1 + h_2 z^{-1} + \cdots + h_M z^{-M+1}$
$Q(x)$	The most convenient function for calculating error probabilities, $$= \frac{1}{\sqrt{2\pi}} \int_x^\infty \exp\left(\frac{-u^2}{2}\right) du = \frac{\text{erfc}(x/2)}{2}$$
$R(z)$	The autocorrelation function of $H(z)$, $= \sum_{m=-M}^{M} r_m z^{-m}$
$S(f)$	The complex frequency spectrum of $s(t)$, $= A(f) \exp j[2\pi f + \theta(f)]$
$S(f,\tau)$	The spectrum of $s(t)$ sampled at $t = nT + \tau$; band-limited to $-f_s < f < f_s$
$\theta(f)$	The phase part of $S(f)$
T	The symbol period, $= 1/f_s$

3. The symbol f is either frequency (usually subscripted), or a time domain transfer function; F is a frequency domain transfer function.

4. The symbols i to n follow the FORTRAN convention and are variable integers. Uppercase letters are fixed integers (e.g., m usually varies from 1 to M).

5. The symbols u to y (and occasionally, under duress, z) are continuously variable signals, progressing as consistently as possible from baseband transmit, through passband transmit and receive, to baseband receive.

6. The symbol z is the variable of the z transform, and z^{-1} is the delay operator; D, the delay operator of coding theory, is not used.

1.3   MODEM STANDARDS AND RECOMMENDATIONS†

Communication between dissimilar entities—persons, organizations, or equipment—is greatly hampered without a complete definition of the interface. The creation of such a mutually established interface is the purpose of any communications standard. Currently there is no complete definition of all the forms of communication that can occur at any but the simplest interface. Nevertheless, standards organizations act as forums for the development of interface standards, and the marketplace acts to test the acceptability of such standards. This combined process has produced sufficient standardization and wide enough acceptance of related assumptions to allow dissimilar manufacturers' communications products to interwork with increasing ease.

† This section was written by Ken Krechmer and John Bingham.

1.3.1   De Facto Standards

The development — either intentionally or incidentally — of standards is the foundation of all forms of communications. The Bell System pioneered the first data communications products and applications, but strictly speaking these did not become standards until somebody else built something that was compatible; then they became de facto standards.

However frustrating and unsatisfactory such unilaterally imposed standards may be, there will always be a place for them in the communications markets. They will survive because a gap exists in the development of de jure standards (of which more anon). The gap may occur because no one recognized a market need; a good example of this is the serial dialing standard popularized by Hayes. Or the gap may persist because the standards committees have not been able to agree on a standard; an error-correcting protocol developed by Microcom (usually called *MNP* for Microcom Network Protocol) is an example of this. No process is perfect, and the communications standards development process is far from it.

De facto standards, while a risk for potential purchasers of modems, may represent a significant marketing advantage for the manufacturer who can lay claim to market preeminence; the Hayes AT command set is an excellent example of this. For this reason manufacturers continually try to have their products accepted as de facto standards; sometimes they are successful.

1.3.2   International (de Jure) Standards

Occasionally a transnational company has been able to establish an international de facto standard (IBM's BISYNC is one of a few), but in the field of communications these standards have generally been de jure, set by either the International Standards Organization (ISO) or the International Telecommunications Union (ITU).

ISO is concerned mainly with data processing standards, but the Open System Interconnect (OSI) model developed by ISO is important to communications designers because it is the best overview of how the functions required

TABLE 1.2   OSI Model (ISO International Standard 7498)

Layer	Major Function[a]	Likely Location[b]
Application	Semantics modification	DTE
Presentation	Syntax negotiation	DTE
Session	Transaction control	DTE
Transport	Channel utilization	Network
Network	Routing	Network
Data link	Error control	Network
Physical	Physical connection	DCE

[a] The functions listed are only the major functions of each layer. See [Vo] for a more detailed discussion.
[b] The likely location is suggested only to help visualization of the model; the physical location of the function of each layer is often dependent on the type of communications equipment used.

for communications relate to processing applications. The model was well described in [Vo] and is summarized in Table 1.2. TC97 is the designation for ISO's overall committee on information processing, and TC97-SC21 is the name of the subcommittee on information retrieval, transfer, and management, which developed the OSI model.

The ITU of today is a descendant of the original International Telegraph Union founded in 1865. The "Telegraph" was changed to "Telecommunications" in 1932, and the organization was administered by the Swiss government until 1947, when it became an agency of the United Nations. It is headquartered in Geneva, Switzerland, and has a membership of 161 nations.

The purpose of the ITU, as defined in its charter, is

to maintain and extend international co-operation for the improvement and rational use of telecommunications of all kinds;

to promote the development of technical facilities and their most efficient operation with a view to improving the efficiency of telecommunications services, increasing their usefulness, and making them, so far as possible, generally available to the public;

to harmonize the actions of nations in the attainment of these common ends.

In 1956 the separate telegraph and telephone groups in the ITU were merged to form the International Telegraph and Telephone Consultative Committee (CCITT in French). CCITT standards, or *Recommendations* — old, amended, and new — are published every four years† after a Plenary Assembly has ratified the work done in the intervening periods by the appropriate Study Group (SG XVII for modems) and its Working Parties.

The V series of recommendations for particular modems are shown in Table 1.3, and for general issues of data transmission on the telephone network (interface to data terminals, dialing on the PSTN, testing, etc.) in Table 1.4. Several of the other related series of recommendations are collected in [Fo4].

In most countries other than the United States, data transmission facilities on the telephone system are provided by the Post, Telegraph, and Telephone (PTT) department of the national government. In the past many of these organizations would procure, and provide for their customers only modems that had been — or were likely to be in the near future — standardized and recommended by the CCITT. This situation is now changing; as in the United States, the ever-widening spread of data communications is creating gaps that only (temporary?) de facto standards can fill, and the increasing privatization of the telephone systems is creating the opportunity for these standards to develop.

Participation in the Work of the CCITT. The CCITT has two classes of membership: operating agencies and scientific and manufacturing organizations. Rep-

---

† Each set is referred to by its identifying color: 1960 — Red, 1964 — Blue, 1968 — White, 1972 — Green, 1976 — Orange, 1980 — Yellow, 1984 — Red, 1988 — Blue.

TABLE 1.3   CCITT Recommendations for Telephone Modems

Rec.	Date[a]	Speed (bit/s)	HDX/FDX	PSTN/Private	Modulation Method
		Voice band			
V.21	1964	200	FDX (FDM)	PSTN	FSK
V.22	1980	1200	FDX (FDM)	PSTN	PSK
V.22 bis	1984	2400	FDX (FDM)	PSTN	16QAM
V.23	1964	1200	HDX	PSTN	FSK
V.26	1968	2400	HDX	Private	PSK
V.26 bis	1972	2400	HDX	PSTN	PSK
V.26 ter	1984	2400[b]	FDX (EC)	PSTN	PSK
V.27[c]	1972	4800	HDX	Private	8PSK
V.27 bis	1976	4800	HDX	Private	8PSK
V.27 ter	1976	4800	HDX	PSTN	8PSK
V.29[d]	1976	9600	HDX	Private	16AM/PM
V.32[e]	1984	9600	FDX (EC)	PSTN	32QAM
V.33[e]	1988?	14400	HDX	Private	64QAM
		60–108 kHz group band			
V.35	1968	48 kbit/s	HDX	Private	SSB
V.36	1976	48–72 kbit/s	HDX	Private	MDB

[a] The dates given are those of the ratifying plenary assemblies; in some cases the recommendation had been finalized and was known to all PTTs and manufacturers who participated in the proceedings several years before.
[b] Generally the CCITT is opposed to having two recommendations for the same speed, but V.22 bis and V.26 ter were both accepted so as to provide compatibility with a lower-speed modem (V.22) and with an HDX modem (V.26 bis).
[c] This used a manual equalizer and is now obsolete.
[d] This used a patented modulation scheme.
[e] These use redundancy in the signal space and convolutional coding to reduce the error rate.

TABLE 1.4   Selected CCITT General Recommendations for Data Transmission over the Telephone Network

V.1	Equivalence between binary notation symbols and the significant conditions of a two-condition code
V.2	Power levels for data transmission over telephone lines
V.3	International Alphabet No. 5
V.4	General structure of signals of International Alphabet No. 5 code for data transmission over public telephone networks
V.5	Standardization of data signaling rates for synchronous data transmission in the PSTN
V.6	Standardization of data signaling rates for synchronous data transmission on leased telephone-type circuits
V.10	Electrical characteristics for unbalanced double-current interchange circuits for general use with integrated circuit equipment in the field of data communications
V.11	Electrical characteristics for balanced double-current interchange circuits for general use with integrated circuit equipment in the field of data communications
V.15	Use of acoustic coupling for data transmission
V.24	List of definitions for interchange circuits between data-terminal equipment (DTE) and data circuit-terminating equipment (DCE)
V.25	Automatic calling and/or answering equipment on the PSTN, including disabling of echo suppressors on manually established calls
V.54	Loop test devices for modems

resentatives from member organizations of either class may submit suggestions for, or comments on, proposed recommendations, and participate in all discussions. However, voting rights on any disputed issues† are restricted to one official delegate from each country.

For most countries the offical delegate is a representative of their PTT. For countries in which the telephone system is not run by a government agency, all Recognized Private Operating Agencies (RPOAs) may belong, and send non-voting representatives, but the single voting delegate is again appointed by the government.

1.3.3 U.S. Standards

For many years the only important data communication standards operating in the United States were the de facto ones established by the Bell System with their 100 and 200 series of modems; each of these usually came close to but never exactly matched the CCITT recommendation for a modem at the same speed. AT&T was not concerned with international modem standards. The USCCITT, an organization sponsored by the State Department, played the role that PTTs from most other countries did, and its appointee was the U.S. voting delegate. However, its designated committee, Study Group D, had too wide an area of responsibility, and the participation of other U.S. modem manufacturers in the CCITT was uncoordinated.

The situation began to change in the mid 1970s when Vadic, Inc. introduced the VA3400, the first 1200 bit/s full-duplex modem, and made a serious challenge to AT&T's dominance. A demand for the service developed in Europe and Asia, and there were requests for a CCITT recommendation. AT&T put forward their recently introduced 212 as a proposed standard, and Racal-Vadic countered with the VA3400. This battle, and several others between rival U.S. manufacturers, were fought out at the CCITT, and it became apparent that with so much dirty linen, the United States needed its own national laundry. In 1978 the U.S. Modem Working Party (MWP) was formed as part of Study Group D to make it easier for independent manufacturers to participate in modem standards activities and to coordinate U.S. participation in all CCITT modem work.

During the next ten years it was not felt that there was a need for any unique national standards, and AT&T and independent manufacturers alike participated enthusiastically, through the MWP, in the framing of V.22 bis, V.32, and V.33. However, the recently developed breadth and diversity of the communications industry, with many independent manufacturers and possibly millions of users, together with the special needs of the United States, seem to be changing this; a few separate U.S. standards may emerge.

The primary U.S. standards body is the American National Standards Insti-

---

† In practice, issues, however hotly debated, are rarely decided by vote; technical discussion and compromise nearly always prevail.

tute (ANSI). Its committees concerned with information processing and data communication are designated X3 and X3S3, respectively; they coordinate the work of the following three accredited organizations and their standards committees and, when appropriate, pass on their recommendations to the USC-CITT:

1. The Electronic Industries Association (EIA) sponsors the TR30 committees, which are concerned with data transmission systems and equipment. Late in 1986 the members of the MWP voted to join the EIA, and all modem standards work is now done in TR30.1.
2. The Exchange Carriers Standards Association (ECSA) sponsors the T1 committees. T1D1 is concerned with ISDN and its subcommittee, T1D1.3, with the next generation "modem"—the ISDN physical layer.
3. The Computer Business Equipment Manufacturers Association (CBEMA) sponsors the ANSI X3 committees, which are concerned mainly with ISO.

1.4   THE VOICE-BAND TELEPHONE NETWORK USED FOR DATA TRANSMISSION

First we must clarify the terminology to be used. It is tempting to call a "complete" telephone connection an end-to-end connection, but this might cause some confusion because the telephone operating companies call these "customer-premises-to-customer-premises" connections, and call local exchanges "end offices." Furthermore, many surveys of the "switched telecommunications network" have, in fact, been concerned only with connections between end offices. We will eschew "end" and "office" because of their other connotations, and, for the sake of brevity, refer to *user-to-user* and *toll* connections.

A user-to-user connection may be either dedicated or dialed. The links in the connection are the same in the two cases, and the only difference for the user is that for some impairments—particularly attenuation and delay distortion—a dedicated (*private* or leased) line is guaranteed to meet certain specifications, whereas a dialed connection can only be described statistically.

In this section we will discuss (1) the network components that might be included in a user-to-user connection, (2) the properties—particularly impairments such as distortion and noise—of those connections, and (3) the constraints that the operating companies place upon modems to be connected to the telephone network.

1.4.1   The Components of a Connection

We will describe the components that a modem designer must be concerned with and briefly discuss the separate effects they have on a data signal. However, for the user it is the properties of the user-to-user connections that are impor-

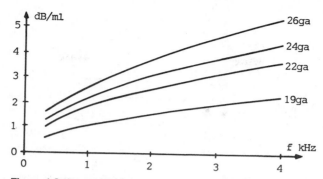

Figure 1.2 Attenuation distortion of unloaded twisted-pair cables.

tant. Although these could be deduced from those of the component parts, it would be a tedious and uncertain process, so as far as possible we will report, in Section 1.4.2, on specifications and surveys of the user-to-user connection.

1.4.1.1 The Subscriber Loop. Access to the local exchange ("end" or Class 5 office) is usually provided by a twisted-pair cable. The attenuation distortion (differential distortion relative to 1 kHz)† of these cables is approximately proportional to the square root of frequency. This is shown in Fig. 1.2 for the four most commonly used gauges. The attenuation can be flattened by breaking the line at regular intervals and inserting *loading coils* in series. These convert the cables into low-pass filters with a fairly sharp cut-off at about 3500 Hz, as shown in Fig. 1.3. The most common loading in the United States is 88 mH every 6000 ft (designated H88).

The modern objective for subscriber loops is to limit the 1 kHz loss to 8 dB (6 dB at 800 Hz in Europe) while using 26 gauge wire without loading coils wherever possible. Sometimes negative-impedance amplifiers are used to help achieve this on long (> 3 miles) loops.

Subscriber loops in the United States were surveyed and reported in great detail in 1969 [Gr]. The changes made since then — motivated mainly by interest in saving copper — were reported on in [Ma2]. They have not significantly reduced the average loss and attenuation distortion of loops but have slightly reduced the variances of those parameters. The noise levels on these loops were surveyed in 1980 and reported in [B&B].

1.4.1.2 Carrier Systems. All calls that traverse more than the obligatory two analog local loops become fairly sharply constrained in bandwidth. The nominal bandwidth of most multiplexing systems is 4 kHz; this is determined by the

---

† Actually, 1004 Hz is used for the measurement in the United States to avoid aliasing effects and beating when the connection includes a PCM system (with an 8 kHz sampling rate); 800 Hz is used for the measurement in Europe.

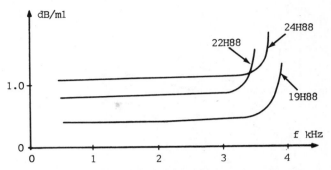

Figure 1.3 Attenuation distortion of loaded twisted-pair cables.

4 kHz spacing of carrier frequencies in FDM systems† and the 8 kHz sampling rate used in Pulse Code Modulation (PCM) systems, but it is reduced by the practically attainable bandwidth of the filters used to separate the signals in FDM, or to band-limit the signal before sampling. Older carrier systems achieved only about 2.5 kHz, but modern carrier and PCM systems are specified to be flat from 0.3 to 3.4 kHz.

Within that band, however, the end-to-end amplitude response may be seriously degraded by misadjusted filters, by the shunt capacitance of any analog switches or, as we have seen, by unloaded subscriber loops.

1.4.1.3 ADPCM Systems. Generally a modem designer need not be concerned with the type of carrier system used; relatively simple descriptions of the noise and linear distortion are usually sufficient to permit calculation of the performance of a modem. However, with the recently deployed 32 kbit/s ADPCM systems the situation is much more complicated. These are designed for voice transmission and assume a strong correlation between successive samples, at an 8 ks/s rate, of the input signal; when there is little or no correlation, as in wide-band modems, serious nonlinear distortion will be introduced.

The compression algorithm is so complex that no theoretical analysis of its effect on data signals has been or is likely to be published; simulation is the only way to check modems. During the discussions leading to the adoption of the V.32 Recommendation it was claimed [AT&T] that the use of trellis coding (see Chapter 10) makes it possible to provide satisfactory service through up to three tandem links. The simulations reported by Kalb [Ka] seem to be less optimistic than the earlier ones for V.32s, and very pessimistic for some "commercially available 9600 bit/s modems" (presumably V.29s); these cannot achieve a satisfactory error rate through only two ADPCM links — even without any linear distortion.

---

† Some high-density submarine cable systems use 3 kHz spacing, and special consideration should be given to modems that may encounter such systems.

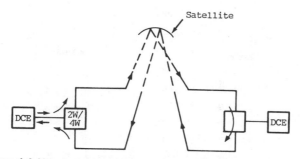

Figure 1.4 User-to-user connection via a satellite, showing talker echo.

Neither of these studies attempted to measure the degradation introduced by the ADPCM as an effective reduction in SNR; perhaps such a simplistic measure is not possible. It should also be noted that both studies were of *toll* connections. For user-to-user connections the attenuation at high frequencies provided by the subscriber loop at the transmitting end will increase the correlation in the data signal and may reduce the distortion introduced by the ADPCM encoders. The deployment of 32 kbit/s ADPCM is not complete, and until it is, it will be difficult to predict the probability of encountering any given number of links. Obviously, more studies of the effects of these systems will be needed.

1.4.1.4 *Echo Suppressors.*   A user-to-user connection via a satellite is shown in Fig. 1.4; the path length of the *talker echo* will be at least $4 \times 22{,}000$ miles, and the delay will be about 0.5 s; this can be very annoying to a talker. The effect was first mitigated by echo suppressors such as are shown, elementally, in Fig. 1.5; ideally, they should be installed at the 4W/2W hybrid, but in practice there may be some more four-wire line on the terrestrial side of the

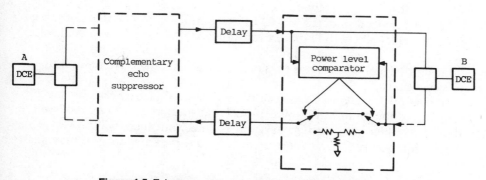

Figure 1.5 Echo suppressors used on a satellite connection.

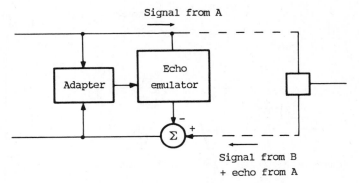

Figure 1.6 A network echo canceler.

suppressor. The device at end B compares the levels at its two input ports, and if it decides that A is talking, it inserts an attenuator in the return (echo) path. Double-talking totally confuses it, and the attenuation may be switched in and out repeatedly.

It is clear that both suppressors must be disabled for FDX data transmission to take place. Each one therefore contains a sharp filter with unity gain in a passband from approximately 2010 to 2240 Hz and a detector that compares the levels at the input and output of this filter; if they are nearly the same — that is, the line signal power is mostly contained within that band — for 250 ms, the suppressor will be disabled and will remain disabled, regardless of the subsequent nature of the signals in the two directions, until the line signal is squelched for 400 ms. This disabling is usually done during the initial handshaking by one modem transmitting an *answer* tone at either 2100 Hz (CCITT standard) or 2225 Hz (modems following the old Bell 103 standard).

1.4.1.5 Echo Cancelers (ECs).

In the 1970s echo suppressors were slowly replaced by ECs, which allow a certain amount of double-talking and do not require a "capture" time for any one talker to assume control of the connection. The basic features of a network EC† are shown in Fig. 1.6, and their characteristics are described in detail in CCITT Recommendation G.165. During periods of single-talking (half-duplex transmission) the canceler learns the characteristics of the echo path, models the path in an echo emulator, passes the "transmit" signal through this emulator, and then subtracts the result from the reflected signal.

The emulator adapts by detecting any correlation between the canceler output and the "transmit" input. If both transmitters of a FDX modem are turned on, then both these signals will be large, but there will be very little correlation between them; consequently, even if the emulator had previously

---

† ECs are also used in some FDX modems, and are discussed in more detail in Section 11.2.

learned the characteristics of the echo, that knowledge will be slowly destroyed. Therefore, the canceler continually compares the incident signals from the two directions, and if it decides that double-talking is occurring, it freezes the emulator taps and continues canceling as it had previously learned.

During the discussions at the CCITT that led to the V.32 Recommendation, it was suggested that since ECs would soon be used on nearly all satellite links, there was no need for the modem to deal with the long-delayed talker echo. However, it was found that:

1. In a few countries the four-wire part of the "tails" (the part of the terrestrial path beyond the EC) may have different (i.e., noncomplementary) frequency-offsets in the two directions. Consequently, the canceler would continually have to adjust its taps to match the rotation of the echo. It can do this well enough to be acceptable for voice communication, but not for high-speed FDX data transmission.

2. The double-talk detection is not completely reliable, and with continuous FDX operation and some configurations of the losses and levels in the two directions, the canceler may often continue to adapt randomly during double-talk, so that in subtracting what it thinks is an echo, it is really adding noise.

Therefore, it was first decided that it is better for a V.32 modem to take responsibility for canceling the long-delayed echo, and for the network EC to be completely disabled. However, it was pointed out that some modems do not include a canceler for the long-delayed echo, and consequently, for satellite connections, they must take the risk of using the network EC, in order to achieve acceptable performance in most cases. Therefore, V.25, the recommendation that defines the calling and answering sequences on the PSTN, was modified to V.25 bis, to allow the option to disable (1) both echo suppressors and cancelers or (2) only the suppressors. This was done by defining two types of answer tone—one a conventional steady tone and the other with periodic reversals of phase; the tone-detect and canceler-disable circuits in the network ECs were modified to respond only to the latter.† The reader is referred to Recommendations V.25 bis and G.165 for details.

1.4.2 Imperfections of a User-to-User Connection

The imperfections that we will be concerned with are (1) flat loss (attenuation), (2) attenuation distortion, (3) envelope delay distortion (EDD) (envelope, or group, delay, is the derivative of the phase with respect to the angular frequency), (4) listener echo, (5) input impedance, (6) noise, (7) harmonic distortion, (8) phase jitter, (9) frequency offset, (10) phase hits, and (11) gain hits.

---

† For FDX modems that use FDM it would be better to have the network ECs disabled, but most V.22 and V.22 bis systems do not send an answer tone with phase reversals. Fortunately, even in the worst cases of misbehavior by an EC, the long-delayed phony "echo" would not add very much to the out-of-band power that must be rejected by the modem's receive filter.

Prior to divestiture, private lines provided by the Bell system were specified by Federal Communications Commission (FCC) Tariff No. 260; those designed for data and voice use were defined as 3002 channels; and there was a basic channel and five types of "C" conditioning. Most operating companies still offer similarly specified lines. The limits for the impairments are given in [BS1] and [Fr4] and are discussed in the following sections.

Information about the U.S. PSTN has been given in [AG&N], [Gr], [BS2], [D&T], [Ma2], and [C . . 2]. Measured impairments have often been grouped in the categories of exchange, short-haul, and long-haul connections, but generally those distinctions are too precise for our purposes; in most of what follows the three categories have been combined. Also, the later surveys have been concerned mostly with either the subscriber loop or the toll connection; these have been combined statistically on the assumption that there is no correlation between the qualities of the two parts of the overall connection.

1.4.2.1   1004 Hz Loss (Circuit Net Loss)

Private Lines. The loss is specified as 16 ± 4 dB over all long-term variations.

PSTN. The average for all connections is 16.0 dB, and the cumulative distribution is shown in Fig. 1.7; the 99 percentile is about 27 dB.

1.4.2.2   Attenuation Distortion

Private Lines. The specifications are shown in Table 1.5. Similar limits, specified by British Telecom for use in the United Kingdom, are shown in Table 1.6. Most other countries conform to the CCITT Recommendations M.1025 and 1020, which are similar to the 3002 C2 and C4 specifications, respectively.

PSTN. The 1959/60 survey revealed that in general the attenuation distortion of user-to-user connections can be approximated by four linear segments, as shown in Fig. 1.8. At that time there was a sharp cut-off below about 280 Hz, a fairly flat section from 300 to 1100 Hz, a linear section of varying slope from 1100 to about 2950 Hz, and then another sharp cut-off above that.

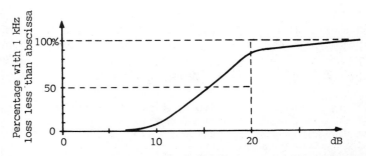

Figure 1.7 Cumulative Distribution Function (CDF) of user-to-user 1 kHz loss on the PSTN.

TABLE 1.5   Line-Conditioning Specifications According to FCC Tariff 260

	Attenuation[a]		Envelope Delay[b]	
	Frequency Range (Hz)	Variation (dB)	Frequency Range (Hz)	Variation (μs)
Basic 3002	300–500	−3 to +12	800–2600	1750
	500–2500	−2 to +8		
	2500–3000	−3 to +12		
C1	300–1000	−2 to +6	800–1000	1750
	1000–2400	−1 to +3	1000–2400	1000
	2400–2700	−2 to +6	2400–2600	1750
	2700–3000	−3 to +12		
C2	300–500	−2 to +6	500–600	3000
	500–2800	−1 to +3	600–1000	1500
	2800–3000	−2 to +6	1000–2600	500
			2600–2800	3000
C4	300–500	−2 to +6	500–600	3000
	500–3000	−2 to +3	600–800	1500
	3000–3200	−2 to +6	800–1000	500
			1000–2600	300
			2600–2800	500
			2800–3000	1500

[a] The loss at 1004 Hz is specified as 16 ± 4 dB, and losses at other frequencies are referenced to the 1004 Hz loss.
[b] Maximum delay variation within specified band.

TABLE 1.6   Line-Conditioning Specifications According to British Telecom

	Attenuation		Envelope Delay	
	Frequency Range (Hz)	Variation (dB)	Frequency Range (Hz)	Variation (μs)
S1	800[a]	0 to +17		
	300–3000	−3 to +16	900–1200	1850
			1200–2300	1100
			2300–2500	1850
S2	800[a]	0 to +10		
	300–3000	−3 to +10	900–1200	1850
			1200–2300	1100
			2300–2500	1850
S3	800[a]	0		
	300–500	−2 to +6	900–1200	1850
	500–2800	−1 to +2	1200–2300	1100
	2800–3000	−2 to +6	2300–2500	1850
T .	800[a]	0		
	300–500	−2 to +6	400–500	3000
	500–2800	−1 to +2	500–1000	1500
	2800–3000	−2 to +6	1000–2600	500
			2600–2800	3000

[a] The loss at 800 Hz must be in the specified range, and losses at other frequencies are references to the 800 Hz loss.

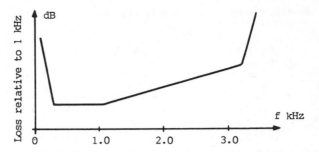

Figure 1.8 Approximating attenuation distortion by linear segments.

The results of later surveys—mostly of separate parts of the network [C . . 2] and [B&B]—have not been presented so graphically, but careful comparison and combination reveal that this model needs to be updated in only two ways: because of the replacement of analog carrier systems by PCM systems, the average midsection slope has been reduced somewhat, and the upper cut-off frequency has been extended to about 3200 Hz. The attenuation distortion of most user-to-user connections—those that use the band only from 300 to 3200 Hz—can therefore still be approximately characterized by just one number—the difference in loss between 1000 and 3000 Hz.† The average value of this difference, ΔA, is 9 dB, and its cumulative distribution is shown in Fig. 1.9. It can be seen that about 90% of all connections meet the basic 3002 line spcification.

It is interesting to note that all the C-conditioning specifications shown in Table 1.5 allow as much roll-off at the low end of the band as at the high. However, although the toll connection (AT&T's end-office connection) studies [D&T and C . . 2] show an average of 2 dB roll-off at 300 Hz compared to 1000 Hz (with a 90 percentile of about 3.5 dB), this is usually compensated by the opposite distortion of unloaded subscriber loops, and the net user-to-user roll-off is not statistically significant. However, most commercially available "worst-case line simulators" do roll off at the low end and are therefore very unrealistic in this respect.

Highly Selective Attenuation: Notches. Some long-distance carriers other than AT&T use a 2600 Hz signaling tone for some bewildering purposes and, to protect their systems from inadvertent operation, insert a very narrow notch filter at 2600 Hz into the connection. This will have very little effect on wideband modems, but would completely cripple one or two channels of a multichannel system (see Section 4.5 on OMQAM systems) if care were not taken to avoid that band.

† This range has been extended from the 2600 Hz used in [AG&N] because modems are now using much more of the (improved) upper part of the band, and 3000 Hz is the most common upper Nyquist (half-power) frequency.

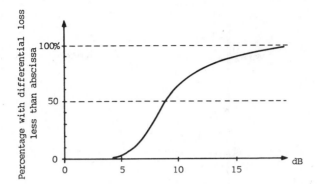

Figure 1.9 CDF of (differential) attenuation distortion between 1 kHz and 3 kHz.

1.4.2.3 Envelope Delay Distortion (EDD). The principal causes of delay distortion in the voice-band telephone system — in descending order of severity — are (1) the sharp low-end cut-off of the filters used in all toll-connecting multiplexing systems, (2) the fairly sharp high-end cut-off of the Voice-Frequency Low-Pass (VFLP) filters similarly used, and sometimes (3) the low-pass filter effect created by loading coils in the subscriber loops.

Private Lines. The 3002 specifications are shown in Table 1.5 and the British Telecom specifications, in Table 1.6. Again, a typical curve within one of these limits must be deduced from the results of the PSTN measurements.

PSTN. Variation of the EDD with frequency is approximately parabolic, with a minimum between 1.4 and 1.9 kHz; although some EDD curves are slightly skewed, the error in describing them all as parabolic, and adjusting the frequency of the minimum accordingly is insignificant. That is,

$$\text{EDD}(f) = \left.\frac{d\theta}{d\omega}\right|_f - \left.\frac{d\theta}{d\omega}\right|_{f_{\min}} = \beta(f - f_{\min})^2$$

The cumulative probability density functions (CDFs) for $f_{\min}$ and the multiplying factor β (in units of milliseconds per square kilohertz) are shown in Figs. 1.10 and 1.11. It can be seen that about 99.5% of all connections meet the basic 3002 specification for EDD.

It should be noted, however, that these parabolas are the result of an averaging process that smooths out the ripples; a plot of an actual, and much more ripply, EDD against frequency is shown in Fig. 1.12. We shall see that the smooth model is adequate for calculating Root-Mean-Square (RMS) and peak intersymbol interference but may make the impulse response look shorter than it really is.

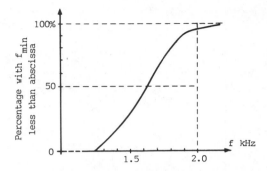

Figure 1.10 CDF of frequency of minimum group delay.

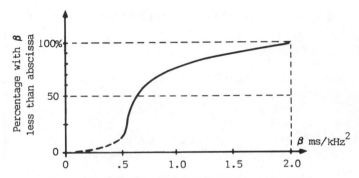

Figure 1.11 CDF of multiplier β in parabolic model of EDD.

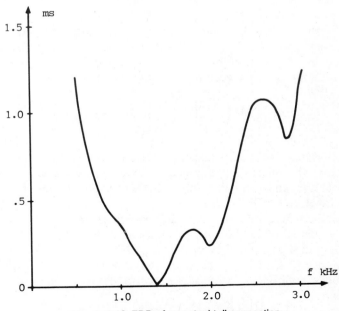

Figure 1.12 EDD of an actual toll connection.

22

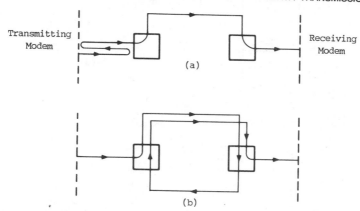

Figure 1.13 Listener echo caused by *(a)* two reflections at ends of two-wire subscriber loop; *(b)* two reflections inside four-wire loop.

1.4.2.4 Listener Echo.
Listener echo is the result of two reflections some- where in a connection; two possible paths are shown in Fig. 1.13. All listener echoes cause amplitude and delay distortion — with the "frequency-period" of the ripples in the frequency characteristic decreasing as the echo delay increases — and short ones are much more easily studied, and compensated for, as distortion. However, medium and long delays (greater than about 5 ms) are beyond the capability of most adaptive equalizers and must be examined sepa- rately.

It can be seen from Fig. 1.13*b* that the level difference between the wanted receive signal and its listener echo is just the loss around the four-wire loop. In the PSTN this is engineered primarily to reduce talker echoes to an acceptable level, and the nominal values for the four-wire and trans-hybrid losses on medium and long connections are 7† and 11 dB, respectively. This would result in a Signal to Listener Echo Ratio (SLER) of 36 dB; since this is much higher than typical SNRs, in most cases listener echo can be ignored.

However, in some cases the four-wire loss may be as low as 4 dB, and, because of the great variability of the input impedance of the two-wire line, the trans-hybrid loss may be as low as 6 dB.‡ This would result in an SLER of only 20 dB! Such a figure would be disastrous for high-speed modems such as V.32s and V.33s. Unfortunately, very little information is available about the tails of the distribution of listener echo, and it is difficult to know whether 20 dB is a 0.1 or a 1.0 percentile.

1.4.2.5 Input Impedance.
The input impedance of the line seen by the user is important for duplex modems because it determines how much of the trans-

---

† This includes the transmission loss of the two hybrid transformers (about 3.5 dB each), so the aim is to use amplifiers to make the simplex line itself have zero loss.

‡ Figures this low would still be acceptable as far as talker echo and singing margin (e.g., see Chapter 2 of [Fr4]) are concerned.

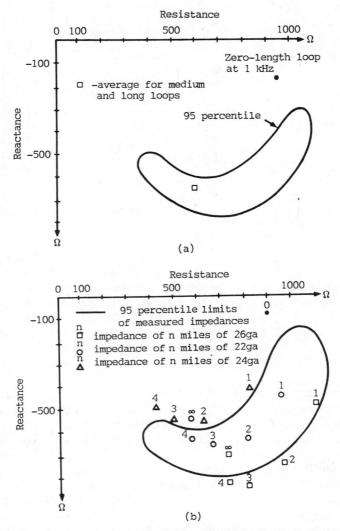

Figure 1.14 95 percentile limits of input impedances: *(a)* at 1 kHz with simulated two-wire termination at the local exchange; *(b)* at 1 kHz with simulated four-wire termination; *(c)* at 3 kHz with simulated two-wire termination; *(d)* at 3 kHz with simulated four-wire termination.

mitted signal is reflected back—as *talker echo*—by the modem's 4W/2W combining network and must therefore be removed by either filtering or echo canceling. The input impedance depends on the gauge and length of the cable, the loading, and the termination at the exchange (another two-wire cable or a four-wire carrier system). Since loading reduces the in-band loss of the cable, and therefore reduces the importance of the matching, we will initially consider only unloaded cables.

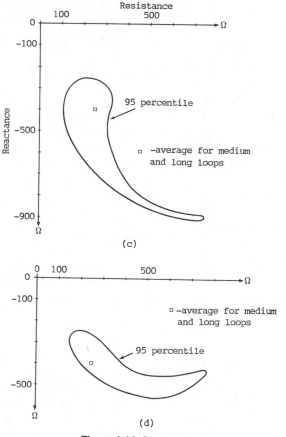

Figure 1.14 Continued

Scatter diagrams of input impedances of cables (at 1 and 3 kHz, and with simulated two and four-wire terminations†) reported by Gresh [Gr] and Manhire [Ma2] are shown in Figs. 1.14a–d. Figure 1.14b also shows the calculated impedance of 1, 2, 3, and 4 mile lengths of 22, 24, and 26 gauge cables; it can be seen that most of the measured impedances are neatly bounded by the curves for the finest and coarsest gauges.

For design of a fixed, compromise 2W/4W hybrid the most important numbers are the average impedances for medium and long loops:

At 1 kHz $Z_{in} \cong 700 - 650j \; \Omega$

At 3 kHz $Z_{in} \cong 250 - 400j \; \Omega$ approximately 3 miles of 24 gauge

† The input impedance of a loaded 22 gauge H88 cable, $(1040 - 177j) \; \Omega$, was used for the 2W termination and 900 Ω in series with 2.16 μF for the 4W.

If these numbers are used for the matching impedance in the hybrid, then a long loop with $Z_{in} = 350 - 500j \ \Omega$ at 1 kHz (around the 95 percentile extreme of the distributions shown in Fig. 1.14) would result in a trans-hybrid loss of 10 dB. For all loops, short and long, loaded and unloaded, the scatter diagrams given in Fig. 39 of [Gr] and Fig. 24 of [Ma2] indicate that 95% of the trans-hybrid losses would be greater than 7 dB.

Effects of Loading, Range Extenders, and Bridged Taps. Loading coils increase the reactive part of the input impedance and make the impedance much less dependent on the length of the loop; they also make the impedance much less predictable† and reduce the effectiveness of any compromise matching impedance. The 1973 survey [Ma2] showed an increase in the use of loading coils on long loops, with the coils typically being added at the user's end.

Range extenders are bidirectional amplifiers (negative resistances connected across the line). In the United States they are usually located at the local exchange, so their large effect on the input impedance of the line is masked for the user by the lossy loop. However, some PTTs reportedly use them further out on the loop, so that the worst-case trans-hybrid loss may be only about 3 dB.

Bridged taps (lengths of open-circuited cables connected across the cable in use) are now rarely used on lines used for data transmission, and need not be considered as a factor in the, necessarily crude, characterization of the line input impedance.

1.4.2.6 Noise. Modem designers are concerned mainly with the SNR at the receiver input, but information gathered from the network is in mixed form; from companded and digitized systems, where the noise is mostly quantizing noise, the information is in the form of an SNR, but from the subscriber loop and carrier systems without compandors it is in absolute noise levels.

Noise levels are expressed as positive numbers of dBrn or dBrnC relative to −90 dBm (1 pW). Two types of weighting network are used: a flat low-pass from 0 to 3 kHz, and a *C-message weighting* filter that has a response as shown in Fig. 1.15. The C-weighting was originally a measure of the relative annoyance to a listener of single interfering tones when using a particular handset. It has nothing to do with data transmission but is nevertheless a useful weighting for most modems because it approximates a wide-band receive filter with −3 dB points at 600 and 3000 Hz. C-message weighted noise is often much lower than 3 kHz flat noise because the latter is dominated by components at the first few harmonics of the power line frequency.‡ C-Message weighted noise figures are adequate only for modems that do not try to use the band below about 450 Hz. Wherever possible, we will quote both noise figures.

---

† The dominating, and random, variable is the distance from the user to the first coil.

‡ The more one learns about noise on the telephone network, the more one realizes that White Gaussian Noise (WGN) plays the role for modem designers that Latin did for medieval clerics — it bears no relation to reality, but it at least allows us to talk to each other!

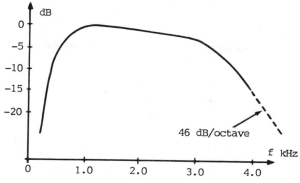

Figure 1.15 C-message weighting.

Noise measured on a quiet line (i.e., with no signal input at the far end) can be misleading if the connection contains a compandor and a T-carrier; most times the measured noise will be unrealistically low because it does not include the PCM system's quantizing noise (the dominant component in most connections). Therefore, measurements are made with a holding tone (1004 Hz at -12 dBm at the far user end), and the SNR is calculated from the ratio of the input to output powers of a narrow-notch filter.

Private Lines. C-message notched noise is guaranteed to be below 50 dBrnC; since the nominal receive level at 1 kHz is -16 dBm (74 dBrnC), this represents an SNR of only 24 dB at 1 kHz (and less at higher frequencies if there is significant attenuation distortion). This is not adequate for high-speed modems, but it is a very conservative specification; most PSTN connections are better than this.

PSTN. The best source of information is the 1982/83 end-office connection survey reported by [C . . 2]. Most of that report was concerned only with toll connections, and Fig. 1.16 shows the CDF of the signal to C-notched noise ratio, averaged over all toll-connection lengths. However, the subscriber loops usually make very little contribution to the total noise seen by the users, and, indeed, because the noise seen at the receiving exchange tends to be flatter than the signal (because of the roll-off of the subscriber loop at the transmitting end), the receiving loop typically filters the noise more than it does the signal and thereby improves the SNR slightly.

The CDF for users' SNR might also be found from the published figures for user-to-user C-message noise levels, but this would require making too many assumptions about the correlations between loop lengths and noise levels and between C-notched noise and C-message noise. Figure 1.16 will have to suffice as a slightly conservative (i.e., low) estimate of the cumulative distribution of signal to C-notched noise ratio as seen by the user.

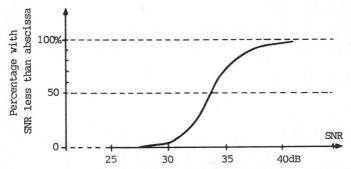

Figure 1.16 CDF of signal to C-notched noise ratio for all toll connections — a reasonable approximation for all user-to-user connections also.

For the signal to 3 kHz flat noise ratio, only noise levels are given in [C . . 2], and we have to try to estimate the SNRs from those. The median SNR can be calculated as 25 dB, but the distribution is much wider than for C-weighted noise.† If we assume that the greatest noise occurs on the loops that have the greatest loss, and that the subscriber loops have little effect on the SNR, then the Circuit Net Loss (CNL) distribution shown in Fig. 1.7 can be combined with Fig. 31 of [C . . 2] to give the CDF for signal to 3 kHz flat noise ratio shown in Fig. 1.17.

Impulse Noise. Impulse noise is caused mainly by coupling from the operation of switches—either mechanical or electronic—on other channels into the observed channel. Its distribution is log-normal rather than Gaussian, and it has been well analyzed and modeled by Fennick [Fe1]. Most of the recent measurements of impulse noise have been made at the local exchange, and although the noise contribution of a local loop is usually negligible, its low-pass filtering effect is considerable; one cannot calculate the magnitude of the pulses seen by the user without more information about the form of those "impulses" at the exchange.

1.4.2.7 Intermodulation (Nonlinear) Distortion. This is sometimes called "harmonic" distortion, but that name obscures the fact that an important effect is the folding back in-band of the product of two tones. Intermodulation distortion is quantified by the ratio, SDR, of the power of a special test signal comprising four tones—fairly representative of a data signal—to that of all the in-band distortion components. The cumulative distributions of SDR for second- and third-order distortions are shown in Fig. 1.18.

---

† Recall that the 3 kHz noise is dominated by tones at the harmonics of the power line frequency, which are much less predictable than the quantizing noise of PCM systems.

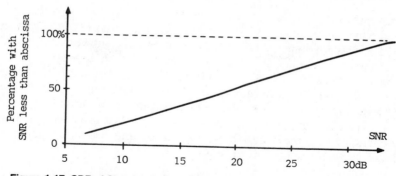

Figure 1.17 CDF of Signal to 3 kHz flat noise ratio for user-to-user connections.

It might appear from these that for about 2% of all connections intermodulation distortion is more serious than noise. However, these measurements were made with a −12 dBm signal at the far exchange; data signals, on the other hand, are typically transmitted by the far user at −9 dBm, and in the worst cases—those for which we are most worried about the SNR at the receiver—will arrive at the exchange at −17 dBm; this reduction of the level of a "vulnerable" signal by 5 dB has the effect of reducing the second- and third-order distortions by 10 and 15 dB, respectively. Consequently, intermodulation distortion is a significant contributor to total "noise" in only a very small number of cases.

1.4.2.8 *Phase Jitter.*

1.4.2.8 Phase Jitter. Measurement of phase jitter is complicated by the fact that its simple manifestation—jitter of the zero crossings of a sinusoidal signal—can also be caused by noise. It is important to try to distinguish between the two effects, because, as we shall see in Chapter 6, a receiver can often be designed to track pure phase jitter, but there is not much it can do about noise.

In the study reported in [C . . 2], both phase and amplitude jitter were measured, and since amplitude jitter per se is very rare, it could be assumed that

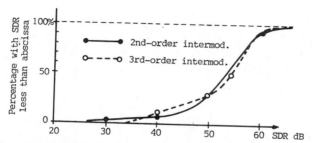

Figure 1.18 CDFs for second and third-order intermodulation distortion.

any that was seen was caused by noise; the implied noise could then be subtracted from the measured phase jitter to give an estimate of the pure phase jitter. This was not done for individual measurements, but we can use the CDFs for both jitters to deduce approximate CDFs for net phase jitter.

The jitters were measured in two bands, 2–300 Hz and 20–300 Hz, and CDFs were shown for short, medium, and long connections. Figure 1.19 shows the CDFs for net phase jitter — after subtracting the approximate contributions of noise — for just medium (average) connections. It is interesting to note the differences between the curves for the two bands, and also to compare them to CDFs for gross phase jiter given in [C . . 2]:

1. With large phase jitters — those for which tracking in the receiver is most important — the wider band shows more than three times as much jitter. This means that the jitter components in the band 2–20 Hz constitute more than 70% of the total jitter power. Since pickup from the 20 Hz ringing current is a major source of phase jitter, it is not clear how much of this the narrower-band filter allowed through, nor what the distribution of the components below 20 Hz was.

2. With large jitters measured in the wider band, the noise and observed amplitude modulation made up less than half of the total observed phase jitter. If the band from 2 to 20 Hz is considered, the contribution of the noise is still smaller.

Comments and a Caveat. It is difficult to correlate the results of the 1982/83 survey with those of the 1969/70 survey [D&T] because the latter used different frequency bands for measuring jitter and did not consider phase and amplitude modulation separately. The total jitter in the band 12–768 Hz was less than the later study showed in the band 2–300 Hz, even though the telephone network had been considerably improved in the intervening 12 years; the only conclusion is that there is considerable jitter in the very low band from 2 to 12 Hz.

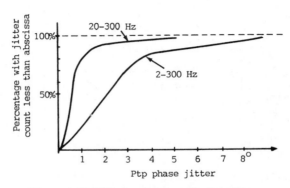

Figure 1.19 CDFs for peak-to-peak phase jitter.

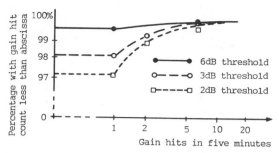

Figure 1.20 CDF for gain hits.

The earlier survey also showed a 90 percentile jitter in the 48–96 Hz band, for medium connections, of 4° ptp. The main contributor in this band would be power-line pickup (60 Hz), and this component would probably account for most of the jitter — the noise contribution in that relatively narrow band would be small. Yet the 1982/83 survey indicated a 90 percentile of only 1.5° in a much wider band! It appears that power-line pickup was greatly reduced in the intervening years.

1.4.2.9 *Frequency Offset.*

CCITT recommendations traditionally call for an ability to deal with a user-to-user frequency shift of 7 Hz, but such shifts can only be caused by very old analog carrier systems; it is not clear how relevant such specifications are in the latter part of the twentieth century. In the past twenty years the U.S. telephone network has been greatly improved, and the 1982/83 survey revealed a shift of 2 Hz on only 0.2% of all connections. All carrier recovery loops are well able to follow this.

A more serious problem is frequency shift seen in an echo. Cancelers typically have a much longer adaptation time than receiver carrier loops and are not able to deal with even 1 Hz of offset. Most connections are made with the same carrier system used for the two-wire line in each direction, and any frequency shift experienced by a signal in one direction will be canceled by an opposite shift of the echo in the reverse direction. However, transatlantic connections are often made by satellite in one direction and submarine cable in the other; the latter systems quite often introduce a shift of about 1 Hz, which is not canceled by a complementary shift.

1.4.2.10 *Gain Hits.*

A gain hit is seen as a rapid change of received signal level.† The CDFs of gain hit counts in a five-minute interval, as reported in [C . . 2], are shown in Fig. 1.20.

---

† Details of the measurement method were given in [BS4], but since the Automatic Gain Controls (AGCs) in high-speed modems are typically very slow acting, any "rapid" change must be considered to be too rapid.

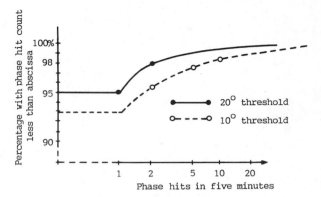

Figure 1.21 CDF for phase hits.

1.4.2.11 Phase Hits. The CDFs of phase-hit counts in five-minute intervals are shown in Fig. 1.21, but these are both questionable and difficult to understand. Careful reconciliation of the conditions of measurement described in [BS3] with the parameters of a proposed carrier recovery loop would be needed to determine whether these hits would cause bursts of errors.

1.4.3 Regulations for Modems Connected to the Telephone Network

These are concerned with safety and noninterference with other users of the medium. In the United States the regulatory body is the FCC, and the regulations are given in the Code of Federal Regulations [CFR]. The rules for direct connection to the PSTN are given in Title 40, Part 68; those for the control of radio frequency emissions (EMI) are given in Part 15. Modems that are to be connected to the PSTN must be tested by an independent registered professional engineer for compliance with the rules in both Part 68 and Part 15.

The situation varies widely in other countries. In some, since all modems for the telephone system are supplied by the PTT, the regulations take the form of procurement specifications. In the United Kingdom the British Approvals Board for Telecommunications (BABT) administers standards published by the British Standards Institute; the major U.K. technical standard for PSTN modems is BS 6305. Compliance Engineering† is a source of regulations — translated into English — from outside the United States.

We will summarize the U.S. regulations as a guide to what might be expected elsewhere.

1.4.3.1 Transmit Level. Some years ago transmit levels in the United States were set for each subscriber loop so that a data signal would arrive at the local

---

† 593 Massachusetts Avenue, Boxborough, MA 01719, Phone (617) 264-4208.

exchange at − 12 dBm. With more and more devices being connected to the telephone system, it became harder for the operating companies to maintain control, and the specification was simplified; the level was defined by a so-called permissive specification at a maximum of − 9 dBm average power transmitted into a 600 Ω load.

Some countries (e.g., the United Kingdom) still specify a "programmed" output level that can be adjusted upon installation between 0 and − 16 dBm.

1.4.3.2 In-Band Signal Spectrum.

The definition of "in-band" varies from country to country, but generally it is from 300 to 3200 Hz. The signal spectrum may have any shape within that band provided it does not interfere with any in-band signaling system that is in use. In the United States, one system is based on a single tone at 2600 Hz, and to prevent inadvertent operation of this system, power in the band from 2450 to 2750 Hz must not exceed the power in the band from 800 to 2450 Hz.

Many systems thoughout the world use a system that is based on 2280 Hz. The British Telecom specification, which is probably typical, defines the forbidden band from 2130 to 2430 Hz and allows power there to be no more than 12 dB above the power in the band from 900 to 2130 Hz.

1.4.3.3 Out-of-Band Power.

PTTs throughout the world use many different specifications for out-of-band (i.e., below about 300 Hz and above about 3200 Hz) power; the permitted levels depend on the ability of the filters in their multiplexing systems, which were originally designed to handle voice, to band-limit the data signal and prevent it interfering with adjacent channels. A composite specification for use in Europe is shown in Fig. 1.22; out-of-band powers below the limits shown should satisfy all PTTs.

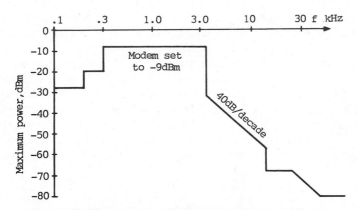

Figure 1.22 Composite out-of-band power limitations for Europe.

1.5   OTHER MEDIA FOR DATA TRANSMISSION

The other media that we will consider very briefly are unloaded twisted pair cables, and radio—in many different frequency bands.† These differ from the voice-band telephone system in several ways:

1. The character of the channel distortion, and the resultant Inter-Symbol Interference (ISI).
2. The character of the interfering noise.
3. Limitations placed upon the form of the signal by the hardware— particularly in repeaters and receivers.

We will consider each factor as it applies to the various media.

1.5.1   Unloaded Twisted-Pair Cables (Subscriber Loops)

These will be used for access to the Integrated Services Digital Network (ISDN) and the eventual requirement will be for FDX communication at 160 kbit/s.‡

   The problems of high-speed transmission on subscriber loops were summarized in [GS&B] and [M&G2].

1.5.1.1   Channel Distortion.   At high frequencies the main contributor is attenuation distortion caused by skin effect in the conductors. As the frequency increases, the depth of penetration of the electromagnetic field into the conductors decreases, and theoretically the attenuation increases as the square-root of frequency; that is,

$$H(f) \simeq e^{-\alpha \sqrt{f}} \tag{1.1}$$

In the frequency band of interest (1 to about 100 kHz) there are many other significant effects; (1.1) is still a reasonable approximation, but the exponent of f should be reduced from 0.5 to 0.35. For 26 gauge twisted pair the loss at 100 kHz is approximately 10 dB/km. These cables are minimum-phase media (see Section 9.1), and delay distortion is not significant.

   A secondary contributor may be *bridged taps*. These are other twisted pairs connected across the user-to-exchange pair to provide party-line service, as shown in Fig. 1.23. When the other telephones are on hook, the bridged taps are open-circuited at the ends. It can be seen that if the cable forming the tap were lossless, its input impedance would be zero at the frequency for which the tap length was a quarter wavelength, which would result in a transmission zero at that frequency. Since the cable is not lossless, the transmission zero moves into

---

† Magnetic tapes and disks used for high-speed, high-density recording also have many of the same characteristics and problems as data transmission media, but I know too little about them.
‡ This is in a medium that had previously been used only for access to 4 kHz voice channels!

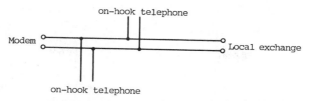

Figure 1.23 Bridged taps on a subscriber loop.

the left-half p plane and causes a minimum-phase attenuation peak around that frequency.

A third minor contributor to the attenuation distortion may be discontinuities—principally changes of gauge—in the cable. These would cause reflections and consequent small ripples in the attenuation/frequency plot.

1.5.1.2 *Noise.*

In most of the frequency band of interest, the cable attenuation is very high, and noise sources at the exchange and beyond are not as harmful as they are for user-to-user voice-band communication. The principal source of interfering signals is now *near-end crosstalk* (NEXT); that is, coupling in the cable sheath from other nearby, similar transmitters whose signals have not yet been attenuated. This noise is far from white; its source is shaped by the same (or similar) transmit coding and filters as the wanted signal, and the coupling transfer function is proportional to $f^{3/4}$.

1.5.2 Radio Links

"Radio Links" seems to be a totally inadequate heading! Since the specific problems depend strongly on those parameters, the reader will ask "What frequency band? What data rate?" However, we can discuss these problems only very generally and refer the reader to [Fr4] for a good description of all the media and to the specialized literature on the different types of modems.

1.5.2.1 *Channel Distortion.*

The main cause of all distortion in radio transmission is *multipath* propagation. This occurs because of tropospheric scatter, ionospheric scatter, or reflection from large objects (buildings, hills, etc.). In most cases "multi" can be simplified to—or at least, well approximated by—two, and then, since the separate paths, with propagation delays T_1 and T_2, are relatively undistorted, the transfer function can be written as

$$H(f) = \alpha \left(1 + \beta \exp -j\omega\tau\right) \exp -j\omega T_1 \qquad (1.2)$$

where the differential delay $\tau = T_2 - T_1$.

Minima of $|H|$ occur when $f\tau = n + \frac{1}{2}$, and it is convenient to define the f_{min}

closest to the carrier frequency as $(f_c + f_0)$. Then the baseband equivalent of (1.2)—ignoring the flat delay—given by Rummler [Ru1 and Ru2] is

$$H(f) = \alpha[1 - \exp -j(\omega - \omega_0)\tau] \tag{1.3}$$

In some media there is a direct path and a delayed path (via a reflection), and the latter is always more attenuated (i.e., $\beta < 1$). As we shall see in Chapter 9, this results in what is called a *Minimum-Phase Fade* (MPF). For others (e.g., tropospheric scatter), both paths are reflected, and either the shorter or the longer may be more attenuated; if the former, then $\beta > 1$, and this results in a *Non-Minimum-Phase Fade* (NMPF).

Variation of Distortion. Any one of the three parameters, α, β, or τ, of a "two-path" radio channel, may change:

1. If the flat gain, α, decreases significantly, this is called a wide-band fade, and some sort of alternative path must be selected.
2. If the relative level of the delayed path, β, changes, for example, from 0.9 to 1.1, the dip will change from a 20 dB MPF, through a perfect notch, to a 20 dB NMPF.
3. If the differential delay, τ, changes, this has the effect of moving the dip (not necessarily a "notch") across the band.

Techniques of dealing with deep wide-band fades using either space or frequency–diversity are beyond the scope of this book, but the other two types of variation can be dealt with by using the adaptive equalizers described in Chapters 8 and 9.

1.5.2.2 *Hardware (Power) Limitations.*

For radio communications via satellite or battery-powered terrestrial repeaters, valuable power savings can be achieved by operating all amplifiers near saturation. This leads to a demand for modulation schemes that generate a nearly constant amplitude signal—thereby achieving power efficiency at the expense of bandwidth efficiency.†
These modulation schemes are described in Section 4.3.

---

† This is in contrast to modems for the telephone system, which make great use of amplitude modulation to increase the bandwidth efficiency.

Chapter 2

Modem Marketing

by Ken Krechmer

2.1 HISTORY OF THE MODEM MARKETS

The growth of the commercial modem marketplace closely parallels expanding communications between commercial data processing systems. Originally data processing was a freestanding application. The tedious nature of manual accounting systems was ideal for automation, so financial applications became the first major commercial use of data processing. Initially communications were confined to terminals with permanent physical connections to the host computer in the same location. With the rapid expansion of population and organizations in the United States in the late 1940s and 1950s came the requirement to interconnect different sites via higher-speed data transfer. The first paper describing an implementation of a modem for this requirement appeared in 1955. The *Bell System Technical Journal* published a paper entitled "Transmission of Digital Information over Telephone Circuits" that described a pulse-modulated carrier system that was capable of supporting data rates up to 650 bit/s [H&V].

The key difference between the earlier telegraph, telephoto,† facsimile, and telex communications and this new site-to-site communications was the use of data rates sufficiently high that a modulation mechanism was necessary to transfer the information through the band-limited telephone channel. Modems

---

† Telephoto transmission was implemented with a forerunner of a modem. It used a form of amplitude modulation (similar to early facsimile systems). Telephoto transmission was used commercially for the first time to send news pictures of President Coolidge's inauguration from Washington to New York, San Francisco, and other major cities in March 1925 [Be2].

enabled the transfer of serial digital data over the existing telephone system at speeds higher than previously thought possible. By February 1958, Dataphone Service, the original designation for the commercial Bell System modems used to transfer data over telephone lines, was first introduced [BLR].

From 1958 until 1968 the Bell System was effectively the only commercial modem supplier for the PSTN; independent modem suppliers did modem development work for the government and manufactured modems for use of leased telephone lines. The first indication of the changes that would eventually occur in the Bell System began in 1957 when the FCC ruled on the use of a simple device called the *Hush-a-Phone*.

The Hush-a-Phone was a plastic cup that fitted over the mouthpiece of a telephone to reduce interference from background noise. The telephone company considered these devices to be "foreign attachment" and therefore prohibited by tariff. The FCC on remand decided, "As we construe the Court's opinion, a tariff regulation which amounts to a blanket prohibition against the customer's use of any and all devices without discrimination between the harmful and harmless encroaches upon the right of the user to make reasonable use of the facilities furnished by the defendants" [A1]. This was the first glimpse of the events to come.

The outcome of the Hush-a-Phone decision paved the way for the development and sale of acoustically coupled modems† by independent firms. In fact, a specialized form of the acoustically coupled modem was first offered by the Bell System in 1965 for transferring electrocardiogram (ECG) information designated the Dataphone 603. The 603 used a half-duplex FM modulation means [CFLP]. Anderson Jacobson, an emerging communications manufacturer, pioneered the use of acoustic coupling for data transmission by showing the first such device in 1967. By September 1969 Bell Laboratories had announced its own accoustically coupled modem, the Data Set 112A. The 112A utilized the standard Bell 100 series FSK modulation and was capable of operating at 300 bit/s half duplex or 150 bit/s full-duplex [D&L].

The emergence of an independent (non Bell System equipment) commercial data communications market began in the late 1960s. The installed base of data processing equipment had become sufficiently large, timesharing was emerging as the first broad communications market application (ideal for acoustically coupled modems), the indifference of the U.S. telephone companies to customer needs was becoming too noticeable, and the beginnings of a change in the regulatory climate in the United States seemed possible. The FCC position in the Hush-a-Phone case would open the market to those products that could be helpful and harmless.

---

† Acoustically coupled modems are a form of modem that employ an acoustic rather than an electrical connection to the telephone system. The digital connection is the same in both cases — usually EIA-232/V.24. The acoustic connection is accomplished through the use of a speaker and microphone held to the telephone handset by rubber cups, thus mimicking the human interface to the telephone handset.

The reason for the telephone companies apparent indifference to customer needs emerged from the effects of regulation, not an inability to understand their customers. Technical development in data communications was occurring continuously. The speed at which data could be transferred over telephone lines and the features available (full-duplex, auto answer, voice alternate, etc.) expanded with each technical development. On the other hand, the operating telephone companies were using equipment depreciation schedules (fixed by regulated tariffs) that extended over twenty years. It is not too surprising that the Bell System presented new communications technology to the market more slowly than did the independent companies that emerged during this period.

As is often the case, the change in the regulatory climate was market driven — users wanted to be able to quickly install timesharing terminals. From the users' view, acoustically coupled modems meant that when the decision was made to install a timesharing application, the installation was immediate. For the Bell System to install a data line and modem often meant several weeks' delay.

Again and again successful communications products utilize technology to provide a market desired service while circumventing an outdated standard or regulation. A small company called Carterfone provided an excellent example of this effect.

In 1967 Carterfone offered mobile radio systems with a device to interconnect the mobile radio system to the PSTN. When the local phone company (Southwestern Bell) found evidence of this "foreign attachment," they threatened to remove telephone service from users of the Carterfone device. Carterfone sued to prevent such action, and the case was remanded to the FCC. Following the direction prescribed earlier in the Hush-a-Phone decision, the FCC found that use of the Carterfone connection to the telephone system did not adversely affect the telephone system and that the tariff prohibiting its use was unreasonable and unlawful [Al].

However, the means to implement the Carterfone decision created a new complexity. The Bell System engineers argued that unrestricted electrical access to the telephone network terminations on the customer premise posed the possibility of damage to the network and suggested that a device termed a *Data Access Arrangement* (DAA) be installed between the telephone line and the customer supplied modem (or other equipment). The FCC agreed with this approach, and a range of different DAAs were designed by the Bell System for different applications.

The use of DAAs opened up the public switched telephone network market to independent modem manufacturers for the first time. Previously they had sold only to the private or leased-line segment of the market. However, the problems of getting the telephone company to install DAAs properly and when requested sometimes seemed to outweigh the advantage of a new market for modems.†

† Personal experience of the author while sales manager at Vadic Corporation in the mid-1970s.

In 1974 one last attempt was made to maintain the telephone monopoly. The State of North Carolina considered requiring that only telephone company-provided telephone equipment could be used for intrastate communications. The Telerent Leasing Company petitioned the FCC for a declarative ruling to the effect that federal law would preempt state law in this matter. The FCC did so, but stated, "Neither the Carterfone ruling nor these tariffs prevent any State from providing *additional* options to customers with respect to interconnection provided that they are alternatives to, rather than substitutes for, the requirements specified in the interstate tariffs, and provided further that such regulations accomplish the protective objective of the interstate tariff regulations and in no way permit interference with or impairment of interstate services" [Al].

With the last issue of jurisdiction resolved, the FCC in 1975 issued the First Order and Report Docket No. 19528. This Order established the nationwide technical standards for the design of ancillary devices and data equipment so that such devices could be connected to the PSTN without the use of protective couplers. A later report dated September 23, 1976 (FCC Docket 20003) was submitted to the chairman of the House Subcommittee on Communications as FCC comments on HR 12323, entitled "The Consumer Communications Reform Act of 1976 [Al].

The first technical requirements that other manufacturers had to meet in order to connect equipment to the PSTN were included in FCC Docket No. CC-79-143, which became the beginnings of Part 68 [Al]. With the publication of the Rules in Part 68, the modem manufacturers could now build "direct-connect" modems that required no DAA to attach to the PSTN. AT&T's share of the modem market has declined ever since.

Now each modem manufacturer had the ability to connect to the PSTN equally. So it became necessary to have standards to ensure that each manufacturer's modem products (and higher OSI layer functions) would be compatible. It was no longer desirable to have the Bell System create de facto standards. This significantly increased the energies directed toward creating de jure standards.

2.2 THE MARKETING IMPACT OF STANDARDS

The Manufacturer's Viewpoint. The standards-making organizations operate by consensus (or at least no strenuous objection), not majority rule. This has the effect of allowing any well-thought-out position (that is not too obviously self-serving) to hamstring the standards committee's ability to create a more generally applicable standard. Conversely, it does prevent major equipment manufacturers, with broad political influence in the committee from swaying the majority.

Once the various technical issues relating to a potential communications standard have been discussed, more pragmatic concerns come into play. Use of certain features may represent a market advantage to certain companies. Even the fact that a certain standard evolved from the work of a specific company is considered of marketing value. Codex and Racal-Vadic both have used advertisements noting that they pioneered specific modulation approaches. (Codex pioneered the modem modulation now specified in CCITT Recommendation V.29. Racal-Vadic produced the first full-duplex dial-up 1200 bit/s modem.)

The range of different organizations involved in the communications standards process and the need to allow many viewpoints to be heard makes the communications standards process painfully slow for the pace of technology. In turn, this slow process requires a significant time investment from key personnel of the communications companies that participate.

Companies make this investment because of the marketing as well as technical importance of the standards process. By maintaining contact with the standards organizations, companies have a window on both the future course of communications standards (and therefore future communications product directions) as well as the ideas and suggestions of their competitors (and thus some view of the competitor's technology and direction). As an increasing number of companies have recognized the market importance of standards activities, it has become decreasingly likely that any one company can have a product of its own making become a formal standard, as Codex did ten years ago. This has not caused any company to stop trying, however.

The User's Viewpoint. Standards are a more pragmatic issue for the user. If the product is not "standard" the buyer has limited the possible communications that can occur. Yet compatibility of communications standards is only one aspect of what the user perceives. The user requires applications that work "properly" with other remote applications. More precisely, all layers of the OSI model (see Table 1.2) must be compatible for two applications to communicate "properly." Given the complexities of achieving compatibility at each layer of the OSI model, the user is often willing to forgo a "standard" if the communications applications can be shown to "work."

Thus the difference between selling standard and nonstandard products is that the nonstandard must be demonstrated in the user's application to "work" to the user's satisfaction. The standard is known to work, and a demonstration is not required. Also, there must be features in the nonstandard product that are sufficiently desirable to make the effort of constructing a demonstration worthwhile.

2.3 MODEM MARKETING

All Product Marketing Has a Four-Stage Cycle

1. *Innovation.* Figure 2.1 shows the conjunction of markets and technology where viable products occur. Innovative manufacturers and users devote re-

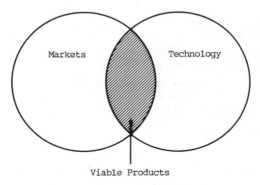

Viable Products

Figure 2.1 Viable products.

sources to finding this area. Technology is developed by a manufacturer that allows a desirable function that previously could not be accomplished. A small percentage of users in any market are innovative and profit from being among the first to put to use the new product or technology. The sales effort required here is to identify the potential buyers who are innovators and present to them the new technology. Since the sale must be initiated by the selling organization, this is termed *demand push.* The Bell System was the manufacturer innovator, and the period up to the mid-1960s was the innovation stage of the original modem market.

2. *Applications Selling.* Most of the possible users are not aware of how the product can be utilized and need to be told of its benefits. When shown the product and its benefits (the sales effort), however, a reasonable percentage of these potential buyers choose to purchase. The term "demand push" is also used to indicate that someone must go out and push prospective buyers to understand and use the new product. The Carterfone decision marked the beginning of the application selling stage of the modem market. Around the time of this decision some of the emerging modem companies began to offer their first products. Codex (now a division of Motorola), Vadic (now Racal-Vadic), General Data Communications, and Timeplex, to name some of the leading companies, all started to provide data communications products within a few years of the Carterfone decision in 1968.

3. *Distribution.* After sufficient users acquire a product and the product has been shown (via demonstration, promotion and advertising) to a large number of prospective buyers, customers begin to contact the manufacturer even before the manufacturer contacts them. This is termed the *demand pull* stage. One definition of the demand pull market stage is: the potential customer in more than 50% of the sales initiates the first contact with the company. By the early 1980s the marketplace had become familiar with the simplicity of the direct connect modem installation. The continuing growth of data processing applications (as example the emergence of the personal computer) dramatically

increased the need to share data between remote sites. Because of this, the low-speed modem market reached the status of being demand pull. One new company positioned itself to take advantage of the different marketing requirements of this stage in the modem market — Hayes Micro Computer Products. The success of Hayes in the demand pull stage of the modem market has been remarkable.

4. *Obsolescence.* When the product has reached a high percentage penetration of the current market and new "better" technology has been developed, the demand for the product starts to decline. This stage of the market cycle is marked by sale prices. Demand may continue as the product price declines. The rapidly falling prices for 1200 bit/s modems when the 2400 bit/s modems were introduced in 1984 was an indication of the obsolescence stage of the market cycle beginning for 1200 bit/s modems.

The Difference between Selling Communications Products and Data Processing Products. Three differences occur as a result of the functional variations between communications systems and data processing systems. Communications always requires three separate systems — the local hardware/software system, the remote hardware/software system, and the transmission system to interconnect them. Data processing applications are implemented on a single system. The greater complexity of selling communications products results from this difference.

1. Commercial applications for communications cannot exist (in any volume) without a defined standard for the interface between the transmission system and the data processing system. The market for communications products grows with the advent of each such standard. This effect modifies the traditional viable product diagram. Figure 2.2 shows that viable *communications* products exist at the conjunction of markets, technology, and standards.

2. Since two physical locations must install equipment (often two technical decision locations, possibly two budgets, as well as two installation locations) for a communications system to exist, the initial rate of insertion (sales) of communications products is slower than the rate of insertion of data processing equipment of similar value.

3. Since half of a communications system's functions are performed remotely, the communications system is more confusing to use than a data processing system of similar hardware/software complexity. This occurs because the local user does not have a simple ability to verify the physical status of the remote system. This complication translates into the need for more sophisticated service procedures, better applications support, and more user training.

Effect of These Differences on the Modem Market Cycle. Markets emerge initially via leading-edge customers who have demanding applications, innovative management, and sufficient capital to risk innovating. In the early com

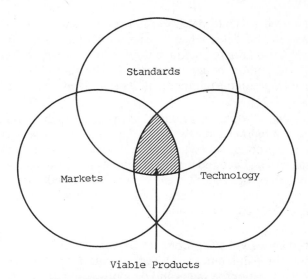

Figure 2.2 Viable communications products.

mercial data communications markets, TWA was a leading edge customer of IBM systems and AT&T Dataphone products. This resulted in one of the first online airline reservation systems—the Planned Airline Reservation System (PARS). Such customers are a small portion of the total market to come. As a rule of thumb, leading-edge customers represent 5–10% of the eventual customers in any market.

When a new modem product (which is not a standard—yet) enters the market (innovation), it must be sold as a system (i.e., at both ends) to the leading-edge customers. This makes the selling and demonstration expense associated with introducing a new modem product significantly larger than a data processing product. Additionally, the time required to create the more complex demonstration and deal with all the associated system interface problems makes the initial sales cycle much longer. A rough estimate would be that a new modem product selling cycle is twice as long as a data processing product selling cycle of similar complexity.

After the leading-edge customers are sold, the applications selling segment of the market cycle emerges. To create a demand requires application selling with a motivated direct selling organization to face and overcome the prospective buyers' resistance to untried solutions. Application selling techniques are employed to instigate customer purchases. Applications examples, reference customer lists, and on-site demonstrations are examples of sales techniques that characterize the applications selling process. In this segment of the marketing cycle the on site demonstration has been refined from previous experience and is not as complex as the demonstrations that took place in the innovation cycle.

During the application selling segment of the market cycle it is desirable to have standards emerge that support this new modem product. The process of creating standards is very slow and not likely to be successful, but it is desirable for the manufacturer to provide to potential buyers a view that a new standard may emerge. The actual existence of a set of standards (both de jure and de facto) is one of the key indicators that the demand pull side of the market cycle is about to begin.

The percentage of the Total Available Market (TAM) that is demand push varies widely depending on the actual size of the TAM, the simplicity of the product, and the existence of standards (that are widely understood) that support the product. When standards exist and the market has been trained regarding the ease of use of the product, the demand pull segment of the market cycle occurs.

The change from demand push to demand pull requires fundamental changes in the distribution strategy for the product. In the application selling segment, direct sales personnel (paid by the company only) or exclusive sales personnel (not allowed to sell competing products) are necessary to explain and demonstrate the product. This activity is quite difficult, and sales personnel who are good at it are usually paid via commissions to reward them for their successes.

When the product reaches the demand push stage of the market cycle, the customer no longer requires convincing of the products usefulness. The customer now requires best price, quickest delivery, and often best payment or credit terms. These attributes of the sales transaction are rarely provided by the manufacturer. At this stage in the marketing process the existing sales organization is repositioned to focus on the major accounts and provide additional service (usually the commission rate is reduced as well), and a new distribution sales organization (usually via separate distribution companies) is developed that can provide the necessary attributes the customer now requires.

Finally, the obsolescence segment of the market cycle appears. Communications standards have the effect of ensuring that modems have a longer obsolescence segment than data processing products. This engenders the requirement to include in the newest modem design the ability to be compatible with the preceding (lower-speed) modem.

The existence of standards or lack thereof lengthens and complicates each segment of the marketing cycle for modems. However, Figure 2.3 indicates that the effort to develop and market an innovative modem design may be repaid by the lengthy distribution and obsolescence segment of the market cycle.

2.4   MARKETS

The markets for modems may be described by each or a combination of the three factors that create them (Fig. 2.2): (1) the technology — the type of implementation; (2) the market — the applications or user requirements; and (3) standards — the Bell, CCITT and/or applicable de facto.

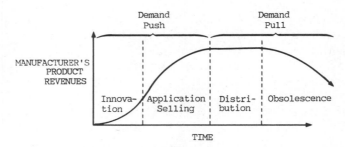

Figure 2.3 Modem product life cycle.

The market may also be described in relation to geography. This is helpful in designating how the sales organization will cover the physical area. With modem products, different countries also represent different interface standards requirements. In this view geography represents, in effect, different standard requirements.

In each segment of the market cycle one of the market descriptions (see Section 2.1.1) tends to predominate: (1) innovation — technology implemented (e.g., V.32 echo canceling), (2) application selling — the user requirements (e.g., facsimile), and (3) demand pull — a standard (e.g., V.22 bis/AT command set).

2.4.1   Markets by Application

The most complex markets to describe are those that relate to application and user requirements. Such markets require an accurate definition if the manufacturer is to properly direct the sales organization to service these applications.

Private-Line Markets. The first markets for modems were private-line markets. Point-to-point private-line applications were the first applications that modem technology could implement. The original Dataphone applications were leased-line point-to-point. Next, polled multipoint private-line applications began to emerge. This service allowed strings of geographically separate locations to be interconnected economically to the host computer. These systems are called *polled* to describe the action of the host computer to control the sequence of communications with each remote station. (See Fig. 2.4 for an example of point-to-point versus multipoint applications.) Because of technical issues it is necessary to use four-wire leased-line service in multipoint configurations.

PSTN Markets. At about the same time as the multipoint applications began to develop, the first major application for dial-up modems emerged — timesharing. Independent modem manufacturers waited until "direct-connect" in 1980 to be able to fully participate in the PSTN market. The original

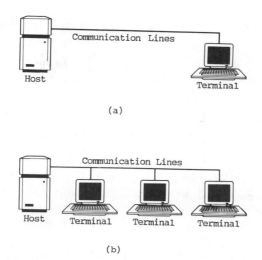

Figure 2.4 Private-line applications: *(a)* point-to-point; *(b)* multipoint.

commercial timesharing systems used 100 series full-duplex modems. In order to speed communications, half-duplex 1200 bit/s Bell 202 series modems were tried. (Full-duplex 1200 bit/s modems did not exist until several years later.) When full-duplex modems are used, the modem can notify the DTE/operator whenever the communications connection was broken. In a half-duplex modem a break in the communications channel can be detected only when the modem is receiving not transmitting. Since a PSTN connection was much more likely to fail than a private-line connection, the use of 202 modems proved to be too unreliable and complicated.

In these early timesharing systems another value of full-duplex communications was to simplify the logic required to perform error detection and correction. In those days logic was expensive and the communication channels inexpensive.

The early timeshare systems used a form of error control called *echoplex* that requires full duplex operation of the modem to accomplish. Echoplex operation is illustrated in Fig. 2.5; a character, *K*, that is input to a keyboard is not immediately displayed (or printed), but is transmitted to the central-site computer, whence it is echoed back by the front-end processor, and only then displayed. When the operator sees the echoed character, he or she makes the decision as to whether the character shown is correct. If it is incorrect, which could be caused by operator or communications channel error, entering the backspace followed by the correct character rights the error.

The importance of the echoplex function for error control of asynchronous communications was partly responsible for the demise of IBM 2741 asynchronous data communications terminals.

By the early 1960s two competing types of asynchronous data terminals were

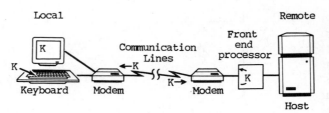

Figure 2.5 Echoplex operation.

being installed for PSTN communications. The telephone companies and Western Union provided Teletype terminals for the early timesharing applications, while IBM provided the IBM 2741 series asynchronous terminals based on the new Selectric printing mechanism. Teletype terminals had no mechanical connection between the keyboard and the printer, while 2741's had a direct physical link between the keyboard and the printer. In the parlance of the period, the 2741 provided "local copy" (i.e., the character that appeared on the printer was locally caused by the operator's key stroke.) The Teletype terminals, on the other hand, could not offer local copy mechanically and instead provided the function electrically with a switch designated "Local" or "Remote" to select whether to echo characters locally.†

Since the IBM 2741 lacked the means to support remote echo, it also lacked any form of error checking protocol on the communications link between the data terminal and the remote computer. And the commonly heard complaint from the users of the 2741 was the lack of reliability. IBM 2741 reliability problems also were also caused by the Selectric printing mechanism, which was not designed for the kind of continuous printing required of a data terminal. The combination of these two reliability problems caused the eventual demise of the 2741 product line, and the Teletype terminal ultimately evolved to become the now ubiquitous ASCII terminal.

The next successful PSTN modem was full-duplex 1200 bit/s. In 1973 Vadic Corporation (now Racal-Vadic Corporation) pioneered PSTN full-duplex 1200 bit/s operation with modems called the VA3400. Three years later Bell introduced the 212. While the VA3400 had an innovative buffer design (offering data rates of 1200 bit/s and 0–300 bit/s without reconfiguration) and a better choice of carrier frequencies for line performance, the 212 had one feature that was significantly better. The Bell 212 included in it a 103 modem also. Because of this, the 212 was automatically downward compatible with the previous PSTN modem. That feature and the size of the Bell System versus Vadic overwhelmed the VA3400 in the marketplace.

---

† Echo locally would be required whenever a half-duplex modem was used, so manufacturers of Teletype replacement terminals often labeled the local copy switch "Half-Duplex" or "Full-Duplex." This incorrect use of the terms "half-duplex" and "full-duplex" has become part of the operating environment. Architects may cover their mistakes in ivy but engineers often ship theirs.

The success of the Bell 212 modem in the late 1970s was followed by yet higher-speed full-duplex modems. The V.22 bis standard for 2400 bit/s operation came next. The desire for 2400 bit/s operation was caused by a major change in the type of information transferred. Realistically 1200 bit/s communications was the highest speed required for interactive applications (no one types that rapidly, and few read that rapidly); 2400 bit/s was needed to support file transfer to and from personal computers. The advent of file transfer requirements changed the requirements for error control. Echoplex is too inefficient a procedure for file transfer error control. This generated the need for a new standard asynchronous error control procedure. Now almost four years after this requirement arose, the standards groups are still discussing and several manufacturers have introduced error control procedures that they hope will become de jure (or at least de facto) standards.

The next modem standard to emerge was the 9600 bit/s full-duplex V.32 modem. It appears that the markets for this higher-speed modem will be quite different from those for the preceding modems. Currently the largest single market for the V.32 modem is automatic dial backup for leased telephone lines. It is likely that dial-up packet switching (based on the CCITT X.32 standard) will also emerge as a market for the V.32. Additionally, the high-speed dial-up capability of the V.32 will be well suited to providing on demand interconnection between multiuser departmental systems and larger mainframe computers.

A new de jure standard for a PSTN modem designated the asymmetrical modem is in the developmental process. A number of manufacturers are also attempting to create their own de facto standards. This modem is described as an asymmetrical modem to denote the fact that it operates on the telephone channel at different data rates in each direction.

2.4.2 Emerging New Technology Markets

Modems were designed to transfer data over an analog telephone network. By the 1950s the high traffic segments of the network between the Central Offices (COs) were being converted to digital for economic and performance reasons. Currently, greater than 90% of the network between central offices in the United States is digital. In many parts of Europe the penetration of digital interconnect is similar. The only remaining segment of the telephone network that is analog is the local loop. The local loop (often termed "the last mile") is the network segment between the end system location and the local central office.

Currently in the United States two digital services have been available in specific areas for several years. These services are described by AT&T as Digital Data Service (DDS) for the private-line version and Switched DDS for the switched AT&T offering. The Regional Bell Operating Companies (RBOCs) offer services equivalent to DDS (although often with a different name). The market success of the DDS services has been sufficient to indicate that more widely available and more powerful digital services will have significant markets.

"Integrated Services Digital Network" (ISDN) is the overall term for all the functions in the network that are necessary to digitize "the last mile." ISDN uses two 64 kbit/s channels for bearer information and one 16 kbit/s for control data at the customer premise. This is termed "2B + D" and described as the basic interface. For terminations into a customer premise switch, 1.544 Mbit/s is the data rate. This data rate is termed "23B + D." In this format the B channels are the same 64 kbit/s; however, the D channel is also 64 kbit/s.

ISDN implementation will require digital central office switches and next-generation digital cross connect switches. Many complex issues of how to integrate ISDN capabilities with existing communications services need to be resolved. Beyond these basic technical choices are hundreds of market-driven feature and function decisions that must be made before ISDN transitions from an innovative market (via DDS and switched DDS) to an application selling market (estimated to start in 1990 or 1991). The first major growth of ISDN will be fueled by applications, the traditional mechanism to generate market demand.

When useful ISDN applications emerge, the marketplace will begin to request ISDN provided capabilities. First, this will occur in narrow business segments and ultimately in homes. During the initial ISDN installation period (at least the first five or more years of the application selling segment) modem use will increase in order to allow non ISDN served locations to participate in desirable ISDN services. Ultimately (like the saga of the punched card), modems will be relegated to use on remaining or older analog facilities such as in developing countries, on nonwire networks (radio or cellular as example), and so on.

Fortunately for modem manufacturers, the growth in technology utilization often overcomes the effect of technology obsolescence. Modem use will expand in underdeveloped countries and will increase in nonwire applications; modems themselves will improve performance and gain features to remain competitive; and, finally, the ISDN interface itself is certainly a form of modem, and many modem manufacturers plan to enter this product area. For these reasons the modem manufacturer will remain in some segment of the modem business at least well into the twenty-first century.

2.4.3 Emerging New Application Markets

Currently the rapid growth in the cellular radio market is opening a new application area for modems. Taxi cabs could accept credit cards by using cellular radio-based communications to verify credit transactions. The emergence of Airphone (which is still in the innovation phase and not yet a proven application) is a potential future area for modem use. One of the larger modem markets is for facsimile transmission. The wide acceptance of Group 3 facsimile that use V.29 modems is creating a large base of compatible machines, over 2,000,000 worldwide in 1986, and growing at 20–25% in 1987. In order to be compatible with this huge installed base, V.29 modems will be required in personal computers, although probably in integrated circuit form.

2.5 THE ASYMMETRICAL MODEM. AN EXAMPLE OF EMERGING TECHNOLOGIES, APPLICATIONS, AND STANDARDS

The asymmetrical modem concept arose out of concerns regarding the complexity and expense of developing and producing full-duplex 9600 bit/s modems such as the V.32. Now that V.32 development is further along, it appears that V.32 modems will be possible at reasonable cost in the next few years. While the original reason for the development of an asymmetrical modem has disappeared, many new and more positive reasons have emerged to indicate the marketability of the asymmetrical modem.

Applications Addressed. Communications applications can be divided into single-thread communications and multithread communications, to borrow terms utilized to describe single and multiple processor environments. In single-thread communications applications, only one communications transaction is occurring at a time. Single-thread applications in systems that do not require echoplex (i.e., any that have a data link layer error checking protocol) do not require a full-duplex modem. The dramatic growth of single-user personal computers is expanding the requirements for single-thread communications.

A major new application for personal computers is termed *desktop publishing* by Apple Computer. Communications applications in desktop publishing are called *desktop communications.* Desktop publishing is an expansion of the original word processing applications that fueled significant personal computer sales in the early 1980s. As such, it will be a large-scale application, and the communications requirements for desktop communications include higher-speed modems to support the higher data rates required. Desktop communications will be one of the major applications of the new asymmetrical modem product.

The desktop communications application has a requirement for very high data rates for communications. The average page size (in character equivalents) for desktop publishing applications is 10,000 to 20,000 characters. This is due to the need to send what is in reality graphics information to the 300 pictel per inch laser printer.

In addition to the graphic information, page description languages must also send sufficient information about where characters are located and where graphics are located so that editing procedures can be used on the page. This header information becomes a significant part of the total characters sent per page.

A normal ASCII page of text contains 1920 characters. Users commonly wish to transfer such data at rates of 2400 bit/s. Users of desktop communications applications who desire the same page transfer rate as they achieved with ASCII pages and 2400 bit/s modems need to transfer page information at data rates between $10,000/1920 \times 2400 = 12,500$ bit/s and $20,000/1920 \times 2400 = 25,000$ bit/s. This indicates one of the emerging reasons for the high-speed requirements of the asymmetrical modem. It also suggests that a data compression capability may be a desirable technology to include in a asymmetrical modem product.

Reviewing further the desktop communications applications, it is apparent that this is a graphics-oriented rather than a character-oriented communications requirement. Facsimile is also a graphic oriented communications requirement. Group 3 facsimile uses pictel densities of approximately 200/inch. Laser printers commonly support pictel densities of 300/inch. The ability to transfer desktop published documents to facsimile machines and/or to transfer documents from a facsimile machine to a personal computer with laser printer without any intermediate steps that decrease document quality is a desirable feature. Also very desirable is the ability, with a personal computer graphics or word processing program, to modify facsimile documents with no document quality loss. By focusing on the existing facsimile market, it is possible to identify and quantify the potential size of these future applications with market research.

Technologies Required. The asymmetrical modem consists of a high-speed main channel and a lower-speed side channel. The term "asymmetrical" is used to distinguish it from full-duplex modems as "full-duplex" is defined as the same data rate in both directions simultaneously. Current modem technology allows the development of a 9600 bit/s modem with a 300 bit/s side channel for costs close to the costs of a V.22 bis modem, yet the asymmetrical modem offers four times the data rate on the main channel.

It is anticipated that a V.33-like modulation could be used for the main channel as an option. This would offer the user a main channel data rate of 14,400 bit/s when the telephone line quality allowed. This offers a 50% performance improvement versus V.32 in applications that do not require full-duplex. The side channel performance possible with a 14,400 bit/s main channel needs to be reviewed.

The implementation of an asymmetrical modem is also predicated on the existence of a number of other new functions in the modem. A data-link layer error control system is required to ensure that telephone channel induced errors are detected and corrected. The performance of this protocol on an asymmetrical physical channel must be reviewed.

Almost all DTEs interface to the DCE via the V.24 or EIA 232 interface using an integrated circuit termed a *Universal Asynchronous Receiver–Transmitter* (UART). Few existing UART implementations support differing data rates on the receive and transmit lines. This forces the asymmetrical modem to include an internal buffer to transfer data to the DTE at high speed in both directions. Since buffering is also required for error control, this does not appear to be an onerous requirement.

The asymmetrical modem is the first modem to require functions that previously were not considered an appropriate part of the DCE. When the asymmetrical modem is sold, it will contain functions above the physical layer of the OSI model. The error control is a data link layer function. Buffering and flow control are transport layer functions. While the existence of such functions in a DCE is causing considerable review at the standards making level, such functions are very desirable for the modem manufacturers. Simplistically, modem manufacturers make more money the more layers of OSI functions they offer.

Physical layer modems currently produce very low profits for modem manufacturers. The addition of a simple error control protocol that is economical to include roughly doubles the price of the modem. Currently a few modem manufacturers are offering buffering in modems for the purpose of electronic mail applications. Again, the buffered modem sells for roughly twice its unbuffered counterpart. The modem manufacturers are very interested in increasing their profits by including higher layer functions of the OSI model, and the asymmetrical modem looks like a modem that will offer this capability.

Finally, there is the issue of what other modulations should be supported. Use of the 212/V.22 and V.22 bis for compatibility with earlier dial-up systems would appear to be vital. It is possible that V.29/V.21 with T30 (T30 is the logical protocol that Group 3 facsimile machines use to establish a data link) to offer compatibility with existing facsimile applications would also be desirable. Since the market and users of facsimile are well known, the modem manufacturer can take advantage of available market research to determine whether the additional modulation and related functions would be cost-effective to include in an asymmetrical modem design.

Standards Issues. Currently, the EIA TR30.1 and CCITT SGXVII WP1 are attempting to develop a standard for the modulation means of the asymmetrical modem. However, the individual modem manufacturers, particularly in the United States, all want to create a de facto standard for their own products, and little progress on a unified standard is occurring.

This rather self-serving approach will likely backfire. When no standard is available, the largest manufacturer has the opportunity to create a de facto one. In this case, Hayes MicroComputer Products is announcing their own version of the asymmetrical modem. Hayes, more than any other company, has the market presence to create a de facto asymmetrical modem standard.

2.6 THE PHYSICAL FORM OF A MODEM

The market for modems is, of course, significantly affected by their physical form.

The first modem market to develop was for standalone modems that included some type of enclosure and power supply. The next market to develop was for modems that consisted of only a printed circuit card assembly that could be inserted in another manufacturers equipment. This market for custom-designed modems is termed *Original Equipment Manufacturer* (OEM) to describe the intermediary company that purchases the card modem from the modem manufacturer.

Eventually a form of modem termed "rack mount" emerged. The independent modem manufacturers were the first to support this market requirement. This configuration was designed to be installed in central-site computer or communications room racks (often termed *relay racks*). This form of product offered the most economical (both in terms of cost and space) means of installing multiple modems at a central-site timesharing computer system.

With the advent of the personal computer market, standard mechanical and electrical interfaces for printed-circuit cards developed. These include the Apple II card, the Macintosh II card, the IBM PC and PS/2 cards, the Multibus formats, and the VME card formats. The standardization of these formats allowed modem manufacturers to sell modems in card form as a standard product in addition to the standalone form.

At this point in the modem market, the semiconductor technology has developed sufficiently to allow the semiconductor manufacturers to create integrated circuits that included all the functions that could traditionally be described as a modem. The only major exception is the telephone line interface circuitry, which remains an elusive semiconductor problem because of high surge voltage and isolation requirements. Current semiconductor technology allows modems as complex as the V.22 bis to be contained in a single semiconductor.

The advent of semiconductor modems creates new markets for modems. Vending machines will call up when they need more goods to vend, or require service, or the change box if full, or to authorize credit card transactions. Large mechanical systems such as commercial air conditioners, or moving sidewalks, or elevators will call when they require service. When the modem is little more than the addition of a chip to a circuit board, such new applications can occur easily.

This emerging application area is termed *integrated modems.* In integrated modem applications the modem is considered an intrinsic part of the product it is in. This affects the design of the interface to the modem chip and the kind of product support that is necessary to sell the semiconductor modem chip. But integrated modem applications promise the kind of volume that the modem semiconductor manufacturers need. In effect, the integrated modem applications replace the older OEM modem application by making the cost of including a modem so low that it is no longer a separate card option — it is embedded into the OEM product.

In the case of standalone modems, the use of modem semiconductors is reducing the cost sufficiently that the traditional modem manufacturers are including many functions that never before existed in the modem. Reviewing some of the applications requirements noted for the asymmetrical modem — error control, multiple modulations, buffers, and compression — gives an indication of the features that are being installed in what once was a simple modem.

Because of these higher OSI layer functions, the traditional modem manufacturer continues to sell "modems" at the same prices as before. The "modems" now do much more than a single form of modulation and demodulation.

The modem manufacturer will continue to produce modems for the emerging applications. Additionally, higher and higher layer communications functions will be included in the "modem." These trends will, of course, be affected by the new communications requirements of the marketplace as they emerge. Successful modem manufacturers will watch the market requirements develop, support the de jure standards activities, and thereby create products that serve emerging market needs.

Chapter 3

Baseband Transmission

The first three sections of this chapter describe, without evaluation or comment, the basic problem of data transfer† and the various baseband solutions of it that have been used. Subsequent chapters show that most passband (modulated) signals have a baseband equivalent, so the results of this chapter will be applicable to nearly all systems.

Section 3.4 discusses the effects of noise on all the baseband systems described, and Section 3.5 begins a comparative evaluation of the systems. This evaluation cannot be complete until we have considered receiver techniques (particularly equalizers), but some initial comparison is possible and indeed necessary.

Operations on Logic and/or Analog Signals. The input serial data stream to the modem that has then been passed through any combination of Asynchronous-to-Synchronous Converter (ASC), encoder and scrambler, is assumed to be in the form of pulses with very fast rise and fall times. The voltage level of these pulses is not important, and they can be considered just as logic zeros and ones. Many of the earlier operations in a transmitter (and, correspondingly, the later operations in a receiver) are most easily considered as being performed on these logic variables, but at some point a Digital-to-Analog (D/A) conversion to physical voltages that are balanced positive and negative is necessary. Many modems use Complementary Metal Oxide Silicon (CMOS) with balanced supplies for analog signal processing; for these, the conversion convention is: Logic

---

† The term "transfer" is used in this book to mean the combined operations of transmit and receive.

$1 \Leftrightarrow +1$ and Logic $0 \Leftrightarrow -1$. This means that the exclusive-OR operation (modulo-2 addition with 0/1 levels) is preserved in the level conversion,
but it must be noted that it is equivalent to *negative multiplication* with ± 1
signals.

Binary Signaling.   Let us assume initially that we have enough bandwidth available in the channel (we will soon see how much that is) to transfer the data bits
singly; that is, with the symbol rate in the channel f_s equal to the input bit rate,
f_b.

The most straightforward way is to convert the data into the signal

$$d(t) = \sum_{k=-\infty}^{\infty} a_k p(t - kT) \tag{3.1}$$

where $\qquad\qquad a_k = \pm 1$

and

$$p(t) = \begin{cases} 1 & \text{for } 0 < t < T \\ 0 & \text{otherwise} \end{cases}$$

The rectangular pulses $p(t - kT)$ typically occupy a bandwidth that is very
much greater than the bandwidth of the channel through which they must be
transmitted, so they will be subjected to some band-limiting, or filtering, in
either the modem or the channel or both. In the receiver the filtered signal is
sampled at the bit rate and decisions about what was transmitted are made bit
by bit (i.e., without considering any past or future received signals or decisions)
as to what was transmitted. If $g(t)$ is the fully filtered signal that results from a
single input bit (i.e., the end-to-end impulse response), then the signal that is
delivered to the decision device is

$$y(t) = \sum_{k=-\infty}^{\infty} a_k g(t - kT) \tag{3.2}$$

3.1   PULSE SHAPING: THE REQUIREMENTS FOR NO INTERSYMBOL INTERFERENCE

3.1.1   End-to-End Shaping

To obtain maximum immunity to noise, each received sample should be dependent on only one transmitted bit; or, stating it another way, there should be
no interference between bits. However, the basic problem for a modem can be
summarized as

$$\begin{aligned} \text{Finite duration} &\Leftrightarrow \text{infinite bandwidth} \\ \text{Finite bandwidth} &\Leftrightarrow \text{infinite duration} \end{aligned} \tag{3.3}$$

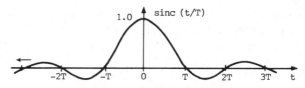

Figure 3.1 Basic band-limited pulse with no ISI.

Therefore, if $g(t)$ is to be band-limited, we cannot constrain it to only one bit period; the best we can do is to arrange that if it is sampled at the bit rate, only one sample is nonzero. That is, for integer values of k and some sampling delay, τ,

$$g(kT + \tau) = \begin{cases} 1 & \text{for } k = 0 \\ 0 & \text{for } k \neq 0 \end{cases} \tag{3.4}$$

A basic pulse that satisfies this requirement with $\tau = 0$ is shown in Fig. 3.1. The transform pair is defined by

$$G_0(f) = T \quad \text{for } -\frac{f_s}{2} < f < \frac{f_s}{2} \tag{3.5a}$$

and

$$g_0(t) = \frac{\sin(\pi t/T)}{\pi t/T} = \text{sinc}\left(\frac{t}{T}\right) \tag{3.5b}$$

This pulse occupies the absolute minimum bandwidth (the subscript 0 indicates zero excess bandwidth), but it requires a "perfect" filter or filter/channel combination with infinitely sharp cut-off and linear phase; both are impossible to achieve. If we are allowed some amount of excess bandwidth, defined as $\alpha f_s/2$,† we can develop a family of pulse shapes and spectra as follows.

The pulse $g_0(t)$, defined by (3.5a), can be multipled by any function $gm(t)$ that has $gm(0) = 1$; the product will still satisfy (3.4), the condition for no ISI. The frequency spectrum of the product is the convolution of $G_0(f)$ and $GM(f)$, the transform of $gm(t)$; that is,

$$G_\alpha(f) = \int_{-\infty}^{\infty} G_0(u)GM(f-u)\,du$$
$$= T\int_{-f_s/2}^{f_s/2} GM(f-u)\,du \tag{3.6}$$

If $gm(t)$ is band-limited to $|f| < \alpha f_s/2$, then $G(f)$ will be band-limited to $|f| < (1 + \alpha)f_s/2$, and it can be shown that

† The value of α is almost always less than unity; that is, there is less than 100% of excess bandwidth.

$$G_\alpha(f) = T \int_{-f_s/2}^{f_s/2} GM(u) \, du$$

$$= T \qquad\qquad \text{for } |f| < (1 - \alpha)\frac{f_s}{2} \tag{3.7a}$$

$$\text{and} \qquad = T \int_{f-f_s/2}^{f_s/2} GM(u) \, du \qquad \text{for } (1 - \alpha)\frac{f_s}{2} < |f| < (1 + \alpha)\frac{f_s}{2} \tag{3.7b}$$

This convolution process is illustrated in Figs. 3.2a and 3.2b.

If the product pulse $g_\alpha(t)$, is sampled at $t = nT + \tau$, the spectrum of the sampled signal is the superposition of $G_\alpha(f)$ and all versions of it shifted by integral multiples of f_s; that is,

$$G_\alpha(f,\tau) = \sum_{k=-\infty}^{\infty} G_\alpha(f + kf_s) \exp[j2\pi(f + kf_s)\tau] \tag{3.8}$$

and in particular,

$$G_\alpha(f,0) = \sum_{k=-\infty}^{\infty} G_\alpha(f + kf_s) \tag{3.9}$$

Since $G_\alpha(f)$ is band-limited to less than f_s, and $G_\alpha(f,\tau)$ is repetitive about multiples of f_s, the only parts of (3.9) that need concern us are

$$G_\alpha(f,0) = \begin{cases} G_\alpha(f) + G_\alpha(f - f_s) & \text{for } 0 < f < \dfrac{f_s}{2} \qquad (3.10a) \\[2ex] G_\alpha(f + f_s) + G_\alpha(f) & \text{for } -\dfrac{f_s}{2} < f < 0 \qquad (3.10b) \end{cases}$$

Substitution of $G_\alpha(f)$ from (3.7) into (3.10a) gives

$$G_\alpha(f,0) = \int_{f-f_s/2}^{f_s/2} GM(u) \, du + \int_{-f_s/2}^{f-f_s/2} GM(u) \, du$$

$$= \int_{-f_s/2}^{f_s/2} GM(u) \, du$$

$$= T \quad \text{for } |f| < \frac{f_s}{2}$$

That is, the spectrum of the sampled signal (obtained by augmenting the original spectrum by one shifted by either $\pm f_s$ as appropriate) is constant across the Nyquist band:

$$G_\alpha(f) + G_\alpha(f - f_s) = T \qquad \text{for } 0 < f < \frac{f_s}{2} \tag{3.11a}$$

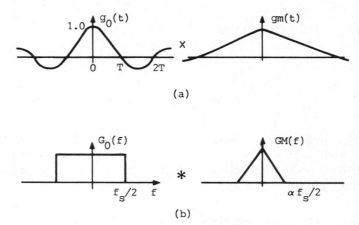

Figure 3.2 Development of general pulse with no ISI: *(a)* multiplication of basic pulse by any pulse band-limited to $<f_s/2$; *(b)* convolution of two spectra.

and
$$G_\alpha(f+f_s) + G_\alpha(f) = T \quad \text{for } -\frac{f_s}{2} < f < 0 \quad (3.11b)$$

This is a necessary condition for $g_\alpha(t)$ to have no ISI; it can be shown [Ny and LS&W] that it is also sufficient. A function that satisfies (3.11) is sometimes denoted by $G_{\text{Nyq}}(f)$.

If its phase $\theta(f)$ were zero, $G_\alpha(f)$ would be real, and (3.11) could be rewritten using just the amplitude, $|G_\alpha(f)|$. However, this is not possible in practice because $g_\alpha(t)$ would then be symmetrical about $t=0$ and would extend to $t=-\infty$; that is, it would be noncausal. In practice, the phase is made as linear as possible ($\simeq 2\pi f\tau$), and $g_\alpha(t)$ is zero for $t<0$ and is approximately symmetrical about $t=\tau$. We will use the term "quasi-real" to describe such a spectrum, $G_\alpha(f)$.

Sampling, Superposition, and Foldover. In (3.8) we showed the spectrum of a sampled signal; since the operation of sampling is fundamental to data transmission, we should briefly review its theory.

If any signal $s(t)$ with transform $S(f)$ is sampled at $t=nT+\tau$, the spectrum of the samples is

$$S(f,\tau) = \sum_{k=-\infty}^{\infty} S(f+kf_s) \exp[j2\pi(f+kf_s)\tau] \quad (3.12a)$$

This spectrum repeats every f_s Hz, so it can be defined by its values over the principal interval $-f_s/2$ to $f_s/2$. Three increasingly specialized cases can now be described.

1. If, like most of the signals considered in this book, $s(t)$ is bandlimited to $|f| < f_s$, then only adjacent segments overlap, and the sampled spectrum (although still infinite) can be defined by

$$S(f,\tau) = S(f - f_s)\exp[j2\pi(f - f_s)\tau] + S(f)\exp[j2\pi f\tau]$$
$$+ S(f + f_s)\exp[j2\pi(f + f_s)\tau] \qquad \text{for } |f| < \frac{f_s}{2} \tag{3.12b}$$

This is the case for most bandpass channels; we will discuss it in more detail in Section 7.5.

2. If, further, the amplitude $|S(f)|$ and phase $\phi(f)$ of $S(f)$ have even and odd symmetry, respectively, about zero frequency, then for f positive (3.12b) can be written as

$$S(f,\tau) = S(f)\exp[j2\pi f\tau] + S^*(f_s - f)\exp[-j2\pi(f_s - f)\tau]$$
$$\text{for } 0 < f < \frac{f_s}{2} \tag{3.12c}$$

where the * represents the complex conjugate. The argument in the second term indicates the folding operation, and the original spectrum is augmented by the *complex conjugate* of the folded spectrum. This is the case for some passband and for all baseband channels.

3. Finally, in the idealized case where $S(f)$ is real and sampling is at $t = kT$, (3.12c) further simplifies to

$$S(f,0) = S(f) + S(f_s - f) \tag{3.12d}$$

which is the way folding is often expressed; this is illustrated in Fig. 3.3.

3.1.2 Special Cases of Shaping for No ISI

We have seen that $G_\alpha(f)$ must be equal to T for $|f| < (1 - \alpha)f_s/2$ and decrease to zero at $f = (1 + \alpha)f_s/2$. To make the filtering as easy as possible, there should

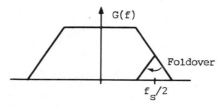

Figure 3.3 Foldover of a symmetrical spectrum by sampling.

be no discontinuities in either $G_\alpha(f)$ or its derivative. One very commonly used shape is called a *raised-cosine;* the transform pair is

$$G_\alpha(f) = \begin{cases} T & \text{for } |f| < (1-\alpha)\dfrac{f_s}{2} \\[3mm] T\dfrac{1 - \sin[\pi(f-f_s/2)/\alpha f_s]}{2} & \text{for } (1-\alpha)\dfrac{f_s}{2} < |f| < (1+\alpha)\dfrac{f_s}{2} \end{cases} \qquad (3.13)$$

so that

$$g_\alpha(t) = \text{sinc}\left(\frac{t}{T}\right) \frac{\cos(\alpha\pi t/T)}{1 - (2\alpha t/T)^2} \qquad (3.14)$$

$G_\alpha(f)$ and $g_\alpha(t)$ are shown in Figs. 3.4a and 3.4b for several values of α. As required, all the $g_\alpha(t)$ are unity at $t=0$ and zero at $t=kT$.

The function $gm(t)$ that multiplies the $\text{sinc}(t/T)$ function can be rewritten as

$$gm(t) = \frac{\cos(\alpha\pi t/T)}{1 - (2\alpha t/T)^2} = \frac{\pi}{4}\left[\text{sinc}\left(\frac{\alpha t}{T} + \frac{1}{2}\right) + \text{sinc}\left(\frac{\alpha t}{T} - \frac{1}{2}\right)\right]$$

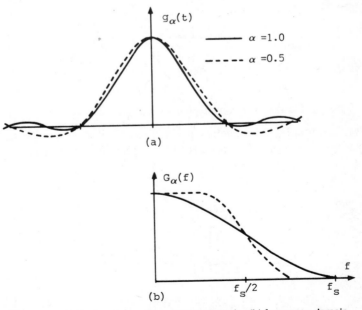

Figure 3.4 Raised-cosine pulses: *(a)* time domain; *(b)* frequency domain.

That is, it is a scaled sum of two offset sinc pulses with period T/α. Its transform has the simple form

$$GM(f) = \cos\left(\frac{\pi f}{\alpha f_s}\right)$$

This function will reappear when we consider duobinary signals in Section 3.3.1.

In an effort to get data through a channel at as high a rate as possible, modern systems use low values of α. This leads to very oscillatory pulses and makes accurate selection of the sampling time very important.

The case of $\alpha = 1$ is of some practical importance, and it will also be helpful later in understanding some other shaping methods. It can be seen that $g_1(t)$ has "extra" zeros at $t = (k - \frac{1}{2})T$ for $k \neq 0$ or 1. If this signal is sampled on the "off-beat," it has two nonzero samples of $2/\pi$ at $t = \pm T/2$.

3.1.3 Generalized Condition for No ISI — Allowing Phase Distortion

Since nearly all band-limiting filters that will be used in transmitters and receivers introduce phase distortion, it is reasonable to ask whether there is some less severe constraint than linear phase that will still result in no ISI. For sampling at $t = kT + \tau$, the condition for no ISI can be generalized to

$$|G(f)| \exp j[\theta(f) + 2\pi f\tau] + |G(f - f_s)| \exp j[\theta(f - f_s) + 2\pi(f - f_s)\tau] = T$$

$$0 < f < \frac{f_s}{2}$$

and a similar equation for negative f.

After splitting this into real and imaginary parts, a little algebraic manipulation gives $\theta(f)$ as a function of the amplitudes:

$$1 - \cos[\theta(f) + 2\pi f\tau] = \frac{|G(f - f_s)|^2 - [T - |G(f)|]^2}{2T|G(f)|}$$

If the phase is to be other than linear, then $1 - \cos[\theta(f) + 2\pi f\tau]$ must be > 0, and since $|G(f)|$ and $|G(f - f_s)|$ are both positive, it follows that

$$|G(f)| + |G(f - f_s)| > T \tag{3.15}$$

This is a negative result but a useful one because if, as recommended in Section 3.1.4, the shaping is split equally between transmitter and receiver, then the amplitudes of the transmitted components at f and $f - f_s$ are $|G(f)|^{1/2}$ and $|G(f - f_s)|^{1/2}$, and (3.15) means that the component of the transmitted power is greater than unity. That is, the use of any phase response other than linear increases the amount of power that must be transmitted for a given magnitude of pulse input to the detector and thereby decreases the eventual SNR of that signal.

3.1.4 Splitting of Filtering Between Transmitter and Receiver

We have now used a requirement in the time domain to derive one in the frequency domain for the pulse that is input to the detector. However, this requirement is for the combination of transmit filter, channel, and receive filter; it does not tell us enough about the individual filters. We will assume initially that the channel is perfect — that is, it has constant amplitude and linear phase across its limited band — and define the requirements for the filters so as to minimize the effect of noise on the sampled signal. If the output power of the transmitter is fixed, and the noise is added somewhere in the channel, then it is intuitively reasonable — and is proved in [LS&W] — that the filtering should be split equally between the transmitter and the receiver. That is,

$$|G_T(f)| = |G_R(f)| = [G_{Nyq}(f)]^{1/2} \tag{3.16}$$

and, since $G_{Nyq}(f)$ is to be real (or, more strictly speaking, quasi-real),

$$\text{Arg}[G_T(f)] = -\text{Arg}[G_R(f)]$$

There is sometimes a small advantage to be gained from allowing the transmit signal to have some delay distortion and then correcting this in the receiver†; however, we will ignore this possibility for the moment and say that $G_T(f)$ and $G_R(f)$, the ideal transmit and receive transfer functions, should each be quasi-real. That is,

$$G_T(f) = G_R(f) = [G_{Nyq}(f)]^{1/2} \tag{3.17}$$

Strictly speaking, $G_T(f)$ given in (3.17) is the spectrum of the transmit signal, and this is produced by the combination of the transfer function of the transmit filter and the shaping of the pulses input to it. In analog transmitters these are usually just rectangular pulses of period T, so the spectrum of the transmit filter (or combination of filters) should therefore be (3.17) divided by the Fourier transform of the rectangular pulse; that is,

$$F_T(f) = \frac{G_T(f)}{\text{sinc}(fT)}$$

In a digital or stored reference transmitter the input pulse will typically consist of a sequence of several rectangular pulses of duration T/N and varying

---

† This may reduce the total complexity of the filters and also help to reduce the effect of impulse noise that is added in the channel.

height. Again, the transmit "filter" must be considered to be the combination of pulse shaper and filter.

3.2 MULTILEVEL SIGNALING

If we wish to increase the bit rate but keep the same symbol rate (and bandwidth), we can group the incoming data in blocks of M bits and, assuming no redundancy, define a more general weighting factor for $g(t)$ in (3.2) that can have any one of $L\ (=2^M)$ equally probable values or levels; for greatest average immunity to noise, these levels should be equally spaced.

It is convenient to retain the form of (3.2) and use the symbol a for the weighting factor for any number of levels, so we define

$$a = \pm 1, \pm 3 + \cdots \pm (L-1); \qquad f_s = \frac{f_b}{M}$$

The average power in this signal is

$$P_s = f_s \frac{2}{L} \sum_{k=1}^{L/2} (2k-1)^2 = f_s \frac{L^2-1}{3} \tag{3.18}$$

It will often be more convenient to avoid having to consider the symbol rate, and to refer to the *energy* in a single pulse, $EN = (L^2 - 1)/3$.

Detection in the receiver is performed by comparing the received signal to a set of decision thresholds that are halfway between adjacent levels. With the addition of noise, the most likely error is that a signal will cross just one (the closest) threshold. Therefore, in order to minimize the error rate, the assignment of levels to each combination of bits is always done according to a Gray code, which ensures that the crossing of one threshold changes only one bit. The code is shown in Table 3.1 for the most common cases of $L = 4$, 8, and 16. As mentioned previously, D/A conversion to balanced positive and negative signals is usually performed after all logic operations (such as duobinary encoding) are completed.

3.3 CONTROLLED ISI: DUOBINARY, OR PARTIAL RESPONSE

As we saw, signaling without ISI at the maximum (Nyquist) rate of f_s symbols/second through a channel that is strictly band-limited to $\pm f_s/2$ requires perfect filtering (i.e., flat passband, infinitely sharp cut-off, and linear phase). The same rate can be achieved more easily if a controlled amount of ISI is created by the filters and then taken account of in the detector. Two equivalent schemes for controlled ISI were invented more or less simultaneously by Lender [Le1] and Kretzmer [Kr1]. The extensive literature was surveyed in [K&P].

TABLE 3.1 Gray Encoding of *M* Bits into *L* (=2^*M*) Levels

Input				Binary Code			
0	0	0	0	1	1	1	1
0	0	0	1	1	1	1	0
0	0	1	1	1	1	0	1
0	0	1	0	1	1	0	0
0	1	1	0	1	0	1	1
0	1	1	1	1	0	1	0
0	1	0	1	1	0	0	1
0	1	0	0	1	0	0	0
1	1	0	0	0	1	1	1
1	1	0	1	0	1	1	0
1	1	1	1	0	1	0	1
1	1	1	0	0	1	0	0
1	0	1	0	0	0	1	1
1	0	1	1	0	0	1	0
1	0	0	1	0	0	0	1
1	0	0	0	0	0	0	0

The nested boxes in the table indicate: *L* = 16 (outermost), *L* = 8, and *L* = 4 (innermost).

The introduction of ISI means that the signal delivered to the detector has more levels; therefore, for the same average energy in the signal, the distance between levels and, consequently, the immunity to noise is reduced. We will analyze this effect for all the different methods together in Section 3.4 and consider methods of mitigating it in Section 9.3.

3.3.1 Duobinary, or Partial-Response Class I

The simplest form of controlled ISI uses an impulse response of . . . 0, 0, 1, 1, 0, 0, ... This *could* be generated by the arrangement shown in Fig. 3.5; the first-order sampled-data filter would have a transfer function of $(1 + z^{-1})$. When this is combined with the perfect Nyquist filter, the end-to-end spectrum is

$$|G_{DB}(f)| = \begin{cases} 2T \cos\left(\dfrac{\pi f}{f_s}\right) & \text{for } |f| < \dfrac{f_s}{2} \\[2ex] 0 & \text{for } |f| > \dfrac{f_s}{2} \end{cases} \tag{3.19}$$

and

$$\theta(f) = \frac{2\pi f T}{2}$$

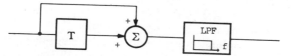

Figure 3.5 Straightforward—but impractical—way of generating a duobinary pulse.

The end-to-end impulse response is

$$g_{DB}(t) = \mathrm{sinc}\left(\frac{t}{T}\right) + \mathrm{sinc}\left(\frac{t}{T} - 1\right)$$

which is unity for $t = 0$ and T and is zero for $t = kT$, $k \neq 0$ or 1.

However, this would have the disadvantages that (1) all the shaping would be in the transmitter and (2) the filter would have to be a perfect "brickwall"—thereby defeating the purpose of the controlled ISI. In practice, the system would be realized with no delay or summer in the transmitter, and the transfer functions of the transmit and receive filters would be

$$F_T(f) = F_R(f) = \left[2T \cos\left(\frac{\pi f}{f_s}\right)\right]^{1/2} \tag{3.20}$$

Since each source impulse generates an impulse doublet at the detector, L equally spaced input levels result in $2L - 1$ levels. These levels are, however, no longer equally probable, and we shall see later that this imbalance can be useful in countering the harmful effects of increasing the number of levels. The levels and their probabilities (in parentheses) are shown in Table 3.2 for the cases of

TABLE 3.2   Levels and Probabilities for Binary and Duobinary

Levels	L = 2		L = 4	
	Binary	Duobinary	Binary	Duobinary
6				
5				(1/16)
4				
3			(1/4)	(2/16)
2		(1/4)		
1	(1/2)			(3/16)
0		(2/4)	(1/4)	
−1	(1/2)			(4/16)
−2		(1/4)	(1/4)	
−3				(3/16)
−4			(1/4)	
−5				(2/16)
−6				(1/16)

$L = 2$ and 4. Since these are actual physical voltage levels, they are shown as both positive and negative.

The energy in a duobinary pulse could be calculated by summing the energies of the different levels weighted by their probabilities:

$$EN_{DB} = \sum_{k=1-L}^{L-1} 4k^2 \frac{L - |k|}{L^2} = 2 \sum_{k=1}^{L-1} 4k^2 \frac{L - k}{L^2}$$

but it is much simpler to realize that the power in the doublet 1 1 is twice that in the single impulse; therefore

$$EN_{DB} = \frac{2(L^2 - 1)}{3} \tag{3.21}$$

Error Propagation in the Detector and Its Prevention. To illustrate the way in which errors can propagate, let us consider initially the simplest, three-level signal. If a zero level is received, the tail of the impulse response resulting from the previously transmitted bit must be subtracted from the present zero in order to determine the polarity of the present bit; if the previous decision was wrong, then the present one will be also. The decision errors will continue, even without noise, until an outer level (± 2) is received; then, since outer levels can result only from two successive bits of the same sign, the polarity of the last received bit is certain, and error propagation stops. For multilevel signals, the propagation will last longer because the probability of an outer level is lower.

This propagation can be prevented by encoding the original blocks of binarily weighted data thus:

$$b_k = a_k - b_{k-1}, \quad \text{mod } L \tag{3.22}$$

and then using the b sequence as input to the D/A and shaping filters. For a two-level signal, the modulo 2 subtraction can be implemented in hardware by an exclusive-OR gate.

The pulse received at time k is then $(b_k + b_{k-1})$—or, more precisely, the balanced equivalent of same—and the original data can be recovered symbol by symbol without error propagation by decoding the $(2L - 1)$ levels modulo L. There is, of course an elegant proof of this, but it need not concern us here.

3.3.2 Modified Duobinary, or Partial-Response Class IV

A duobinary, or partial-response Class I, signal has a spectrum that extends down to zero frequency just like any uncoded "full-response" signal. If a baseband signal is to be single-sideband-modulated (see Section 4.4) or if the "baseband" channel does not allow d.c. transmission (see Section 11.3) then a shaping that allows transmission at the full Nyquist rate and has a zero at d.c. is desirable.

TABLE 3.3   Bit and Signal Sequences for AMI and $(1 - z^{-1})$ Shaping

a_k	1	0	0	0	1	1	0	1	0	1	1	1
AMI[a]	+1	0	0	0	−1	+1	0	−1	0	+1	−1	+1
b_k[a]	1	1	1	1	0	1	1	0	0	1	0	1
$b_k - b_{k-1}$		0	0	0	−1	+1	0	−1	0	+1	−1	+1

[a] Note that the initial values are arbitrary.

Bipolar or Alternate Mark Inversion (AMI).   A very simple way to eliminate the d.c. component in a two-level signal is to combine the encoder and the D/A Converter (DAC) so that input zeros are transmitted as a zero level and ones as alternating $+1$ and -1. This method was originally called bipolar in the United States and AMI in Europe; AMI is now the preferred term — presumably because of the many other possible meanings of bipolar.

Although AMI predated duobinary, it can be understood most readily in the way we have used for duobinary; that is, as a digital precoder followed by a DAC and a shaping network. Table 3.3 shows a random data sequence a_k and its AMI conversion; it also shows a second sequence b_k related to the a_k by

$$b_k = a_k + b_{k-1}, \quad \text{mod } 2 \tag{3.23}$$

If this sequence is passed through a filter with the transfer function $(1 - z^{-1})$, the sequence of $(b_k - b_{k-1})$ is generated; it can be seen that this is the same as the AMI sequence. AMI can therefore be regarded as a type of partial response† with a different shaping and precoding.

If, however, this shaping is used to transmit at the full Nyquist rate through a bandwidth of $\pm f_s/2$, the very impractical spectrum shown by the dashed line in Fig. 3.6 would be produced. In practice, this spectrum is always relaxed by allowing some amount of excess bandwidth (often up to 100%), but that rather defeats the purpse of the encoding and shaping.

It was realized in the early 1960s that greater bandwidth efficiency could be achieved by splitting the input bit sequence into two sequences, and applying the inverting rules to each separately; this was called *Interleaved Bipolar*. Lender [Lei] showed how to avoid error propagation in the receiver by precoding and shaping the single sequence according to

$$b_k = a_k + b_{k-2}, \quad \text{mod } 2$$
$$x_k = b_k - b_{k-2} \tag{3.24}$$

Kobayashi [Ko3] showed that interleaved bipolar (or interleaved AMI) and modified duobinary (MDB) are equivalent, and this can be seen from the

---

† The term "partial response" can be used quite generally; Kretzmer [Kr2] defined four classes, and Kabal and Pasupathy [K&P] extended this to nine classes.

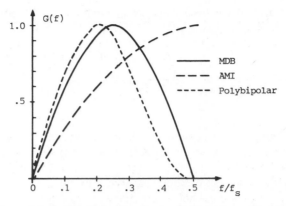

Figure 3.6 Amplitude spectra of pulses with no d.c. component.

example shown in Table 3.4. However, MDB is more specific because it refers only to the most useful case of no excess bandwidth, whereas IAMI is concerned only with the coding, and its pulse shaping must be specified separately. The spectrum of MDB is also shown in Fig. 3.6.

Modified duobinary can also be used with multilevel signals; then the precoding addition of (3.24) would be performed modulo L.

3.3.3 Generalized Duobinary: Polybinary

The transfer function $(1 + z^{-1})$ in tandem with the perfect filter in Fig. 3.5 has the effect of "rolling off" the spectrum. Lender [Le2] suggested that this be generalized to a class of polynomial functions $(1 + z^{-1} + \cdots + z^{-N})$. These cause faster roll-off by putting extra zeros in the spectrum; however, the polynomials with an odd number of terms do not have a zero at $z = -1$, and the discontinuity of the spectrum at $f = f_s/2$ makes them very impractical [Kr2]; the polynomial with four terms ($N = 3$) is probably the only one that might be useful.

TABLE 3.4 Bit and Signal Sequences for Interleaved AMI and Modified Duobinary

a_k	1	1	0	0	0	1	0	1	0	1	1	1
a_k (k even)	1		0		0		0		0		1	
AMI	+1		0		0		0		0		−1	
a_k (k odd)		1		0		1		1		1		1
AMI		+1		0		−1		+1		−1		+1
IAMI	+1	+1	0	0	0	−1	0	+1	0	−1	−1	+1
b_k	1	1	1	1	1	0	1	1	1	0	0	1
$b_k - b_{k-2}$			0	0	0	−1	0	+1	0	−1	−1	+1

The precoding to prevent error propagation must be generalized to

$$b_k = a_k - b_{k-1} - \cdots - b_{k-N}, \mod L$$

although it is very unlikely that this generalized method would be used for anything other than binary input ($L = 2$).

3.3.4 Generalized Modified Duobinary: Polybipolar

The same principle of introducing extra zeros can also be applied to MDB, but, as with polybinary, three zeros seems to be the limit of usefulness. The polynomial $(1 + z^{-1} - z^{-2} - z^{-3})$ produces the spectrum shown in Fig. 3.6, and requires the precoding

$$b_k = a_k - b_{k-1} + b_{k-2} + b_{k-3}, \mod L$$

Again, it is very unlikely that this would be used for $L > 2$.

Polybipolar has been proposed recently for high-density magnetic recording [T&P] and high-speed data on the subscriber loop [B..2], where it was renamed modified modified duobinary.

3.4 NOISE AND ERROR RATES

Throughout most of this book it will be assumed that the noise μ added to a signal has a Gaussian distribution with variance σ^2. That is,

$$p(\mu) = \frac{1}{\sigma\sqrt{2\pi}} \exp\left(\frac{-\mu^2}{2\sigma^2}\right). \tag{3.25}$$

Although there are other important types of noise (e.g., impulse noise), they are much harder to model and their effects are much harder to analyze. The justification for mostly restricting ourselves to Gaussian noise is that if systems are ranked on the basis of their performance with Gaussian noise, the ranking (and the approximate decibel differentials) are usually valid for other types of noise even though the absolute performances may change.

3.4.1 Noise at the Input of the Detector

If the noise is added at the input of the detector, and if the distance from any level to its nearest threshold is normalized to unity†, it can be seen from (3.25)

---

† Lucky et al. [LS&W] generalized this by using d as the distance, but this would lead to confusion when we use d as the "distance" between sequences in Chapter 9; if there were no ISI, that d would be twice LS&W's.

that the probability of any particular level being detected as one of its nearest neighbors is

$$\mathscr{P}(\mu > 1) = \frac{1}{\sigma\sqrt{2\pi}} \int_1^\infty \exp\left(\frac{-x^2}{2\sigma^2}\right) dx$$

Since each level of an L-level Pulse Amplitude Modulation (PAM) system has a probability of $1/L$, and each, except for the two outside levels, can err in two directions, the total probability of symbol error is

$$\mathscr{P}_E = \frac{2(L-1)}{L} \frac{1}{\sigma\sqrt{2\pi}} \int_1^\infty \exp\left(\frac{-x^2}{2\sigma^2}\right) dx \qquad (3.26)$$

which, using the terminology of [LS&W], is equal to $2(1 - 1/L)\,Q(1/\sigma)$, where

$$Q(v) = \frac{1}{\sqrt{2\pi}} \int_v^\infty \exp\left(\frac{-x^2}{2}\right) dx$$

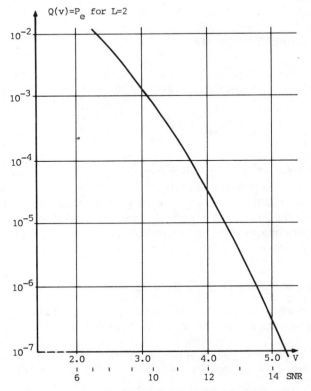

Figure 3.7 $Q(v)$, the probability of error for binary signals.

Since $\mathscr{P}_E$ is the symbol error rate, and since each symbol conveys $\log_2 L$ bits, the bit error rate is

$$\text{BER} = \frac{2(1 - 1/L)}{\log_2 L} Q\left(\frac{1}{\sigma}\right) \tag{3.27}$$

This is plotted in Fig. 3.7 for the basic case of $L = 2$; for other systems [PAM, duobinary, Quadrature Amplitude Modulation (QAM), etc.] the main difference will be in a scaling of the argument of the Q function†; this scaling can be expressed as a loss of SNR relative to the basic case:

$$\Delta \, \text{SNR} = 10 \log\left[\frac{(L^2 - 1)}{3}\right]$$

It should be noted that nothing we have said so far depends on the frequency spectrum of the noise; each decision is made independently of the others, and any correlation between successive noise samples is irrelevant.

3.4.2 Noise at the Input of the Receiver

In practice, of course, the noise is measured at the input of the receiver, and the design of the receive filter is a factor in the error/noise performance. If the noise density at the input of the receiver is defined as N_0 W/Hz, then

$$\sigma^2 = N_0 \int_{-f_s}^{f_s} |F_R(f)|^2 \, df$$

If the system is "full-response" with filters designed according to (3.13) and (3.16), then

$$\sigma^2 = N_0 \int_{-f_s}^{f_s} G_\alpha(f) \, df = N_0 \tag{3.28}$$

The signal power needed at the input to the receiver to produce the normalized sampled levels of $\pm 1, \pm 3, \ldots$, is

$$P_s = \frac{3f_s}{L^2 - 1} \int_{-f_s}^{f_s} |G_T(f)|^2 \, df$$

$$= \frac{3f_s}{L^2 - 1}$$

† The multiplier of $Q(1/\sigma)$ in (3.27) will vary from 1 to about $\frac{1}{2}$ as L varies from 2 to 16 (the maximum feasible), but with a high SNR, such a halving of the error rate could be caused by a change of SNR of just a few tenths of a decibel; the effects of the multipler can be reasonably ignored.

so that the noise variance at the detector is related to the SNR at the receiver input by

$$\sigma^2 = \frac{3}{L^2 - 1} \frac{P_S}{P_N}$$

For a duobinary system

$$\sigma^2 = 2TN_0 \int_{-f_s/2}^{f_s/2} \cos\left(\frac{\pi f}{f_s}\right) df = \frac{4N_0}{\pi} \tag{3.29}$$

and a similar integration for modified duobinary gives the same result. Thus the effective noise has been increased by a factor of $4/\pi$, which is equivalent to a 2.1 dB SNR penalty. This penalty is the result of increasing, by the shaping, the number of levels from L to $(2L - 1)$. Since this is almost a doubling for large values of L, one might have expected the 3 dB that results when the number of levels of a full-response PAM system is doubled, but the smaller penalty is due to the fact that the receive filter has been redesigned to "match" the received signal. This principle of matching will be referred to frequently in later chapters.

It should be noted that this penalty is incurred only if *the detection is done on a symbol by symbol basis*. In Chapter 9 we will consider several methods of "retrieving" the lost 2.1 dB.

3.5 COMPARISON OF THE VARIOUS METHODS OF BASEBAND ENCODING AND SHAPING

Duobinary encoding and shaping were originally proposed as a way of avoiding the sharp cut-offs needed for efficient bandwidth utilization with full-response PAM. However, the apparent SNR penalty was a deterrent to their use, and as the art of filter design improved in the past ten years, full-response systems became predominant in voice-band modems.

Modified duobinary is especially suitable as the baseband prototype for single-sideband (SSB) modulation, because the zero at d.c. in its spectrum greatly facilitates the filtering out of the unwanted sideband. However, SSB systems were mostly developed before QAM systems were fully understood; they are much less used nowadays.

The spectral shaping of both types of duobinary and, especially, the polybipolar extension, offers a very convenient way of matching a data signal to channels and media such as wide-band subscriber loops, and magnetic tapes and disks. We will discuss these possibilities in Chapters 10 and 11.

Chapter 4

Passband Transmission and Modulation Methods

Most channels that are used for data transmission do not allow any transmission at frequencies below some lower band edge or above some upper band edge. Consequently, the baseband signal, which may have been encoded, amplitude-modulated or filtered, or any combination of these, is used to modulate some property of a carrier (or carriers) in order to shift the frequency band of the signal.

Modulation or Keying? If the unfiltered, fast rising and falling pulses that have been scrambled and/or encoded are used to modify some property of a carrier, that property is said to be "keyed" or "shift-keyed," and the resultant signal has a wide (theoretically infinite) bandwidth. If, on the other hand, the data signal is filtered to constrain its bandwidth before modifying the carrier, the modification process is called *modulation*. Sometimes a keyed signal can be postfiltered to achieve the same effect as modulation; if this is done without distortion, the effect prevails over the method in deciding the terminology, and the process is called *modulation*. However, one must be very careful in analyzing such systems; nonlinear distortion can often creep in unobserved!

Any of the properties of a carrier—amplitude, phase or frequency—can be either shift-keyed or modulated: ASK, PSK, and FSK (Amplitude, Phase and Frequency Shift-Keying) and AM, PM, and FM (Amplitude, Phase, and Frequency Modulation) have all been used.

Linear and Nonlinear Modulation. If the modulation is linear (i.e., it obeys the superposition rule), the original baseband spectrum is merely shifted; if it is nonlinear, the spectrum is invariably widened as well as shifted. Linear (i.e., amplitude) modulation is generally to be preferred because (1) the bandwidth required is less, (2) the error rate in the presence of noise is less, and (3) an adaptive equalizer can be used in the receiver to alleviate the effects of distortion.

As the demand for bandwidth conservation has intensified, linear modulation has been used more and more, and the supposedly nonlinear frequency and phase modulation methods have been made only as nonlinear as is needed to achieve a desired effect. The most notable examples of these "quasi-linear" methods are those that strive for a compromise between bandwidth conservation and constancy of the envelope. Amplitude modulation will be discussed in detail in the first two sections of this chapter, and then, in Section 4.3, the quasi-linear methods will be described and analyzed using the linear approach.

4.1 DOUBLE-SIDEBAND (SUPPRESSED CARRIER) MODULATION

If a carrier signal $\cos \omega_c t$, is multiplied by a baseband signal $x(t)$, the product can be represented in the time domain as

$$s(t) = x(t) \cos \omega_c t = \frac{x(t) \exp j\omega_c t + x(t) \exp - j\omega_c t}{2} \tag{4.1}$$

Multiplications in the time domain correspond to convolutions in the frequency domain, so that if $x(t)$ is band-limited to $|f| < f_c$, it follows that

$$S(f) = X(f + f_c) + X(f - f_c)$$

The baseband and passband spectra are as shown in Fig. 4.1.

If $x(t)$ is not band-limited (rectangular pulses are typically used), superposition or aliasing will occur in $S(f)$; this method is called *binary PSK*.

It can be seen that the double-sideband (DSB) modulation process has doubled the bandwidth of the original signal; it is therefore rarely used.

4.2 QUADRATURE MODULATION AND KEYING

The redundancy (and consequent spectral inefficiency) of DSB modulation can be avoided by modulating two orthogonal carriers with two independent half-speed data signals and then adding these modulated carriers to form a composite signal. The two carriers have the same frequency but are 90° apart in phase and can be thought of as defining a two-dimensional signal space. The axes in this space are represented by the carriers $\cos \omega_c t$ and $- \sin \omega_c t$† and the coordi-

† The minus sign is necessary to preserve the convention of the x_q coordinate being positive upward.

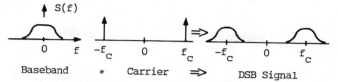

Figure 4.1 Spectrum shifting by DSB modulation.

nates by the in-phase and quadrature components of the baseband signal, $x_p(t)$ and $x_q(t)$. Thus

$$s(t) = x_p(t) \cos \omega_c t - x_q(t) \sin \omega_c t \qquad (4.2)$$

Before we develop the theory of quadrature modulation further, it is worthwhile to establish the credibility of the method by showing that the two original data signals can be recovered in a receiver. If an undistorted $s(t)$ as defined by (4.2) is multiplied by two carriers $\cos \omega_c t$ and $-\sin \omega_c t$, generated in the receiver, the products are

$$
\begin{aligned}
y_p(t) &= x_p(t) \cos \omega_c t - x_q(t) \sin \omega_c t \cos \omega_c t \\
&= \frac{x_p(t)(1 + \cos 2\omega_c t) - x_q(t) \sin 2\omega_c t}{2}
\end{aligned}
\qquad (4.3)
$$

and

$$
\begin{aligned}
y_q(t) &= -x_p(t) \cos \omega_c t \sin \omega_c t + x_q(t) \sin \omega_c t \\
&= \frac{x_p(t) \sin 2\omega_c t + x_q(t)(1 - \cos 2\omega_c t)}{2}
\end{aligned}
\qquad (4.4)
$$

The sidebands of the second harmonics of the carrier ($x_p(t) \cos 2\omega_c t$, etc.) are then removed by low-pass filtering, and the receiver baseband signals y_p and y_q are then equal (within a factor of 2) to the originals.

A prototype of all two-dimensional transmitters is shown in Fig. 4.2; the low-pass filters may be replaced by any combination of pulse-shaping networks and filters; they are omitted altogether in all "keying" systems. The Serial-to-Parallel converter (S/P) is discussed in the sections on input encoding.

Complex Notation. Later discussions of QAM (particularly of adaptive equalizers) will be easier if we define the baseband signal $x(t)$ as the complex signal $x_p(t) + jx_q(t)$, and a carrier as $\exp j\omega_c t$. Then the transmitted signal of (4.2) can also be written as

$$s(t) = \text{Re}[x(t) \exp j\omega_c t] \qquad (4.5)$$

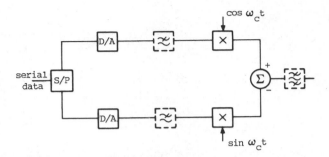

Figure 4.2 Generic QPM transmitter.

and the demodulation process as

$$y_p(t) = \text{Re}[s(t) \exp(-j\omega_c t)]$$

and

$$y_q(t) = \text{Im}[s(t) \exp(-j\omega_c t)]$$

Magnitude and Phase Notation. The signal of (4.2) can also be written in a magnitude-and-phase form:

$$|s| = (x_p^2 + x_q^2)^{1/2} \cos(\omega_c t + \phi) \tag{4.6}$$

where

$$\phi = \tan^{-1}\left(\frac{x_q}{x_p}\right)$$

The length and angle of the baseband vector (x_p, x_q) modulate the amplitude and phase of the carrier, respectively. If the length (magnitude) is kept constant, the modulation is pure PM; systems with 4, 8, and 16 phases have been used. For the higher numbers of phases, the transmit signal is not usually generated by this two-dimensional baseband method — constraining the length of the vector to be constant becomes too complicated — but these systems can still be considered as special cases of quadrature modulation and analyzed accordingly.

A very simple special case is if x_p and x_q can have only the values $\pm 1/\sqrt{2}$; the phase can then be $\pm \pi/4$ or $\pm 3\pi/4$. If the signals are not filtered before modulating, then this is called "QPSK"; if they are filtered, it can be considered as either two-level (four-point) QAM or four-phase modulation (QPM).

Error Rate Resulting from Noise. It can be shown that the error rate of a properly configured QAM system is the same as that of a baseband system that uses the same bandwidth, but it is more convenient to postpone the proof until we consider the general case of multipoint systems in Section 4.2.3.

4.2.1 Baseband Equivalent of a QAM Signal

If appropriate filtering is used to remove all extraneous products, the processes of modulation and demodulation can be considered as simple frequency translation up to a passband (multiplication by $\exp j\omega_c t$) and down again [multiplication by $\exp(-j\omega_c t)$]. If the signal $s(t)$ is not distorted by the channel, and the modulating and demodulating carriers have the same phase, then the end-to-end transfer function of the system is equivalent to that of two independent undistorted baseband systems. If, however, the channel has a transfer function $C(f)$, then demodulation with a carrier with phase angle θ_c [i.e., by $\exp j(-\omega_c t + \theta_c)$] translates $C(f)$ to the baseband as $C(f+f_c)\exp(j\theta_c)$ as shown in Fig. 4.3a. Then the impulse response of the end-to-end system is

$$
\begin{aligned}
h(t) &= \int_{-\infty}^{\infty} G(f)\, C(f+f_c)\, \exp j(2\pi ft + \theta_c)\, df \\
&= \int_{-\infty}^{\infty} G(f)\, |C(f+f_c)|\, \exp j(2\pi ft + \phi(f+f_c) + \theta_c)\, df
\end{aligned}
\tag{4.7}
$$

where $\phi(f) = \text{Arg } C(f)$. Therefore

$$
\begin{aligned}
h(t) &= \int_0^{f_s} [G(f)\,|C(f_c+f)|\, \exp j(2\pi ft + \phi_+) \\
&\qquad\qquad + G(-f)\,|C(f_c-f)|\exp j(-2\pi ft + \phi_-)]\, df \\
&= \int_0^{f_s} \{G(f)\,[|C(f_c+f)|\cos(2\pi ft + \phi_+) + |C(f_c-f)|\cos(2\pi ft - \phi_-)] \\
&\qquad + jG(f)\,[|C(f_c-f)|\sin(2\pi ft + \phi_+) - |C(f_c-f)|\sin(2\pi ft - \phi_-)]\}\, df
\end{aligned}
\tag{4.8}
$$

where
$$
\phi_+ = \phi(f_c+f) + \theta_c
$$
and
$$
\phi_- = \phi(f_c-f) + \theta_c
$$

and it is assumed that the ideal base-band end-to-end response $G(f)$ is real and is band-limited to $|f| < f_s$.

It can be seen from (4.8) that in general $h(t)$ is complex and can be defined as $h_p(t) + jh_q(t)$. This means that a signal that is input to one channel in the transmitter appears (distorted of course) at the output of both channels. This relationship for a *linear* channel can be summarized as follows:

From x_p to y_p the transfer function is $\quad h_p$
.... x_p ... y_q .... h_q
.... x_q ... y_p .... $-h_q$
.... x_q ... y_q .... h_p

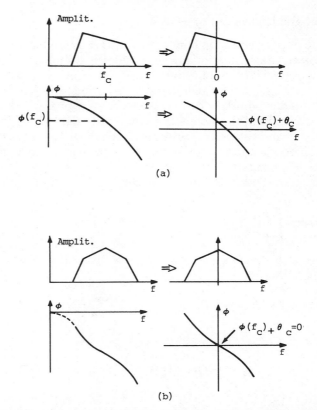

Figure 4.3 Translation of channel characteristics to baseband: (a) general channel; (b) symmetrical channel.

These are, of course, just the equations that define a complex multiplication; they are emphasized this way to establish firmly the two ways of considering QAM — complex and two-dimensional "real."

The amount by which the real part $h_p(t)$ differs from the ideal represents *intrachannel* distortion; the entire imaginary part $h_q(t)$, which is supposed to be zero, represents *interchannel* distortion†; both contribute to ISI. However, it must be noted that not all intrachannel and interchannel impulse responses can be treated as the real and imaginary parts of a complex response; we shall see in Section 4.3.3 that *nonlinear* interchannel distortion is sometimes deliberately introduced.

† We will have to be careful in using the term "interchannel" because it will also be used to describe the signal leaking from one channel into another using an adjacent frequency band in FDM systems (see Sections 4.5 and 11.1). This latter will be called *interference*.

A "Symmetrical" Channel. An important special case occurs when the amplitude of $C(f)$ is symmetrical about f_c and the sum of the phases of $C(f)$ and of the demodulating carrier is antisymmetrical about f_c; that is,

$$|C(f_c - f)| = |C(f_c + f)|$$

and

$$\phi(f_c - f) + \theta_c = -[\phi(f_c + f) + \theta_c]$$

This will be called a *symmetrical bandpass* channel (with an implication that the appropriate carrier phase is to be used) from now on (see Fig. 4.3b).

It is interesting to note that in this case

$$\theta_c = -\phi(f_c)$$

and the demodulating carrier counteracts any rotation introduced by the channel, so that the phase of the d.c. component in the baseband is zero. In the more general case of an asymmetrical channel θ_c is sometimes chosen in this way, but it will probably not be the phase that minimizes the ISI.

4.2.2 Differential Phase Encoding

Quadrature Modulation (QM) signals are always transmitted without a carrier pilot because one would interfere with the d.c. component of the data signals; therefore, it is difficult for the receiver to obtain an absolute phase reference. Without such a reference, and using a constellation that looks just the same after it has been rotated by any integer multiple of $\pi/2$ (it has "quadrantal invariance"), the phase of the demodulating carrier that is used for coherent detection can only be found modulo $\pi/2$, and the receiver cannot decide which quadrant a particular point is in; it can only detect changes of quadrant.

For a four-phase [Quadrature Phase Modulation (QPM)] system, this means that the four input "dibits", 00, 01, 11, and 10, should cause a *change* of carrier phase of 0°, 90°, 180°, or 270°(−90°); to minimize the number of errors, the changes should be assigned to the dibits according to a Gray code. Phase changes were originally defined as occurring from one end of one signal element to the beginning of the next. This was unambiguous as far as the phase shift was concerned, but it implied that the phase was to be changed abruptly (i.e., keyed). The new wording (*CCITT Yellow Book,* vol. VIII.1, 1980) refers to a "phase shift from the center of one signal element to the center of the next"; this may be misleading if there are not an integer number of carrier cycles in one bit period. It must therefore be interpreted as phase *in addition* to that which would be experienced by an unmodulated carrier.

Early voice-band modems (e.g., V.26) used a straightforward Gray code, but all recent ones (V.22, V.22 bis, V.32, etc.) have standardized on a shifted and reversed code; both codes are shown in Table 4.1.

TABLE 4.1   Phase Changes for Differential QPM

	Phase Change (°)	
Dibit	V.26	V.22, etc.
00	0	90
01	90	0
11	180	270
10	270	180

An Argument for Sometimes Not *Using Differential Phase Encoding.* As we shall see in Section 5.7, synchrodyne demodulation and decoding of a signal that has been differentially phase encoded causes bit errors to occur in pairs, because if noise causes a received point to move into another quadrant, then both that symbol and the next will have one bit in error. Some error correcting methods are not able to deal with such correlated errors, and differential phase encoding must be eschewed. Then very special techniques are needed to establish a phase reference in the receiver.

4.2.3   Multipoint Constellations

In the one-dimensional baseband case, the symbol rate can be reduced by a factor of M by encoding M bits into L $(= 2^M)$ equally probable levels. Similarly, M bits can now be encoded into L^2 points in a two-dimensional space.

Much work has been done to devise multipoint constellations [TW&D] that are optimum in some specified conditions. If added Gaussian noise is the only impairment, then for L large, the optimum constellations are based on an equilateral–triangular grid, which packs a set of points into a minimum area (minimum total power) while maintaining a given distance between neighbors. To deal with other impairments such as phase jitter (see Sections 1.4.2.8 and 6.4.2.1) more complicated constellations, such as that of V.29 [F&G], have been devised. However, each of these constellations has some or all of the following disadvantages: they (1) are difficult to generate and even more difficult to detect optimally in the receiver, (2) do not satisfy the quadrantal invariance requirement discussed in Section 4.3.2, or (3) are patented.

Consequently, if M is even, then square constellations, in which $M/2$ bits are used independently for each dimension, are now most often used.

Constellations for M Odd.   In these cases a simple square is not possible, and the constellations become more complicated. For $M = 3$, three different ones have been used: the eight-phase configuration in Fig. 4.4a and the two "stars" in Figs. 4.4.b and 4.4.c. The star in Fig. 4.4b is theoretically preferable because its total power is less for the same minimum distance between points. However, the transmit signal is more difficult to generate by analog techniques because the x_p and x_q coordinates are not integrally related; furthermore, optimum decisions

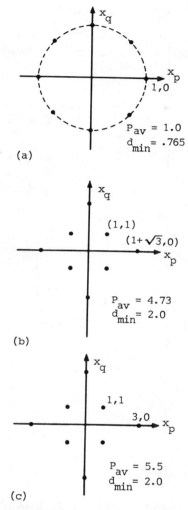

Figure 4.4 Eight-point constellations.

in the receiver cannot be made independently in each dimension. Consequently, the star in Fig. 4.4c was chosen for the 4800 bit/s fall-back mode of V.29 and is recommended when constant amplitude (pure phase modulation) is not needed.

For $M = 5$ and 7, the almost square "crosses" in Fig. 4.5 were proposed and codified by Ungerboeck, [Un4].

4.2.3.1 Practical Constellations.
For modems that may have to transmit at more than one speed, it is desirable that all the points of each constellation lie on

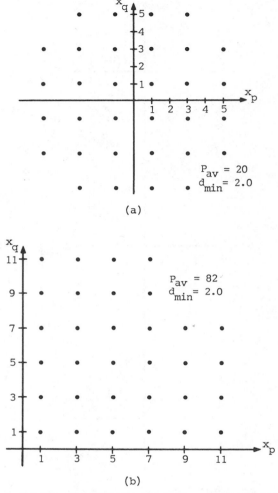

Figure 4.5 Nonsquare (*B* odd) constellations: (*a*) 32-point "cross"; (*b*) one quadrant of a 128-point cross.

a simple integer grid and that each set have approximately the same average energy. These desiderata can be almost achieved by rotating the sets with *M* odd by 45°, expanding the basic four-point set so that the minimum distance between points is 8, and then reducing this distance by a factor of $\sqrt{2}$ each time *M* is increased by one. One quadrant of each of the resultant sequence of constellations is shown in Figs. 4.6*a* – 4.6*e*.

These preferred constellations all have quadrantal or rotational invariance, and the two bits out of each block of *M* bits (by convention, the first two) that define the quadrant of the transmitted signal must be differentially encoded as described in Section 4.2.2. The other bits must then be assigned to the points in each quadrant in such a way that they can be detected correctly when the constellation is rotated by any integer multiple of $\pi/2$. The 16-point constellation — shrunk back again to its usually quoted size — used in V.22 and V.32 is shown in Fig. 4.7. Note that there are three possible amplitudes ($\sqrt{2}$, $\sqrt{10}$, and $\sqrt{18}$) for this signal which correspond to the dibits 00, 01 or 10, and 11, respectively.

4.2.3.2 Error Rates for Square Constellations.

The total power is twice the power in each dimension; that is, from (3.18):

$$P_S = 2 f_s \frac{L^2 - 1}{3} \tag{4.9}$$

and since for low error rates, the *symbol* error rate is approximately twice the error rate in each dimension,

$$\mathcal{P}_e = 4 \left(1 - \frac{1}{L} \right) Q \left[\left(\frac{3}{L^2 - 1} \frac{TP_S}{2N_0} \right)^{1/2} \right]$$

The factor $2N_0/T$ is the input noise power in the Nyquist bands $(-f_c - f_s/2)$ to $(-f_c + f_s/2)$ and $(f_c - f_s/2)$ to $(f_c + f_s/2)$; as in the one-dimensional case, this can be denoted by P_N. Furthermore, since each symbol now conveys $2 \log_2 L$ bits, and, with Gray encoding, each symbol error causes one bit error, the bit error rate is

$$\text{BER} = \frac{2}{\log_2 L} \left(1 - \frac{1}{L} \right) Q \left[\left(\frac{3}{L^2 - 1} \frac{P_S}{P_N} \right)^{1/2} \right] \tag{4.10}$$

Recognizing that the number of levels for QAM is the square of the number for a one-dimensional baseband system, it can be seen that the BERs given by (3.27) and (4.10) are the same. Thus the intuitive feeling that a bandwidth-efficient QAM system should have the same error rate as an optimum baseband system that uses the same bandwidth is indeed correct.

4.2.3.3 Error Rates for General Constellations.

For well-designed constellations — those for which the minimum distances between each point and its nearest neighbor(s) are equal — the error rate in conditions of high SNR is mainly determined by that minimum distance† and by the number of points that are thus separated. The error rate for a general constellation can be approximated by

$$\mathcal{P}_e = K_1 Q \left[\left(K_2 \frac{P_S}{P_N} \right)^{1/2} \right]$$

---

† Or conversely, by the average power of the signal if the minimum distance is fixed.

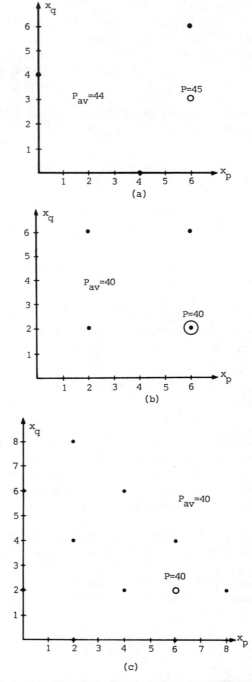

Figure 4.6 One quadrant of standard constellations, expanded and — for B odd — rotated by 45° showing training points: (a) to (e) M = 3 to 7.

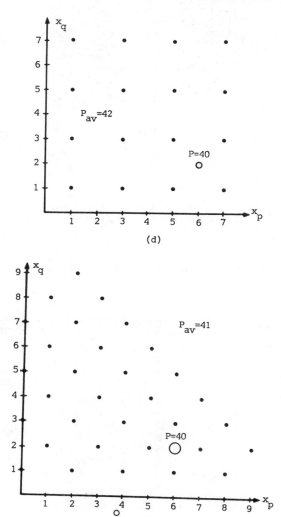

Figure 4.6 Continued

The values of K_1 and K_2 and the approximate decibel increase in SNR needed to maintain the same (low) error rate as with a basic two-level baseband system are given in Table 4.2 for all the recommended constellations. This table can be used together with Fig. 3.7 to find the error rate for any multipoint signal.

4.2.3.4 Four-Point Subsets.

Nearly all hand-shake sequences involve the transmission of a 4QAM signal—using both repeated patterns (see Section 4.2.5) and random data. It is desirable that the power of this signal be nearly the same as that of the eventual multipoint signal, and that the four points form a

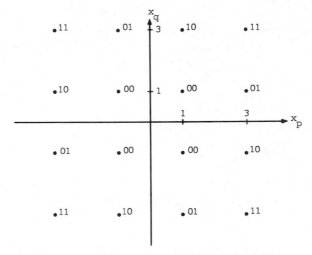

Figure 4.7 V.22 bis constellation showing assignment of bits 3 and 4.

TABLE 4.2 Error Rate Multipliers and SNR Penalty for Multipoint Constellations

Constellation	B	L^2	K_1	K_2	ΔdB
2×2	2	4	1.0	1.0	0
8ϕ (Fig. 4.4a)	3	8	0.67	0.54	5.3
Star (Fig. 4.4b)	3	8	0.67	0.65	3.7
Star (Fig. 4.4c)	3	8	0.33	0.60	4.4
4×4	4	16	0.75	0.45	7.0
32-pt cross	5	32	0.65	0.32	10.0
8×8	6	64	0.58	0.22	13.2
128-pt cross	7	128	0.52	0.16	16.1
16×16	8	256	0.47	0.11	19.3

subset of the full set. In most cases these two cannot be achieved together, and four new points — still conveniently on the integer grid — must be used. Those defined in various CCITT recommendations† are shown as circles in Fig. 4.6.

It should be noted that the absolute phases of these points are not important — they can be easily taken into account in the receiver; as always, it is differential phase that conveys the information.

4.2.4 Implementation of QAM

An implementation of a V.22 bis transmitter that uses a mixture of CMOS digital logic and switched-capacitor analog circuitry is shown in Fig. 4.8. Some

---

† It is puzzling why (6,2) was chosen for $M = 6$ and 7; (5,4) is closer to the average in both cases.

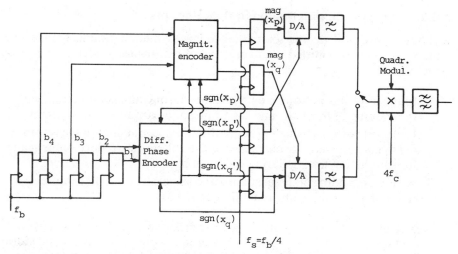

Figure 4.8 CMOS and switched-capacitor implementation of a V.22 bis transmitter.

comments — some particular (P) to this modem, and some generally applicable (G) to all QAM modems — may be helpful:

G. The arrangement of the basic serial-to-parallel converter — an M-bit serial input register clocked at f_b, some static digital logic, and an M-bit parallel output register clocked at f_s — is typical of all QAM transmitters.

G. The lines labeled $\text{sgn}(x_p, x_q)$, $b_{1,2}$, and $\text{sgn}(x_p', x_q')$ carry the two-bit definitions of the old phase, the phase change, and the new phase, respectively.

P. The DAC serves as the input network to the low-pass filters. The "mag" input controls the size of the input capacitor (1.0 or 3.0), and the "sgn" input controls the phasing of the two input switches so as to make the first integrator of the filter inverting or noninverting [M&S2].

G. The inputs to the filters that are shown here are at the symbol rate f_s, but the output sign and magnitude registers could be clocked at any integer multiple of f_b and more weighting elements appropriate to the technology (resistors, capacitors, controlled gain amplifiers of D/A converters) used in order to do some or all of the baseband filtering using sampled-data Finite Impulse Response (FIR) filter techniques. The specifications of these filters in order to prevent nonlinear foldover distortion in the modulators are derived in Section 4.6.

P. The switched-capacitor filters must be switched at some integer multiple of the symbol rate f_s.

P. The stringent filtering requirements of a V.22 bis can be met by any combination of baseband and passband filters.

P. The quadrature modulator timeshares all its capacitors [Bi4 and Bi5] in order to maintain nearly perfect balance between the two channels.

4.2.5 QAM Signals with Repeated Data Patterns

Although most modems use scramblers to ensure that continuously repeated patterns are *not* transmitted, there are several reasons for studying such patterns:

1. Many modems use simple data patterns during initialization to speed up the training of the AGC and the timing and carrier recovery circuits.
2. Repeated patterns, which produce discrete spectral lines in the transmitted signal, are useful for characterizing the filters during development and for simplifying testing during production.
3. An understanding of these patterns and signals is essential to the development of an intuitive understanding of QAM.

Nearly all useful repeated patterns are based on the four-point constellations described in Section 4.2.3.4. For a 4QAM system with 100% excess bandwidth the four simple patterns of $\Delta\phi = 0°$, $90°$, $180°$, and $270°$ result in discrete frequency components as shown in Fig. 4.9; with lesser bandwidths, the outer components will be attenuated or even eliminated altogether.

0° Phase Change. For QAM, this is not interesting because it results in d.c. in the two baseband channels. The single tone at the carrier frequency could be used for carrier recovery, but it obviously contains no timing information, and, because it is in the middle of the band, it might give unreasonably optimistic information to the AGC.

For Minimum Shift-Keying (MSK) signals, however, it can be very useful. The baseband signals are as shown in Fig. 4.10; they have components at every harmonic of the symbol frequency. The components of the transmitted signal all occur at the peaks of the spectral lobes shown in Fig. 4.11; therefore, it is a very useful signal for testing output filters and checking compliance with the FCC regulations on adjacent channel interference. As we shall see in Section 4.3.2, it is generated by an input of alternating ones and zeros.

+90° or −90° Changes. These are not of much practical use, but they help in understanding the basic modulation process. The baseband signals contain

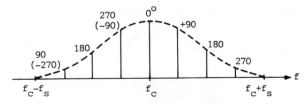

Figure 4.9 Discrete spectral components resulting from repeated phase changes.

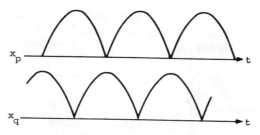

Figure 4.10 MSK baseband signals for $\Delta\phi = 0$.

components at $f_s/4$ and $3f_s/4$, and their phases are such that each component results in a single sideband as shown in Fig. 4.9. For systems with less than 50% excess bandwidth, the outer components vanish, and if either of these signals persists for a long time, the single tone (offset from the carrier by $f_s/4$) can play havoc with both timing and carrier recovery.

180° Changes. This signal was proposed in [Bi1] for initialization because the tones at the two Nyquist frequencies $(f_c \pm f_s/2)$ are ideal for fast carrier and timing recovery, as is discussed in Sections 6.4.1.3 and 7.3.3. It is also very

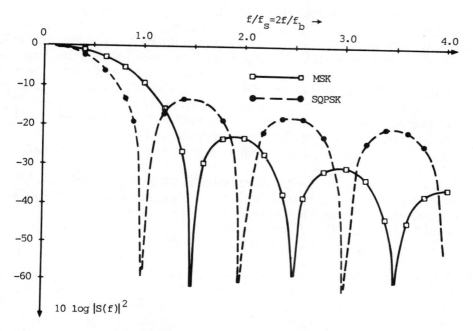

Figure 4.11 Power spectra of MSK and SQPSK.

useful for diagnosing problems in the circuits used for these functions. However, for systems with a small excess bandwidth, because the tones are near the edges of the band, they may give unduly pessimistic information to the AGC.

Alternate + 90° and − 90° Changes. This is the initializing signal adopted by the CCITT in its recent recommendations. One of the baseband channels has a d.c. signal, and the other has a sinewave at $f_s/2$. The resultant mixture of carrier and band-edge tones is ideal for AGC training and is adequate for carrier and timing recovery.

4.3 CONSTANT ENVELOPE MODULATION

Many radio systems use Class C amplifiers in their repeaters in order to economize in power, and repeaters in satellites, in order to achieve the utmost economy, nearly hard-limit the input signal. If the signal has any significant amount of amplitude modulation, these nonlinear operations will cause a spreading of the spectrum of the signal, which would cause interference with similar signals in adjacent bands. The signal can sometimes be refiltered, but this distorts it (causes intersymbol interference) and thereby reduces the effective SNR. Moreover, hardware constraints sometimes preclude refiltering, and then spectrum spreading could be disastrous.

Power Conservation versus Bandwidth Conservation. The original way of avoiding amplitude modulation was to use frequency modulation, but this was very wasteful of bandwidth. As data traffic and the consequent demand for communication channels have increased, there has been a parallel (and fortunate) reduction in the cost of "power," so that the trend has been toward more bandwidth-efficient linear or "quasi-linear" modulation methods. The compromises between bandwidth conservation and constancy of envelope (leading to power conservation) in any particular medium are beyond the scope of this book; we can only discuss the bandwidth required, and the envelope variation produced by various methods that have been devised, and let the designer choose.

We will consider only signal constellations in which all the points have the same amplitude and where the information is contained in the phase. This is not sufficient to ensure a constant envelope of the modulated signal, but it is certainly necessary. We will deal mainly with four-phase systems, but sometimes they will look like eight-phase systems.

Finite Bandwidth/Zero ISI/Constant Envelope. For linear modulation methods, finite bandwidth means that the impulse response is of infinite duration, and only achieves zero ISI at the sampling times; between these times the nonzero and data dependent values of the signal vary the envelope. Conversely, the nonlinear phase and frequency modulation methods that achieve constant envelope all have infinite bandwidth. Thus, although each of the listed properties is desirable for data transmission by radio, all three cannot be achieved in the same system.

Many attempts to achieve acceptable compromises between these three requirements have been reported. We will consider overtly linear modulation methods and those other methods, traditionally considered nonlinear, that can be studied and implemented by linear techniques. The very large class of nonlinear methods called *Continuous Phase Modulation* (CPM) has been treated in detail in [AR&S], [A&S1], [A&S2], and [AA&S]; it is beyond the scope of this chapter.

The methods that will be considered are:

1. *Staggered QPM (SQPM).*† This achieves zero ISI in a finite bandwidth but has considerable amplitude modulation; it is discussed in Section 4.3.1.

2. *Minimum Shift-Keying (MSK).* This achieves zero ISI with a constant envelope, but has significant "out-of-band" power. It was originally described in [D&H] and is discussed in Section 4.3.2. SQPM and MSK were comparitively evaluated in [ACHM].

3. *Tamed Frequency Modulation (TFM).* This achieves a constant envelope in a theoretically infinite, but practically finite bandwidth; however, it has considerable ISI, which may result in a reduction of the effective SNR. It was originally described in [J&D] and is discussed in Section 4.3.3. Generalized TFM, as originally described in [Ch3], reduces both the ISI and the bandwidth by flattening the passbands of the filters (both real and implied) used in TFM. It is discussed in Section 4.3.4.

4. *Modified Staggered QPM (MSQPM).* For reasons that we will discuss in Section 4.3.2, most TFM and GTFM transmitters are implemented as variations of SQPM, with the result that the signal can be strictly band-limited; however, a small variation of the envelope ensues. TFM can be considered as a quasi-linear modulation method that is very closely approximated by a linear modulation method that we will call MSQPM.

Conventional and Staggered Four-Phase Systems. The eye pattern of a conventional QPM system with 100% excess bandwidth is shown in Fig. 4.12a. If we normalize the signals so that the sum of the powers of the two sampled signals is unity, we can see that the continuous sum, and therefore the power of the modulated carrier, can vary between its maximum of 1.0 and a minimum of zero.

If, however, the timing of the data in the two channels is staggered by $T/2$, the two base-band signals can no longer be zero at the same time; this is shown in

† This has also been called *offset QPM*, but the term "staggered" is preferable because the initial "O" has also been used for "orthogonally multiplexed QAM" systems, which we will consider in Section 4.5.

Fig. 4.12b. It can be seen that, at the sampling times, the signal on one channel is $\pm 1/\sqrt{2}$ and on the other 0 or $\pm 1/\sqrt{2}$; therefore, $P_{max} = 1.0$ and $P_{min} = 0.5$; this is 3 dB of amplitude modulation.

It should be noted that staggering the timing on the two channels has no effect on either the baseband or the passband spectra.

Differential Encoding of SQPM Systems. Application of differential phase encoding to staggered systems can get complicated. It can be seen from Fig. 4.12b that from, for example, the point $(1,1)$, the possible paths in one symbol period may lead to $(1,0), (1,1), (-1,1)$, or $(-1,0)$ for a phase change of $-45°, 0°, +90°$, or $+135°$.

However, because of the staggering of the data on the two channels, they will now be distinguishable in the receiver, and it is adequate to assign the input data to the transmitter a_{2k} to each channel on an alternating basis (S/P conversion, or "demultiplexing") and then separately differentially encode each half-speed stream a_{2k} and a_{2k+1} according to

$$b_{2k} = a_{2k} \oplus b_{2k-2}$$
$$b_{2k+1} = a_{2k+1} \oplus b_{2k-1}$$

(4.11a)

It will be more convenient from now on to let the as and bs have the values of $+1$ or -1 instead of 1 or 0; modulo two addition then becomes negative multiplication. Then, if the two parallel encoders are combined into one encoder operating on the serial input data, the encoding becomes

$$b_n = -a_n b_{n-2}$$

(4.11b)

and the pulses that are input to each channel are

$$x_{p,2k} = b_{2k} = -a_{2k} x_{p,2k-2}$$

and $\qquad x_{q,2k+1} = b_{2k+1} = -a_{2k+1} x_{q,2k-1}$

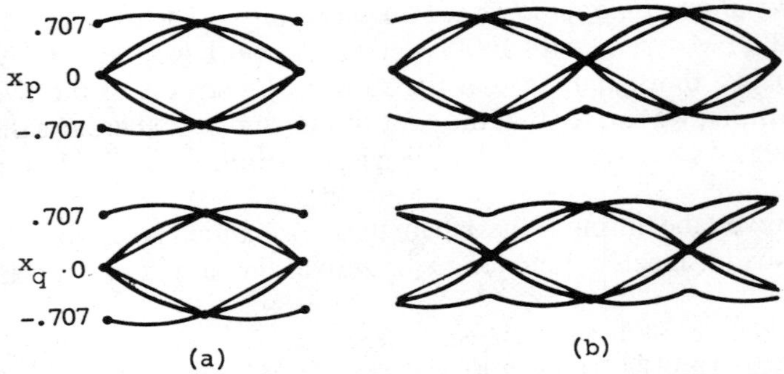

Figure 4.12 Eye patterns for $\alpha = 1$: (*a*) conventional; (*b*) staggered (offset).

4.3.1 Finite Bandwidth and Zero ISI: SQPM

As shown above, an SQPM system using 100% excess bandwidth has 3 dB of amplitude modulation. If bandwidth conservation is paramount, this is the preferred modulation method; however, it must be noted that if the bandwidth is reduced below the 100% excess the signals on each channel will overshoot between sampling times so that P_{max} and the amount of amplitude modulation will be increased.

All the filtering and modulation techniques developed for conventional synchronized QPM are applicable here; the only difference occurs in the timing recovery in the receiver; this is discussed in Section 7.4.2.1. The method is of interest also because it will serve as a starting point for the development of MSQPM.

4.3.2 Zero ISI and Constant Envelope: MSK

In order to maintain constant amplitude of the carrier, the baseband signals $x_p(t)$ and $x_q(t)$ must satisfy

$$x_p^2(t) + x_q^2(t) = 1. \tag{4.12}$$

The simplest solution to this is, of course, $x_p, x_q = \pm 1/\sqrt{2}$; this is staggered QPSK (sometimes, euphoniously but redundantly, called OK QPSK). The spectrum of this is shown in Fig. 4.11; we will compare this to the spectra of the other systems later.

If we wish to shape each pulse in order to reduce the out-of-band power, then (4.12) can be satisfied by

$$x_p(t) = \pm \sin\left(\frac{\pi t}{T}\right) \quad \text{and} \quad x_q(t) = \pm \cos\left(\frac{\pi t}{T}\right)$$

or vice versa. Since the data signals on the two channels are staggered by $T/2$, the responses to input pulses (generally called the *impulse response* even though the input is rarely an impulse) can be the same — a half-sinusoid. Since one half-sine pulse is zero at $t = 0$ and T, and the other at $t = T/2$ and $3T/2$, each can be modulated by the data without discontinuities.

The possible signals on the two channels during the period $t = kT$ to $(k + 1)T$ are shown in Fig. 4.13; in the first half of the period from kT to $(2k + 1)T/2$ the modulated output signal is

$$\begin{aligned}
s(t) &= b_{2k} \sin\left[\frac{\pi(t - kT)}{T}\right] \cos \omega_c t - b_{2k-1} \cos\left[\frac{\pi(t - kT)}{T}\right] \sin \omega_c t \\
&= \sin(\omega_c t + \phi_{2k})
\end{aligned} \tag{4.13a}$$

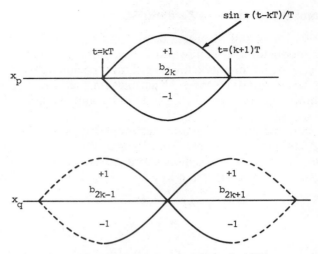

Figure 4.13 Baseband pulse waveforms for MSK.

where

$$\phi_{2k} = \tan^{-1} \left\{ -\frac{b_{2k}\sin[\pi(t - kT)/T]}{b_{2k-1}\cos[\pi(t - kT)/T]} \right\}$$

Similarly, from $(2k + 1)T/2$ to $(k + 1)T$,

$$s(t) = b_{2k} \sin[\pi(t - kT)/T] \cos \omega_c t$$
$$- b_{2k+1} \cos[\pi(t - kT - T)/T] \sin \omega_c t \quad (4.13b)$$
$$= \sin(\omega_c t + \phi_{2k+1})$$

where

$$\phi_{2k+1} = \tan^{-1} \left\{ \frac{b_{2k}\sin[\pi(t - kT)/T]}{b_{2k+1}\cos[\pi(t - kT)/T]} \right\}$$

It should be noted that this terminology differs from that in other writings on MSK [G&M1, Pa, and Z&R], in that to emphasize the similarity between MSK and all other four-phase systems, (1) T is the *symbol* period ($= 1/f_s$) which is *twice* the input bit period; and (2) the basic unmodulated pulse in each channel and in each symbol period is a positive half-sinewave—the relative polarity from period to period is controlled only by the encoded data.

The signal of (4.13) is "frequency shift-keyed" between two frequencies $f_c \pm 1/2T$, and if it is arranged that the carrier frequency is an integer multiple of one fourth of the bit rate, this shifting can always take place when the modulated signal is zero; this is an example of Continuous Phase FSK (CPFSK). It should be noted that MSK generated in this way is a special case (albeit a very common one); a more general case [A&S1] is CPFSK with the same frequency shift but any center (carrier) frequency.

Because the shift of $1/2T$ is the minimum that allows the two FSK signals to be coherently orthogonal [Pa] this method is called *minimum shift-keying;* because the modulation index of $\frac{1}{2}$ is much less than that often used, it is also called *Fast FSK* (FFSK). We thus have the rather surprising result that what is often regarded as the most nonlinear of modulation methods, FSK, can, with some very mild constraints, be implemented by linear techniques.

4.3.2.1 Encoding of Input Data for MSK.
If it is certain that an MSK signal will be demodulated only by linear, "quadrature" methods, then the input data could be encoded as shown in (4.11b). To allow the receiver the option of using a simple frequency discriminator, however, the encoding must be such that an input of repeated ones or zeros generates a single tone at $(f_c + f_s/2)$ or $(f_c - f_s/2)$. This is not achieved by the simple differential encoding of (4.11b); some further encoding is needed, and none of the writings on MSK has made this clear. It is much simpler to use differential phase encoding.

Since, when each channel is sampled, the signal on the other is zero, and since the power of an MSK signal is constant, the constellation and possible paths between sampled points is as shown in Fig. 4.14. The values of x_p and x_q at their appropriate sampling times uniquely define the phase of the signal. This is in contrast to the case of SQPM, where, as shown in Fig. 4.12, the phase is not uniquely defined by just one sampled value.

The differential encoding is therefore binary at the *bit* rate. Each input bit causes a phase change of $\pm\pi/2$, and the polarities of the pulses on the two channels are given by

$$x_{p,2k} = -a_{2k} x_{q,2k-1}$$

$$x_{q,2k+1} = a_{2k+1} x_{p,2k}. \tag{4.14}$$

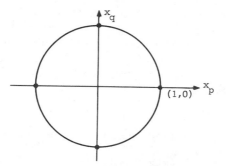

Figure 4.14 Constellation for MSK.

That this encoding does indeed obey the "FSK" rule can be checked by considering input data of repeated ones. The baseband signals would be two sinewaves at $f_s/2$, and the phase relationship between them would ensure that quadrature modulation produces only an upper sideband at $(f_c + f_s/2)$.†

An interesting insight into MSK can be obtained by considering an input of alternating ones and zeros. The half-sinewave pulses on each channel are all of the same sign — resulting in strong d.c. components — and the *main* component of the transmitted signal is at f_c. The sidebands at $(f_c \pm nf_s)$ do allow the instantaneous frequencies to be observed alternately as $(f_c + f_s/2)$ and $(f_c - f_s/2)$, but if these sidebands are seriously attenuated, the "average" frequency of f_c prevails, and the output of a discriminator would be near zero. Band-limited MSK, therefore, begins to look like a duobinary FSK signal [Le3].

4.3.2.2 Spectrum of MSK.

The baseband impulse response of an MSK transmitter (shifted so as to be symmetrical about $t = 0$) is

$$x(t) = \begin{cases} \cos \dfrac{\pi t}{T} & \text{for } |t| < T/2 \\ 0 & \text{otherwise} \end{cases} \tag{4.15}$$

The Fourier transform of this (normalized to unity at d.c.) is

$$X(f) = \frac{\cos \pi f T}{1 - (2fT)^2} \tag{4.16}$$

This is plotted in Fig. 4.11.

4.3.2.3 Implementation of MSK.

Strictly speaking, an MSK signal could be generated by a VCO with *any* center frequency *if* the difference between the upper and lower frequencies (those produced by steady input ones and zeros) could be synchronized with half the bit rate. In practice this is very difficult to do, and the slightly specialized, but much more easily realizable, form of MSK represented by (4.13) and (4.14) is almost always used.

An MSK signal is often generated in this way, with the shift between two intermediate frequencies thereby locked to the bit rate, and then the signal is frequency translated (heterodyned) up to the transmission band. In this case the final upper and lower frequencies do not individually have to be integer multiples of $f_b/4$.

A simple implementation of a transmitter is shown in Fig. 4.15a; a few explanations may be helpful:

---

† The instantaneous frequencies in the two bit periods, obtained by differentiating (4.13a) and (4.13b) with respect to time, are $(b_{2k}/b_{2k-1})/2T$ and $-(b_{2k+1}/b_{2k})/2T$, and since the sinewave on the p channel leads that on the q channel, both of these are equal to $1/2T$.

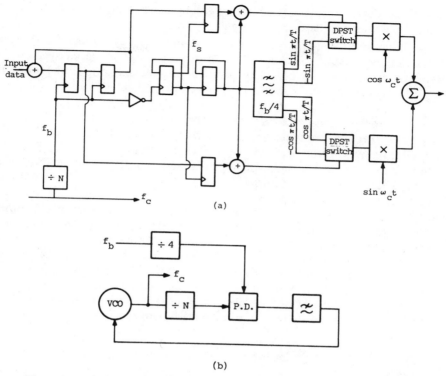

Figure 4.15 QM implementation of MSK: (a) transmitter; (b) PLL to lock f_c to $f_b/4$.

1. The bandpass filter is centered at $f_b/4$ and is shown with two pairs of outputs (positive and negative, sine and cosine); the data modulation is then performed by selecting one of each pair with a Double-Pole, Single-Throw (DPST) switch. If a multiplying modulator is used, only the positive sine and cosine outputs would be needed. In either case, the method of generating the outputs will depend on the hardware technology. For example, at low bit rates, when using a switched-capacitor or active-RC, state-variable implementation, two or three (respectively) of the outputs are available from the same filter; at higher speeds it may be simpler either to add a single-frequency 90° phase shift network or to use two identical filters and drive them from quadrature outputs of the divide-by-N counter.

2. The zero crossings of the input and the sine output of the bandpass filter should coincide with the rising edge of the input of the divide-by-N counter; at high data rates and carrier frequencies it will be necessary to use a synchronous counter—or at least to resynchronize the output—in order to ensure this.

3. If modulators of the "square-wave $\times$ continuous" type are used, a simple output filter to remove the sidebands of odd harmonics of the carrier will be needed. However, some further economy of circuitry can be achieved by com-

bining the data and carrier modulating functions using digital logic, and then using only one modulator per channel.

4. Because there are no narrow passband filters and all the signal shaping is done in the baseband, the channel center frequency can be varied over a wide range of integer multiples of $f_b/4$ by using a Phase-Locked Loop (PLL) containing a frequency divider as shown in Fig. 4.15b.

4.3.2.4 *Performance of MSK.*

As we shall see in Chapter 5, MSK can be detected using synchrodyne demodulation† and a matched filter in the time domain; its noise/error performance is then the same as optimally detected QAM. It can also be detected, with approximately a 3 dB penalty in SNR, by a simple frequency discriminator.

4.3.2.5 *Other Zero ISI and Constant Envelope Pulse Shapes.*

Another pulse shape that was used many years ago [B&D] because of its ease of generation using a sinusoidal oscillator, is the raised cosine in the *time* domain, $(1 + \cos 2\pi t/T)/2$. It has been proposed for systems requiring a constant envelope [Fe3] and given the name of NLF-OK-QPSK‡, but its amplitude modulation is almost exactly the same as that of the pulse with the raised cosine ($\alpha = 1$) frequency spectrum that we have already discussed, and its out-of-band power is higher than that strictly band-limited pulse; it will not be considered further.

4.3.3 Finite Bandwidth and Almost Constant Envelope: TFM, GTFM, and MSQPM

The third and fourth combinations that we proposed earlier, constant envelope with almost finite bandwidth (TFM) and nearly constant envelope with strictly finite bandwidth (MSQPM)—both achieved at the expense of introducing ISI—will be considered together. TFM was developed and originally explained as a nonlinear phase modulation [J&D], but, in order to maintain a precise relationship between the carrier and data clock frequencies, it is often implemented as a variation of the by now familiar SQPM.

If 100% excess bandwidth is used, and the impulse response of each staggered low-pass channel is . . . , 0, 0, $1/\sqrt{2}$, 0, 0, . . . , then, as shown in Fig. 4.12b, the eye pattern of each channel passes through $\pm 1/\sqrt{2}$ at its appropriate sampling time and $\pm 1/\sqrt{2}$ or 0 at the "half-times"; thus the constellation of SQPM (Fig. 4.16) is a square. In order to keep the sum of the powers in the two channels constant at all sampling times, we have to do the reverse of what medieval mathematicians strove to do; we must circle the square. If the sampled value on one channel is zero the sampled value on the other must be increased to 1.0.

---

† Often, but misleadingly, called *coherent demodulation*—see Section 5.4.1.

‡ The search for constant envelope modulation methods has produced many systems with long, exotic initials; there is even a rumor that the Russians have developed "Offset MSK" and "Tamed Offset MSK" (OMSK and TOMSK)!

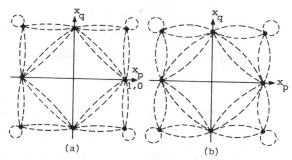

Figure 4.16 Signal-space diagram of SQPM at sampling times and signal paths between sampling times: (a) $\alpha = 1.0$; (b) $\alpha = 0.5$.

This can be done by modifying the basic QPM transmitter shown in Fig. 4.2 so that at even and odd multiples of $T/2$ the inputs to the x_p and x_q channels are augmented by correcting pulses whose magnitudes are defined by

$$x_{p,2k} = \frac{b_{2k}|b_{2k-1} - b_{2k+1}|(\sqrt{2} - 1)}{2\sqrt{2}}$$

$$x_{q,2k+1} = \frac{b_{2k+1}|b_{2k} - b_{2k-1}|(\sqrt{2} - 1)}{2\sqrt{2}} \tag{4.17}$$

where the even and odd subscripted bs are the encoded data input to the x_p and x_q channels, respectively.† These pulses cause ISI that is nonlinear — because it depends on a nonlinear function of the data — and that furthermore, cannot be described in complex notation — because, in contrast to the rules for complex multiplication given in Section 4.2.1, the expression for the interchannel pulse has the same sign from the "real" channel to the "imaginary" as vice versa.

These small correction pulses must have an effect only at the time they are introduced (the amplitude at later sampling times will either be correct or will be corrected by subsequent pulses); that is, the response to one of these pulses at the input to the modulator, sampled at the $2f_s$ rate, must be nonzero at only one sampling time. Therefore, the low-pass filter for these pulses must have a Nyquist frequency of f_s, which is twice that used for the main pulses, and for practicality it must have an excess bandwidth α. A possible implementation using two filters is shown in Fig. 4.17.

However, it is better to put both the main and the correcting pulses through the same filter; therefore, the response of each channel to the main pulses, $x_{p,2k}$ and $x_{q,2k+1}$, must be considered as $\ldots, 0, 0, 1/2\sqrt{2}, 1/\sqrt{2}, 1/2\sqrt{2}, 0, 0 \ldots$ at

---

† These equations and (4.18) may look a little strange because the signal x_p at time $2k$ depends on the input at time $(2k + 1)$! In practice a delay would be included, but the explanation of the method is clearer without the delay.

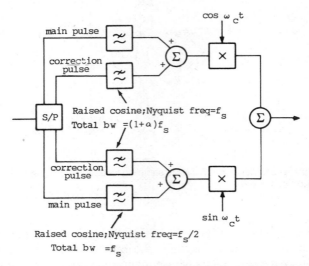

Figure 4.17 Possible, but undesirable, implementation of TFM.

the $2f_s$ sampling rate. Then the composite pulses input to the x_p channel are given by

$$x_{p,2k} = \frac{b_{2k}(1 + |b_{2k-1} - b_{2k+1}|(\sqrt{2} - 1))}{2}$$

$$x_{p,2k+1} = \frac{b_{2k} + b_{2k+2}}{2}$$

(4.18)

The inputs to the x_q channel are defined by similar expressions offset by $T/2$.

4.3.3.1 Encoding of Input Data for MSQPM. The phase encoding for TFM is defined in [J&D] as

$$\phi_n - \phi_{n-1} = (b_n + 2b_{n-1} + b_{n-2})\frac{\pi}{8},$$

(4.19)

and then the pulses on the two channels are defined as $\cos \phi$ and $\sin \phi$. However, the coding given in (4.11b) is equivalent and is much simpler to implement when a QM type of transmitter (as opposed to an FM type) is used.

4.3.3.2 Power Spectrum of MSQPM (and TFM). The voltage spectra of the intrachannel impulse triplets that are input to the low-pass filters in each channel are raised cosines [viz., $(1 + \cos \pi f/f_s)T/2\sqrt{2}$], and those of the single inter-channel impulses are flat with a magnitude of $(1 - 1/\sqrt{2})T$. The two signals are independent, so their power spectra must be added; however, the average probability of occurrence of the interchannel impulses is only $\frac{1}{2}$. Therefore, the power spectrum at the input to each low-pass filter is given by

$$|S_i(f)|^2 = \left[\frac{(1 + \cos \pi f/f_s)^2}{4} + (0.293)^2\right] T \qquad (4.20)$$

and that at the outputs by

$$|S(f)|^2 = \begin{cases} |S_i(f)|^2 & \text{for } |f| < (1-\alpha)f_s \\ |S_i(f)|^2 \left\{1 - \sin\left[\dfrac{\pi(f-f_s)}{2\alpha f_s}\right]\right\} & \text{for } (1-\alpha)f_s < |f| < (1+\alpha)f_s \\ 0 & \text{for } |f| > (1+\alpha)\dfrac{f_s}{2} \end{cases} \qquad (4.21)$$

This is plotted in Fig. 4.18 for the case of $\alpha = 0.5$, along with the power spectrum of TFM [J&D]; it can be seen that the two agree very closely. If these are compared to the spectra of MSK and SQPSK in Fig. 4.11, it can be seen that a dramatic reduction of out-of-band power has been achieved. One simple (perhaps too simplistic?) measure of the "compactness" of a spectrum is the bandwidth (relative to the bit rate) that contains 99% of the total power. For SQPSK the ratio is 8.0, for MSK it is 1.2 [G&M1], and for MSQPM ($\alpha = 0.5$) and TFM it is approximately 0.82.

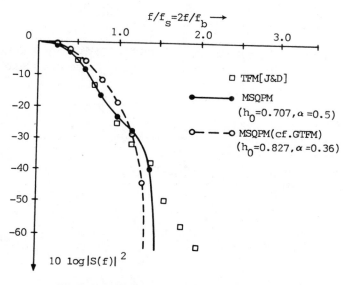

Figure 4.18 Power spectra of MSQPM and TFM.

4.3.3.3 Implementation of an MSQPM Transmitter. An implementation based on the use of digital logic and an analog filter is shown in Fig. 4.19; some comments may be helpful:

1. The analog switches may not be appropriate components with many hardware technologies; they are drawn thus only to show how the x_p and x_q signals are composed. The controls for the first two are the logic signals shown; the four steering switches are controlled complementarily by the two phases of the symbol clock.
2. The gain devices are one-bit D/A converters; if the logic is implemented in CMOS with accurate complementary supplies, single resistors will probably be adequate. If switched-capacitor filters are used, then complementary supplies are not necessary; the polarity of the input signals can be controlled by the phasing of the input switches of the filters.
3. The input pulses to the filters are rectangular with a duration of $T/2$; their spectral shaping of $\text{sinc}(f/f_s)$ must be allowed for in calculating the frequency response required of the filters.
4. The stop-band attenuation required of the filters depends on limits put on adjacent channel interference — usually by the FCC — and discussion of it would be beyond the scope of this book.

4.3.3.4 Performance of TFM. The comment that was made about MSK — that detailed discussion of the performance must await analysis of receiver structures and algorithms — is even more applicable here, because we have now introduced ISI, and we do not know yet whether we can correct for it. A claim has been made [J&D] that TFM is only 1.2 dB worse than optimum QPM; this needs to be examined closely.

4.3.3.5 Amplitude Modulation of TFM and MSQPM. The addition of the interchannel correction pulses to an SQPM signal makes the squared magni-

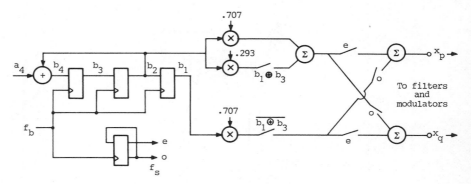

Figure 4.19 TFM encoder.

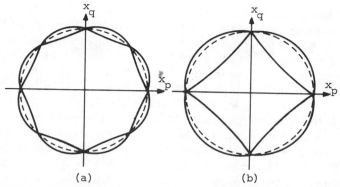

Figure 4.20 Showing the deviations (exaggerated) from a circle of TFM signal paths: (a) $\phi = 45°$; (b) $\phi = 90°$.

tude of the signal, when sampled at $t = nT/2$, equal to 1.0; that is, all the sampled points lie on a circle of unit radius. However, between sampling times the magnitude will vary. For phase changes of $\pm \pi/4$ the paths of the points lie within the bounds shown (exaggerated) in Fig. 4.20a; the deviation from the circle is very slight. However, for phase changes of $\pm \pi/2$ (shown in Fig. 4.20b), the deviation will be considerably greater. This can be calculated as follows.

Using a low-pass filter with 50% excess bandwidth, the response to the intrachannel and interchannel pulses can be truncated with negligible error to

$2t/T$:	$-5/2$	-2	$-3/2$	-1	$-1/2$	0	1/2	1	3/2	2	5/2
Intrachannel:	-0.028	0	0.133	0.354	0.594	0.707	0.594	0.354	0.133	0	-0.028
Interchannel:	0.005	0	-0.035	0	0.176	0.293	0.176	0	-0.035	0	0.005

For a phase change of $+\pi/2$ from the point $(1,0)$ at the time kT to the point $(0,1)$ at the time $(2k + 1)T/2$, the data sequence and signals on the x_p and x_q channels that result in the greatest deviation of the signal power at $(2k + 1/2)T/2$ are

$2t/T$	$2k-2$	$2k-1$	$2k$	$(2k+1/2)$	$2k+1$	$2k+2$	$2k+3$
b	$+1$	-1	$+1$		$+1$	-1	$+1$
x_p	0.707	0.707	1.0	(0.609)	0	-0.707	-0.707
x_q	-0.707	-0.707	0	(0.609)	1.0	0.707	0.707
$x_p^2 + x_q^2$	1.0	1.0	1.0	(0.742)	1.0	1.0	1.0

Thus the *peak* amplitude modulation is $10 \log(0.742) = 1.3$ dB. The average, of course, is much less than this, but careful simulation of the nonlinearities of each particular channel would be needed in order to decide whether this amount of modulation is tolerable.

4.3.3.6 Generalized TFM. The amplitude modulation and the ISI of TFM can be reduced by making the pulses look more like those of MSK. In the frequency domain this can be done by reducing the excess bandwidth of the filter and thereby flattening its passband. In the time domain, we want to reduce the magnitude of the impulse response at $t = \pm T/2$; that is, we want to change the main impulse response from . . . , 0, $1/2\sqrt{2}$, $1/\sqrt{2}$, $1/2\sqrt{2}$, 0, . . . to . . . , 0, h_1, h_0, h_1, 0, . . . where $h_0^2 + (2h_1)^2 = 1.0$ and $h_0 > 2h_1$.

Chung [Ch3] found that an excess bandwidth of 36% and phase changes of 0.31 and 0.19 are in some sense optimum. These phase changes correspond to $h_0 = 0.827$ and $h_1 = 0.281$ and therefore require a correction pulse (and result in a peak ISI) of 0.173 (compared to 0.293 for TFM). By applying the methods of analysis described in Sections 4.3.3.2 and 4.3.3.5, the output power spectrum is found to be as shown in Fig. 4.18; the 99% power bandwidth is approximately 0.82, and the peak amplitude modulation is 0.4 dB. It appears that the only price that must be made for the reduction of ISI and amplitude modulation is the need for a sharper filter, but GTFM is still quite new, and needs more analysis, simulation, and testing [L&K].

4.4 SINGLE-SIDEBAND MODULATION

A double-sideband signal is redundant, and one of the sidebands could be eliminated without any loss of information; the spectral efficiency that is achieved with QAM can also be achieved with single sideband. However, if $G(f)$ were to extend to d.c., the filtering would have to be perfect, and even then there would still be some doubt about what happened to the frequency component at d.c. (now at f_c). To avoid this, $x(t)$ must be filtered or encoded in some way to remove its d.c. component; a common way of doing this is to use the modified duobinary (MDB) coding and shaping described in Section 3.3. The frequency domain representation of the generation of a lower sideband from an MDB baseband signal is shown in Fig. 4.21. The spectrum is exactly the same as that that would be obtained by quadrature modulation of a duobinary (DB)-shaped baseband; in fact, if stable carriers were available in the receiver for demodulation, the performance of the two systems would be exactly the same.

The unwanted sideband can be removed by filtering after modulation, or by either the Hartley [Ha1] or the Weaver [We1] method. These latter have the advantage that they are baseband operations and do not need a bandpass filter; the same circuit or software program can therefore be used for many different

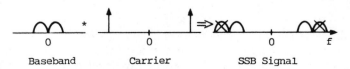

Baseband   Carrier   SSB Signal

Figure 4.21 Generation of an SSB signal.

carrier frequencies. The Hartley method (the more useful for data transmission) of generating a single sideband can be summarized as

$$s_1(t) = x(t) \cos \omega_c t + \hat{x}(t) \sin \omega_c t$$

or
$$s_u(t) = x(t) \cos \omega_c t - \hat{x}(t) \sin \omega_c t \tag{4.22}$$

where $\hat{x}(t) = H[x(t)]$, the Hilbert transform of $x(t)$.

The precise mathematical definition of a Hilbert transform need not concern us here; for our purposes it can be considered as an operation that preserves the amplitude spectrum of the original and delays the phase by 90° at all frequencies. An ideal Hilbert transformer is physically unrealizable, and the best that can be achieved is two all-pass networks that produce a phase difference between their outputs of approximately $-90°$ across the band of the signal. The specification of these networks is discussed in Section 11.1 and their design in Appendix I; their use in a Hartley modulator is shown in Fig. 4.22.

It can be seen by comparing (4.22) to (4.2) that SSB generated in this way is a special case of QAM where

$$x_p(t) = x(t)$$
$$x_q(t) = \pm \hat{x}(t)$$

and the + or − sign is used to generate an upper or a lower sideband, respectively. That is, the same data is input to the quadrature channel, and the shaped and/or filtered signal is then shifted by 90°. In the receiver the quadrature channel is ignored because it contains the same information.

The main disadvantage of SSB modulation is that recovering or regenerating a carrier of the correct frequency and phase for demodulation in the receiver is a difficult task. It is possible to generate this carrier from information contained in the signal, but such methods have not been used much. A more common

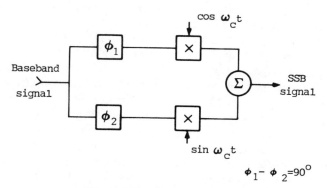

Figure 4.22 Hartley modulator.

technique is to add to the SSB signal a low-level "pilot" tone at the carrier frequency and use a narrow-band PLL in the receiver to extract this tone. However, as we shall see in Chapter 6, this method is appropriate for only certain transmission media.

In general, it seems that SSB modulation for *data* transmission has waned in popularity as quadrature modulation has become more widely understood.

4.4.1 Baseband Equivalent of a Single-Sideband Signal

With an SSB signal we are only interested in the real part of the impulse response (the quadrature channel contains redundant information); therefore, the third and fourth terms in the integrand of (4.8) can be discarded. Furthermore, there is only one sideband to contribute to the integral, so that for the case, for example, of an upper sideband:

$$h_p(t) = G(f) |C(f_c + f)| \cos(2\pi ft + \phi(f_c + f) + \theta_c) \, df$$

Again, it is important to note the role of the phase of the demodulating carrier; if there were no distortion in the channel, but θ_c were $\pi/2$, then $h_p(t)$ would be the Hilbert transform of the desired response. In the QAM case an adaptive equalizer would be easily able to perform the derotation, but for an SSB signal the equalizer is only one-dimensional, and implementation of the necessary transfer function (equivalent to a further Hilbert transform) may be difficult. If there is distortion the effect is not so dramatic, but θ_c must still be chosen carefully; criteria for this choice will be discussed in Chapter 6.

4.5 ORTHOGONAL MODULATION OF MULTIPLE CARRIERS

If a single multilevel (many bits per symbol) signal is transmitted through a distorted channel, some form of adaptive equalization will be required in the receiver. If a linear equalizer is used, noise enhancement will occur and the SNR of the signal delivered to the detector will be reduced.

Some early modems [DH&M and F&L1] used the technique of dividing the band into many subbands and transmitting the high-speed serial data stream as a multiplexed set of low-speed data streams modulated onto a set of carriers evenly spaced throughout the band. Holsinger [Ho1] proved that the method results in the lowest possible error rate if the channel has amplitude distortion, and a linear equalizer is used in the receiver. Chang [Ch1] showed how the carrier phases and data timing of the separate channels can be arranged to maintain a flat power spectrum of the composite signal and to permit separation of the overlapped channels in the receiver without interchannel interference; Saltzberg [Sa2] developed the special case where each subchannel uses SQAM, and considered the effects of channel distortion on such a modulated signal.

Since the development of powerful DSP capabilities Orthogonally Multiplexed QAM (OMQAM) has been used by Gandalf, Telebit (see [Ba] and [TC]), and NEC for modems that transfer between 9.6 and 19.2 kbit/s in a telephone voice channel, but it has never been what one might call "mainstream." This is because it requires complicated, and usually proprietary, DSP techniques in the receiver, and because the trade-offs of improved performance (if any) against complexity of implementation have been difficult to assess. We will attempt that assessment here.

Performance with Gaussian Noise. If a distorted channel is divided into many subchannels, the variation of attenuation and delay across each subband is greatly reduced. Therefore, the impulse responses of these subchannels will be much less dispersed, and adaptive equalization may not be needed. Consequently, it is often claimed that the method has the advantage that the noise is not enhanced and the SNR is not reduced. However, a type of piecewise amplitude equalization may be performed by the separate AGCs in each subband. They make the average attenuation of all the subchannels appear to be equal† and, in the process, amplify the noise in the bands that had been attenuated. We will examine this effect carefully in the next two sections.

We shall find that a system with a predetermined bit loading of the bands — a nonadaptive system — may perform worse than a conventional single-band system and that the full potential of this modulation method is realized only if the transmitter can be configured to "match" the channel. This requires measurements of the SNRs in the receiver and feedback to the transmitter. Therefore, even though the other sections of this chapter are concerned only with transmitters, for this method we will have to consider also those parts of the receiver that affect the transmitter.

Performance with Impulse Noise. We shall find that a multichannel modem is more sensitive to narrow-band spikes of noise‡ than is a single-channel modem, which averages the noise across the whole band. Conversely, it is *less* sensitive to narrow time spikes (impulses) of noise than is a single-channel modem because it averages them over a long symbol period.

Complexity of Implementation. If the number of bands is very large (the Telebit modem can use as many as 512), each band will be very narrow and essentially undistorted; adaptive equalization will not be needed. If the number of bands is only moderate (the NEC modem uses 14), some simple equalization will be needed for the highest data speeds.

However, the fact that no equalizer — or only a very simple one — is needed does not necessarily mean that the implementation is simple; the multichannel

---

† Similarly, if separate timing recovery is performed for each subband, this is a type of piecewise delay equilization because it makes the average delay of all the subchannels appear to be equal.
‡ Assuming that it has not avoided them by adaptively configuring the bands.

filters or the DFT needed in the receiver to separate the subchannels may be as complex as a single-channel adaptive equalizer.

4.5.1 Ideal Performance of a Fixed System with Distortion and Noise

This method is very different from single-channel modulation, so, in order to generate motivation for the study of it—the "why" before the "how"—we must first calculate its theoretically ideal performance in the presence of distortion and noise.

The ideal performance can be achieved only if (1) the number of bands is large enough that the effects of *intrachannel* distortion can be ignored and (2) the reciever is able to completely separate the signals in the different bands so as to eliminate *interchannel* interference (ICI). However, calculation of the error rate for a system with so many bands would be very tedious, so we will do it only for the 14-band system described by Hirosaki [Hi . .]—ignoring, at least at first, the need for an adaptive equalizer [Hi1].

The symbol rate is 200 s/s, and the total number of bits per symbol B_{tot}, is 96, so that the aggregate bit rate is 19.2 kbit/s. Twelve of the fourteen QAM transmitters used in the subbands encode seven bits per symbol (i.e., $B_i = 7$ for $i = 2 - 13$) and the two in the bands at the edges of the 400 – 3200 Hz overall band encode just six ($B_1 = B_{14} = 6$).

If the channel is a so-called worst-case 3002 line as shown in Fig. 4.23,† its amplitude response can be approximated by 14 flat segments centered about equally spaced carrier frequencies as shown ($f_{c,i} = 300 + 200i$ for $i = 1 - 14$); with ideal filtering each subchannel may be considered to be undistorted but attenuated by an amount A_i. If the power transmitted in each band is normalized to unity, the total received power is

$$P_{tot} = \sum_{i=1}^{14} 10^{-A_i/10}$$

and with the noise input to the receiver assumed to be white, the SNRs in the subbands are given by

$$SNR_i = SNR_{tot} - A_i + 10 \log \left(\frac{14}{P_{tot}} \right) \qquad (4.23)$$

For an input SNR_{tot} of 30 dB, the SNR_i and the resultant error rates are as shown in Table 4.3. It can be seen that the error rate is dominated by the errors in bands 12, 13 and 14 and is greater than $(0.004 + 0.002 + 0.014 + 0.037 + 0.036)/14 = 0.0066$. This could be obtained in an undistorted channel with an

---

† This channel is unusual in that its attenuation rises sharply at the low end (see Section 1.4.2.2), but that wil not affect the results very much; it is the greater attenuation at the high end that will mostly concern us.

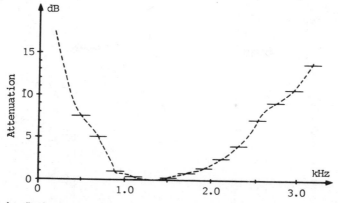

Figure 4.23 Amplitude response of a "worst-case" 3002 channel divided into 14 flat segments.

SNR of 24 dB, so the deterioration due to distortion is at least 6 dB. In practice, in order to support a 128-point constellation, the variation of amplitude response across the subbands would have to be partially equalized in order to reduce both ISI within each channel and interference between channels (ICI), and there would probably be a further noise enhancement of about 1 dB. The total degradation of about 7 dB is *worse* than would be experienced by a conventional wide-band modem using linear equalization!

4.5.2 Performance of an Adaptive System with Distortion and Noise

A multiband modulation system has many more degrees of freedom than a single-band system, and we should consider whether they can be used to improve what is, at least so far, a very unpromising predicted performance. The overall error rate can be reduced by adjusting separately for each band the proportion of the total power that is transmitted in that band and the number of bits that are encoded per symbol. Thus the transmitter can be partially "matched" to the channel. Let us consider how this might be done.

If the constellations in each band are assumed to be square,† the aggregate probability of error from the fourteen channels is

$$\mathcal{P}_e = \sum \left(\frac{B_i}{B_{\text{tot}}}\right) \mathcal{P}_{e,i}$$
$$= \sum \left(\frac{B_i}{B_{\text{tot}}}\right) 2 \left(1 - \frac{1}{L_i}\right) Q(x_i) \tag{4.24}$$

† This is not, of course, possible if B_i is odd; however, the power in the cross constellations with $B_i = 5$ or 7 is reduced from that of the hypothetical square constellations with $L = 4\sqrt{2}$ or $8\sqrt{2}$ (and therefore the effective SNR is increased) by less than 0.2 dB.

TABLE 4.3 Bit Assignments, Attenuations, SNRs, and Error Rates for a 14-Band OMQAM System on a Worst-Case 3002 Line

Channel no.	1	2	3	4	5	6	7	8	9	10	11	12	13	14
Bits per symbol	6	7	7	7	7	7	7	7	7	7	7	7	7	6
Attenuation	7.5	4.5	1.0	0.4	0	0.2	0.7	1.3	2.5	4.0	7.0	9.0	10.5	13.5
SNR	25.4	28.4	31.9	32.5	32.9	32.7	32.2	31.6	30.4	28.9	25.9	23.9	22.4	19.4
$\mathscr{P}_{e,i}$	0.004	—	—	—	—	<0.001	—	—	—	—	0.002	0.014	0.037	0.036

TABLE 4.4 Optimum Bit Assignments and Power Distribution for the 14-Band OMQAM System

Channel no.	1	2	3	4	5	6	7	8	9	10	11	12	13	14
B_i	5.84	6.84	8.00	8.20	8.33	8.27	8.10	7.90	7.50	7.00	6.01	5.34	4.84	3.86
B_{iq}	6	7	8	8	8	8	8	8	8	7	6	5	5	4
γ_i	0.079	0.079	0.071	0.062	0.056	0.059	0.066	0.076	0.100	0.071	0.070	0.056	0.079	0.078

where $L_i = 2^{B_i/2}$, the number of levels in each dimension in the ith band,

$$x_i = \left| \frac{3N}{(L_i^2 - 1)} R_i \gamma_i \right|^{1/2}$$

where γ_i is the proportion of the total transmit power allotted to the ith band, R_i is the linear signal to noise ratio in each band ($= 10^{SNR_i/10}$), the B_i are constrained by $\Sigma B_i = B_{tot}$, and all summations and products in this section are taken over i or $j = 1$ to N.

This expression for the overall error rate can be simplified slightly by making the approximation that $L_i \gg 1$; hence

$$\mathcal{P}_e = \Sigma \left(\frac{B_i}{B_{tot}} \right) Q(x_i) \qquad (4.25)$$

where now
$$x_i = \frac{(3N \, R_i \gamma_i)^{1/2}}{L_i}$$

Nevertheless, minimizing $\mathcal{P}_e$ with respect to all the γ_i and L_i would still be a very complicated task. We can approach it in two stages.

Optimum Power Allocation per Band. The power allocated to each band could be calculated by the optimal "water-filling" method of Gallager [Ga1] and [LS&W]; this is usually depicted as shown in Fig. 4.24a. In our example, the full band extends from $(f_{c1} - f_s/2)$ to $(f_{c14} + f_s/2)$—that is, from 400 to 3200 Hz; it is assumed that the signal must be limited to that band—that is, the walls of the "bowl" are vertical. The term $N(f)$ is the noise power spectral density at the receiver—conveniently expressed in our example in picawatts per hertz—and

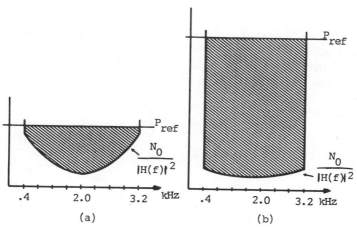

Figure 4.24 Transmit power allocation by water-pouring analogy: (a) usual depiction; (b) more realistic situation.

$|H(f)|^2$ is the dimensionless power transfer function of the channel. The power to be transmitted at frequency f is defined by

$$P(f) = P_{ref} - \frac{N(f)}{|H(f)|^2} \tag{4.26}$$

so that the total transmitted power is represented by the shaded area. However, as we shall see, this is usually a very unrealistic portrayal.

The following sequence of calculations could be performed with normalized powers, considering only the signal power spectrum and the SNR at the receiver, but the argument will be clearer if we use actual power levels. In order to simplify comparison with the wide-band and nonadaptive multiband cases, we will assume that the noise is white, but this is not necessary in practice.

Transmit level	-9 dBm	(126 μW, or 0.045 μW/Hz)
Attenuation at 1 kHz	21 dB	
Receive power at 1 kHz	357 pW/Hz	
Average receive power across full band	184 pW/Hz	
Receive SNR	30 dB†	
Noise power per hertz, $N(f)$	0.184 pW/Hz	
$N(f)/|H(f)|^2$ at 1 kHz	23.1 pW/Hz	

The shaded area in the bowl contains 1.13 μW of power. However, the total transmit power needed to achieve the 30 dB receive SNR is 126 μW, so the concave part of the "bowl" is insignificant compared to the part with vertical sides; Fig. 4.24b is a much more realistic portrayal of the theory. It can be seen, therefore, that in (4.26) $P_{ref} \gg N(f)/|H(f)|^2$, and $P(f)$ should be essentially constant across the full band.

Optimum Bits per Symbol (Loading) in Each Band. It can be shown — and it is intuitively reasonable — that the overall error rate, $\mathcal{P}_e$ in (4.25), is almost minimized if the error rates in all the bands are equal. If we ignore the small effect of the multiplying factors, B_i/B_{tot}, this means that with all the γ_i now equal to $1/N$,

$$x_i = x_{av} = \frac{(3R_i)^{1/2}}{L_i}$$

Then, since

$$\prod_{i=1}^{N} L_i = 2^{B_{tot}/2}$$

† Very optimistic for such a line, but we'll use it anyway.

it follows that

$$x_{av}^2 = \frac{3R_{av}}{2^{B_{av}/2}}$$

where

$$B_{av} = \frac{B_{tot}}{N}$$

$$R_{av} = [\Pi R_j]^{1/N}$$

Then the optimum number of levels and bits per symbol in each band is

$$L_i^2 = 2^{B_{av}} \frac{R_i}{R_{av}} \tag{4.27a}$$

$$B_i = B_{av} + \log_2 R_i - \log_2 R_{av} \tag{4.27b}$$

In practice, of course, the L_i and B_i cannot be continouosly variable but must be quantized by either rounding or truncating the B_i. This quantization will probably again unbalance the error rate in the separate bands and thereby increase the overall error rate. Therefore, the equality of the error rates should, ideally, be restored by adjusting the γ_i so that the x_i are equal when the quantized numbers of levels L_{iq} are used. This means that

$$\gamma_i = \frac{L_{iq}}{R_i \, \Sigma(L_{jq}^2/R_j)} \tag{4.28}$$

If these calculations are performed for the 14-band system transmitted through the channel shown in Fig. 4.23, the optimum bits per symbol and power for each band are found to be as shown in Table 4.4. With an input SNR of 30 dB, the error rate is 2×10^{-5}. Compared to an undistorted ideal of 10^{-6}, this represents a deterioration of only 1.2 dB. If the calculation is simplified by keeping the transmitted powers the same in all bands ($\gamma_i = \frac{1}{14}$), there is a further deterioration of 0.3 dB.

Sounding of a Channel† to Establish the Optimum Loading. During the initialization of communication through an unknown channel the transmitter should send a signal to the receiver that will enable the latter to calculate the SNR in each band; the receiver should then calculate the γ_i and B_i, store the results for its own use, and also send the results back to the transmitter. If N is large, distortion and noise that change slowly across the whole band can be ade-

---

† The metaphor of navigation fits well here; a wide region of low SNR is a shoal and a single-frequency interfering tone is a mast of a sunken ship.

quately identified by sending signals in only a subset of all the bands (e.g., one out of every eight), but the single-frequency tones that are sometimes encountered can be detected and avoided only by checking the quality of signals in every band.

It is important to note that the receiver should measure SNRs rather than only receive levels. This is because some noise sources encountered in the telephone network (particularly in ADPCM systems) generate colored noise, and even noise that was originally white will be colored by the amplitude response of the part of the channel that comes after its source.†

Furthermore, we shall find that, particularly with the simpler methods of implementation, the signals within each band may be significantly contaminated by ISI, ICI, and noise. Therefore, the sounding signals that are sent by the transmitter should be such that the receiver can measure the combined effects of all three; the calculation of the optimum loading should then be based on these rather than on noise alone. The format of these signals has not been disclosed by Telebit (the only manufacturer to use adaptively loaded OMQASK so far), but we may be able to deduce some more about them after we have considered receiver methods in Chapter 5.

Adaptive Adjustment of the Data Rate. In the 14-band example it was a coincidence that rounding of the B_i resulted in a B_{tot} equal to the nominal 96; in general, it will not. Therefore, all modems that use this modulation method with adaptive configuring of the bands must have some way of controlling the input data rate. Since neither modem (DCE) nor terminal (DTE) is usually capable of dealing with a variable clock frequency, the data rate should be controlled by accepting data at the nominal rate and using some form of flow control (see Chapter 12) to tell the DTE when the DCE's input buffer is full and it can accept no more data.

Then, instead of striving to achieve a nominal total bit rate at a minimum but unspecified and unpredictable error rate, a more reasonable approach is to set the number of levels in each band so as to achieve a specified error rate with the measured SNR in that band. In this case the modem throughput would be continuously variable up to some maximum.

4.5.3 Conventional Implementation of a Transmitter: OMQAM

The transmitters described by Saltzberg [Sa2] and Hirosaki [Hi2] use $2N$ parallel data signals, each with a symbol rate of f_s, to modulate N carriers with frequencies given by

$$f_{c,n} = f_{c,1} + (n - 1)f_s \qquad n = 1 \text{ to } N$$

---

† For this reason the standard laboratory practice of testing modems by adding white noise at the input to the receiver is unrealistic. This will be discussed in more detail in Section 4.5.5.

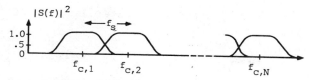

Figure 4.25 Power spectra of partially overlapping multichannels.

The signal within each band is band-limited to no more than 100% excess bandwidth (i.e., each spectrum overlaps only its immediate neighbors) and uses conventional Nyquist spectral shaping (shared equally between transmitter and receiver). The individual power spectra are shown in Fig. 4.25, and it can be seen that the aggregate power spectrum is flat across most of the band. The lower and upper "Nyquist" frequencies are $f_{c,1} - f_s/2$ and $f_{c,1} + (2N - 1)f_s/2$, and the equivalent excess bandwidth ratio is α/N.

Orthogonality is achieved (i.e., ICI is eliminated) by using SQAM within each band and alternating, from one band to the next, the assignment of the cosine and sine carriers to the undelayed and delayed data streams; this is illustrated in Fig. 4.26.

The filtering of the baseband signals could be done in any one of several ways, but Hirosaki showed that the amount of processing (usually measured by the number of multiplications needed) can be greatly reduced by using FIR filters and implementing the complete set by an FFT. If the composite signal is

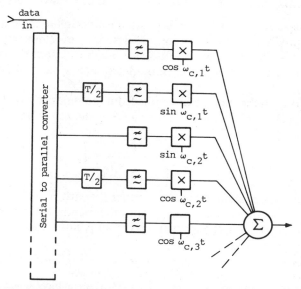

Figure 4.26 OMQAM transmitter: showing alternate staggering of subbands.

"wide-band" (i.e., $Nf_s \gg f_{c,1}$), then the number of multiplications per output sample is given by

$$N_{\text{mult}} = \frac{2M}{N} + 2 \log_2 N + 2$$

where M is the order of the FIR filter. A further economy can be achieved if the first carrier frequency $f_{c,1}$ is an odd-integer multiple of $f_s/2$†; then

$$N_{\text{mult}} = \frac{M}{N} + 2 \log_2 N$$

Because the baseband filters must select a band from 0 to $f_s/2$ using a sampling frequency of at least Nf_s, their required order M is approximately proportional to N. Hirosaki calculated that for an ICI to signal ratio of -30 dB, $M = 14N$, but for the multilevel signals for which OMQAM is often used, the ICI should be less than this, and $M = 20N$ is probably preferable. Then

$$N_{\text{mult}} = 20 + 2 \log_2 N$$

The theory of an FFT implementation of an OMQAM transmitter is described very well in [Hi2]; further details would depend very much on the characteristics of the particular signal processor used.‡

4.5.4   Implementation of a Transmitter Without Filters: OMQASK

The $2N$ low-speed data streams can be used to modulate the carriers directly without any prefiltering. This method was described by Weinstein and Ebert in [W&E] and has been used in the Telebit modem. Because the carrier amplitudes are keyed instead of being modulated, the system should be called *OMQASK*.

The spectrum of the signal that results from the input of a rectangular pulse $p(t)$ $(= 1/T$ for $0 < t < T$, and $= 0$ elsewhere) to the nth band is

$$S_n(f) = \text{sinc} \left[\frac{f - f_{c,n}}{f_s} \right] \tag{4.29}$$

and since the carrier frequencies are spaced f_s apart, it can be seen that the spectrum is zero at the center of each of the other bands. Nevertheless, it is clear that there will still be ICI; the spectrum is certainly not zero across the whole of

---

† The symbol rate is usually low enough that this is not a serious constraint; the 14-band system already considered is an example with $f_s = 200$ Hz and $f_{c,1} = 500$ Hz.

‡ Some processors (for example, Texas Instruments' TMS320) multiply almost as fast as they do the other simpler operations; for these, it becomes more important to simplify the structure of the whole program than to minimize just the number of multiplications.

the other bands, and, as shown in Fig. 4.11, it decays so slowly outside its passband that it would seem that about 60 channels on either side might contribute interfering signals at a level higher than the permitted -40 dB. Let us consider the question of ICI in more detail.

If the signal with a spectrum given by (4.29) is demodulated by $\cos 2\pi f_{c,n} t$, then the original baseband signal, $p(t)$, is recovered. If, on the other hand, it is demodulated into another band that is offset by mf_s, then the baseband signal that may cause ICI is

$$p_m(t) = \int_{-\infty}^{\infty} \frac{\sin \pi(f - mf_s)T}{\pi(f - mf_s)T} \exp(j2\pi ft) \, df = p(t) \exp(j2\pi mf_s t) \quad (4.30)$$

Thus the ICI appears in both the in-phase and quadrature channels of the other bands as a burst, of duration T, of sine (or cosine) waves of the offset frequency.

All the detection techniques that may be used in a receiver are equivalent to integrating this signal from 0 to T; because of the spacing of the carriers, this integral is zero for all values of m. Thus, at least for undistorted channels, orthogonality is ensured.

Use of a Guard Space to Combat the Effects of Channel Distortion. The main effect of channel distortion is to disturb the integration of $p_m(t)$ in (4.30); the contributing terms no longer cancel exactly, and ICI results. A method of remedying this was suggested in [W&E]; we will describe the method here, but must postpone an analysis of its effectiveness until we have studied receivers in Chapter 5.

The basic pulse duration of T and symbol rate of f_s are retained, but the spacing of the carriers is increased from f_s to $f_s(1 + \epsilon)$; consequently, the number of bands is reduced to $N/(1 + \epsilon)$. Orthogonality on an undistorted channel is restored by integrating the demodulated signals only over time intervals of $T/(1 + \epsilon)$ or, approximately, $(1 - \epsilon)T$. The part of each baseband pulse that is transmitted from $t = 0$ to ϵT is used to allow the transient response of each channel to decay, and data detection is achieved by integrating from ϵT to T, as shown in Fig. 4.27*b*.

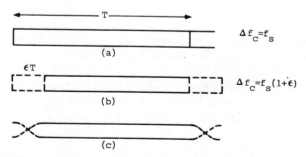

Figure 4.27 Block of data: (*a*) with no guard space; (*b*) with guard space but no shaping; (*c*) with guard space and shaping.

Weinstein and Ebert also suggested shaping the pulses in the time domain as shown in Fig. 4.27c in order to increase the rise-time of the baseband pulse, or, more specifically, to reduce the time derivatives of the signal; if the amplitude response of the channel is modeled as a polynomial of the frequency variable, this reportedly reduces ICI caused by both amplitude and delay distortion. However, the mechanism by which this occurs is not clear, and no quantitative analysis has been published; moreover, since the shaping cannot be implemented without a lot of extra signal processing, we will not consider the method further.

4.5.5   A Tentative Comparison of Single-Band and Multiband Modulation

The relative merits of single-channel and multichannel modulation (with or without adaptive loading) depend on the types of distortion and noise encountered on the channels.

If there is a lot of unpredictable amplitude distortion and the noise is white, then, as we saw in the example, a fixed configuration multiband system will probably perform considerably worse than a single-band system. This is because the error rate in a single-band system is a function of the average noise across the whole band, but in a fixed multiband system it is dominated by the error rate in the bands with the *lowest* SNR. Therefore, a fixed system should be considered only for those channels for which the amplitude response is accurately predictable.

Performance over Narrow Voice-Band Channels with Gaussian Noise. For communication over telephone channels whose effective bandwidth is limited by amplitude distortion to much less than the nominal 3.1 kHz, the noise enhancement of a conventional single-channel modem that uses a linear equalizer will typically be about 3–4 dB; with a decision-feedback equalizer it will be about 1–2 dB†; with the multichannel approach it will typically be 1 dB. Therefore, if decision-feedback equalization can be used, the few tenths of a decibel of improvement in performance possible from a multiband modem certainly do not justify the complicated hand-shaking protocol needed to configure the transmitter and the large amount of signal processing needed in the receiver.

Wider (Only Mildly Distorted) Voice-Band Channels. All high-speed single-channel modems designed for the PSTN confine 99% of their signal power in a bandwidth of about 2.6 kHz (usually centered around 1.8 kHz) under the presumption that the voice channel outside that range will often be useless; this is an example of designing for the worst case. Because the signal spectrum is fixed, the modem cannot take advantage of channels that are significantly wider than this.

---

† We are anticipating here the results of Chapters 8 and 9, but that is unavoidable; decisions about modulation methods must consider the capabilities of receivers.

If the channel is fairly wide and has little amplitude distortion (achieved with loaded local loops and high-quality carrier systems), a multichannel modem designed to transmit 19.2 kbit/s in a total bandwidth of, say, 3.2 kHz would have a very significant 4.8 dB noise advantage over a single-channel modem with only a 2.4 kHz Nyquist bandwidth.

Channels with Constant Attenuation and Impulse Noise. Hirosaki [HH&S] described an OMQAM modem for use in a 60–108 kHz groupband. The amplitude response is very flat, so all the bands can be loaded equally, and no adaptive configuring is needed. However, if the number of bands is not very large, delay distortion in the outer channels becomes significant; this must be anticipated and the adaptive equalizers for the outer channels made larger.

With a flat channel, the performance of an OMQAM system with white Gaussian noise is theoretically the same as that of a single-channel system,† but the sensitivity to impulse noise is much less; therefore, since this is sometimes the predominant impairment in groupband transmission, an OMQAM (or OMQASK) system is an attractive alternative to the more traditional approaches.

4.6 SQUARE-WAVE MODULATION AND STORED-REFERENCE TRANSMITTERS

So far we have considered four different types of modulators in which the following conditions exist:

1. The input baseband signal is band-limited and the carrier is a pure sinusoid. This is an "ideal" situation that is rarely achieved, or even attempted, in practice.

2. The baseband signal is band-limited, but the carrier is a square-wave; some simple output filter must be used to reject the sidebands of odd harmonics of the carrier. Most QPM and QAM transmitters are implemented in this way.

3. The baseband signal has infinite bandwidth, but the carrier is a single-frequency sinusoid; QPSK and SQPSK are examples of this. Nonlinear distortion occurs in this case when those components of the baseband signal at higher frequencies than the carrier are folded over into the passband as shown in Fig. 4.28. However, most QPSK signals are "narrow-band" (i.e., the ratio of carrier frequency to symbol rate is high), and the folded over components are either removed by a mild-output bandpass filter, or just ignored.

---

† The comparison in [HH&S] with a very suboptimum V.37 modem ignores the fact that the 3 dB "loss" of the partial-response Class IV shaping can be easily retrieved in the receiver (see Chapter 9).

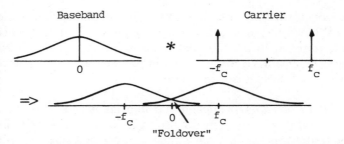

Figure 4.28 Modulation of baseband signal with bandwidth greater than f_c.

4. Both the baseband signal and the carrier have infinite bandwidth. The only example of this that we have considered so far is MSK, and the previous comments about narrow-band signals usually apply to MSK also. However, for voice-band modems the modulation is usually "wideband," and because it is sometimes advantageous to use infinite bandwidth inputs, we must study the phenomenon of foldover carefully.

4.6.1 The Theory of Fold-Over

If any (analog or derived from data) baseband signal $f(t)$ is used to modulate carriers $\cos \omega_c t$ and $\sin \omega_c t$, the spectra of the outputs at positive frequencies can be expressed most informatively as shifted versions of the baseband spectrum:

$$F_c(\omega + \omega_c) = \frac{F(\omega)}{2}$$

and

$$F_s(\omega + \omega_c) = \frac{F(\omega)}{2j}$$

If now $f(t)$ has infinite bandwidth and is used to modulate square-wave carriers $c2(\omega_c t)$ and $s2(\omega_c t)$† that are normalized so that the amplitudes of their fundamentals are unity, the products are given by

$$f_{c2}(t) = f(t)\, c2(\omega_c t) = f(t)\, [\cos \omega_c t - \tfrac{1}{3} \cos 3\omega_c t + \tfrac{1}{5} \cos 5\omega_c t - \cdots]$$
$$f_{s2}(t) = f(t)\, s2(\omega_c t) = f(t)\, [\sin \omega_c t + \tfrac{1}{3} \sin 3\omega_c t + \tfrac{1}{5} \sin 5\omega_c t + \cdots]$$

Again, the spectra of these can be expressed as the sums of shifted versions of the baseband spectrum $F(\omega)$:

---

† The "2" part of the symbols indicates that they are two-level carriers; later we will use $c3$ and $s3$ to denote three-level carriers.

$$2F_{c2}(\omega_c + \omega) = F(\omega) + [F(\omega + 2\omega_c) - \tfrac{1}{3}F(\omega - 2\omega_c)]$$
$$- [\tfrac{1}{3}F(\omega + 4\omega_c) - \tfrac{1}{5}F(\omega - 4\omega_c)] \qquad (4.31a)$$
$$+ [\tfrac{1}{5}F(\omega + 6\omega_c) - \tfrac{1}{7}F(\omega - 6\omega_c)] \ldots$$

and $\quad 2jF_{s2}(\omega_c + \omega) = F(\omega) - [F(\omega + 2\omega_c) - \tfrac{1}{3}F(\omega - 2\omega_c)]$
$$- [\tfrac{1}{3}F(\omega + 4\omega_c) \text{ h } \tfrac{1}{5}F(\omega - 4\omega_c)] \qquad (4.31b)$$
$$- [\tfrac{1}{5}F(\omega + 6\omega_c) - \tfrac{1}{7}F(\omega - 6\omega_c)] \ldots$$

Some interesting observations can be made from these equations:

1. The components of the wideband baseband signal that, when multiplied by square waves, fold into the passband are in the frequency regions around integer multiples of $2\omega_c$.
2. Those components around even multiples of $2\omega_c$ fold over and add to the baseband spectrum $F(\omega)$ in the same way in both channels, but those around odd multiples add in opposite ways.
3. The modulated spectra are not symmetrical about ω_c; the foldover causes a tilt in the passband. Furthermore, the tilt is different in the two channels.

The infinite bandwidth of the baseband signal occurs whenever sampled-data processing of the signal is used. An example of this is shown in Fig. 4.29, where the carrier and data samples are synchronized. So that the number of different samples of the carrier that have to be generated will be finite, the argument of the cosine must repeat after a finite number of samples. That is, the ratio of carrier frequency to symbol rate must be a ratio of integers.

This is a necessary condition, but it is not sufficient; the sufficient condition depends on how much filtering is done before the modulator. We will consider two extreme examples: one where the input data signal is not filtered at all and one where very precise filtering is performed (but the advantages of square-wave midulation are nevertheless retained).

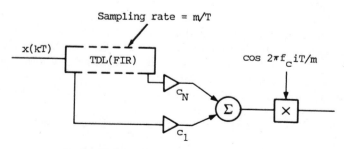

Figure 4.29 Sampled-data filter and modulator.

4.6.2 Square-Wave Modulation of Square-Wave Data Signals

If the baseband signal is an unfiltered data signal (i.e., rectangular pulses with the information contained in the sign), the modulation can be performed very simply by an exclusive-OR gate.

If two baseband data signals are to modulate two quadrature "square-wave" carriers, the most straightforward way is to use two exclusive-OR gates and algebraically sum the outputs; however, this requires balancing the outputs of the two gates. A more tolerant circuit results from considering the carriers as three-level signals that take the values $+1, 0, -0$, and 0 successively as shown in Fig. 4.30a; then the modulator (exclusive-OR gate) is timeshared between the two channels as shown in Fig. 4.30b.†

This type of modulation (one- or two-dimensional, with two- or three-level carriers) produces the greatest foldover effects, and we must examine them very carefully.

Consider a repetitive data signal of period nT. The input to the modulator will have a fundamental frequency of f_s/n and, if not strictly band-limited, harmonics at mf_s/n; the foldover terms in (4.31) will generate an equivalent set of baseband components at $(mf_s/n) \pm 2if_c$. In order for these to coincide with the unmodulated baseband components (i.e., for the foldover not to introduce any extra frequencies),

$$\frac{mf_s}{n} \pm 2if_c = \frac{m'f_s}{n}$$

That is

$$f_c = \frac{m - m'}{ni}\frac{f_s}{2}$$

Since this must be true for all data patterns, and for all harmonics of the fundamental data frequency and of the carrier (i.e., for all values of n, m, and i), it follows that, as first shown by van Gerwen and van der Wurf [G&W1],

$$f_c = \left(\frac{K}{2}\right)f_s \tag{4.32}$$

If the carrier frequency and symbol rate are related thus, the effect of foldover is only to change the magnitude of the baseband components and thereby to change the pulse shaping and spectrum in a linear way that could, if necessary, be corrected by a subsequent bandpass filter.

This ratio holds for the following medium-speed voice-band modems:

Bell 212, V.22 and V.22 bis: $f_c = 1200$ or 2400 Hz, $f_s = 600$ s/s and $K = 4$ or 8
Bell 201 and V.26: $f_c = 1800$, $f_s = 1200$ s/s and $K = 3$

---

† Ignoring, for the moment, the signals f'_c and f'_s, and the circuitry within the dotted lines.

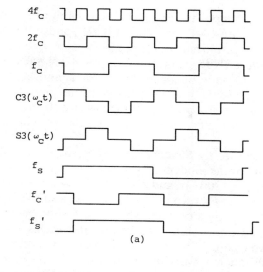

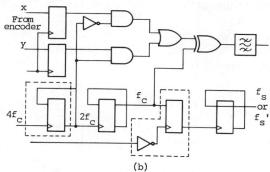

Figure 4.30 Square-wave times square-wave modulation: (*a*) timing diagram; (*b*) schematic— dotted lines enclose circuitry for syncopated timing.

However, these all use quadrature modulation and, as noted previously, the effects of foldover would be different in the two channels. Consequently, if the two modulated signals were added directly, the passband signal would be non-linearly and irreversibly distorted.

There are several ways to correct for this:

1. Use two separate bandpass filters with different slopes in their passbands for the two channels; this is too wasteful of hardware to be acceptable.

2. Stagger the data timing on the quadrature channel so that the relationship between the data and the carrier, and therefore the effects of foldover, are the same on both. This occurs in all SQPM and SQPSK modulators, but it is not applicable to voice-band modems, which, in order to facilitate timing recovery in the receiver, use simultaneous timing in the two channels.

3. Offset the timing of the carrier by one-eighth of a period. All previous descriptions of square-wave modulation [C&P] and [C&N], have shown the carrier changes synchronized with the data changes (f_s in Fig. 4.30 for the case of $K = 4$). However, the pairs of terms in (4.31) such as $[F(\omega + 2\omega_c) - \frac{1}{3}F(\omega - 2\omega_c)]$ that lead to inbalance of the two channels can be partially canceled by delaying the data by one eighth of a carrier period (f_s' in Fig. 4.30). We will call this timing *syncopated* to distinguish it from *staggered* or *offset* that describe the relationship between the data on the two channels.

Reduction of Foldover by Syncopated Timing. The principle of syncopation can be illustrated by the modem on which it was first used — the Bell 212 [Pe]. In the low-band $f_c = 1200$ Hz, and $K = 4$; the carrier waveforms that are modulated by the data are as shown in Fig. 4.30a, and a simple (unpatented) implementation of the modulator is shown in Fig. 4.30b — now including the circuitry within the dotted lines. As before, the exclusive-OR is timeshared between the two channels, but the timing of the data is now delayed by resampling the carrier square wave with the inverse of the $4f_c$ clock. The effects of the foldover (in both nonlinear distortion and shaping the common passband) can be calculated as follows.

The Fourier transforms of a basic rectangular pulse $f(t)$ modulated by the three-level carriers, $c3(\omega_c t)$ and $s3(\omega_c t)$, can be written as functions of u, the passband angular frequency:

$$F_{c3}(u) = 2\text{Re}\left\{ \frac{\sin(uT/32)}{u} \exp\left(\frac{-juT}{32}\right) \right.$$
$$\left. \left[1 - \exp\left(\frac{-j3uT}{16}\right)\right]\left[1 - \exp\left(\frac{-juT}{4}\right)\right]\right\} \quad (4.33a)$$
$$= 8\left[\frac{\sin(uT/32)}{u}\cos\left(\frac{uT}{4}\right)\sin\left(\frac{uT}{8}\right)\sin\left(\frac{3uT}{32}\right)\right]$$

and

$$F_{s3}(u) = 2\text{Im}\left\{ \frac{\sin(uT/16)}{u} \exp\left(\frac{-juT}{8}\right)\left[1 - \exp\left(\frac{-juT}{4}\right)\right]\right\}$$
$$= 8\left[\frac{\sin(uT/32)}{u}\cos\left(\frac{uT}{4}\right)\sin\left(\frac{uT}{8}\right)\cos\left(\frac{uT}{32}\right)\right] \quad (4.33b)$$

These can be transformed to shifted versions of the baseband spectra by replacing u by $(\omega_c + \omega)$ and recognizing that $\omega_c T = 4$. Then two modulator transfer functions can be formed from the ratios of half the sum and half the difference of $F_{c3}(\omega_c + \omega)$ and $F_{s3}(\omega_c + \omega)$ to the baseband transform, $F(\omega)$; the average term is the common transfer function for the two channels, and the difference term is a measure of the nonlinear distortion introduced. A lot of tedious trigonometry results in

$$H_{av}(x) = \frac{F_{c3}(\omega_c + \omega) + F_{s3}(\omega_c + \omega)}{F(\omega)}$$

$$= \frac{\pi}{4} \frac{x}{1+x} \left[\sqrt{2} + \frac{1}{\sin(\pi x/4)} \right] \tag{4.34a}$$

and

$$H_{nl}(x) = \frac{F_{c3}(\omega_c + \omega) - F_{s3}(\omega_c + \omega)}{F(\omega)}$$

$$= \frac{\pi}{4} \frac{x}{1+x} \left[\sqrt{2} - \frac{1}{\cos(\pi x/4)} \right] \tag{4.34b}$$

where $x = \omega/\omega_c$.

These two transfer functions are plotted in Fig. 4.31. Several interesting points about these functions should be noted:

1. For $x \ll 1$, the narrow-band case, the average is approximately unity, and the nonlinear part is very small, as would be expected.

2. At $x = -0.25$, the lower bandedge for a low-band 212 or V.22 modem, the difference transfer function has a magnitude of 0.103. This means that at this frequency the components that cause nonlinear distortion are only 19.7 dB down—just tolerable for a four-point signal space. By comparison, the more conventional arrangement in which the carriers and data change simulta-

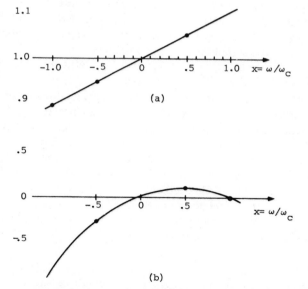

Figure 4.31 Modulator transfer functions: (a) linear (average); (b) nonlinear.

neously, results in "nonlinear" components that are only about 14 dB down at the bandedges — intolerable even for a cheap 212.

3. The shaping that is common to both channels increases with frequency (by 0.48 dB from the lower to the upper bandedge) — the opposite of what would be expected from the fact that the output signal is composed of rectangular pulses of duration $T/8$ that have a $\text{sinc}(uT/16)$ shaping.

4.6.3 Square-Wave Modulation of Partially Filtered Baseband Signals

For all values of K less than 4 (i.e., for signals with a band wider than the 212 low-band one just considered), the amount of nonlinear distortion introduced by square-wave quadrature modulating two square-wave data signals would be intolerable, and some filtering to remove the components of the baseband signals around odd multiples of $2\omega_c$ is needed. The basic method is to process both channels as shown in Fig. 4.29; since four samples of the carrier per period are needed, the most common sampling frequency for the data filter is $4f_c$ ($= 2Kf_s$). For modems such as a V.26 ($K = 3$), very simple transmitters comprising CMOS digital logic and a few elementary D/A converters (resistors for active RC, or capacitors for switched-capacitor implementation) can be designed.

4.6.3.1 A Stored-Reference Transmitter for a V.32 Modem.   As an example of a much more complex design, we will consider a V.32 modem. This uses echo canceling to achieve full-duplex operation, and in order to reduce the complexity of the canceler, the echo — and therefore the transmitted signal itself — must be a linear function of the transmitted data.† We will discuss the required attenuation of the echo in more detail in Chapter 11; at this point it is sufficient to say that the nonlinear components of the transmit signal must be at least 60 dB below the signal level in the passband.

As shown in Table 1.3, $f_c = 1800$ Hz and $f_s = 2400$ s/s. Since $K = 1.5$, this does not even satisfy the basic requirement of (4.32) for a one-dimensional sampled-data transmitter. However, filtering out the components of the baseband signal around $2\omega_c$, $6\omega_c$, and so on will ensure that the square-wave modulation introduces no extra frequency components *and* will make the modulator transfer functions for the two channels the same.

The baseband spectrum uses approximately 16% excess bandwidth and extends to about 1400 Hz; the passband spectrum therefore extends from about 400 to 3200 Hz. We could, theoretically, continue to use the three-level carriers; this would simplify the design of the digital part of the transmitter, but it would mean that the sampling frequency would be 7200 Hz ($4f_c$), and the output smoothing filter would have to pass 3200 Hz and reject 4000 Hz (the

---

† The cancellation of nonlinear echoes has been described by Agazzi [AM&H], but not enough information about telephone channels is available yet to determine whether such sophisticated processing is necessary.

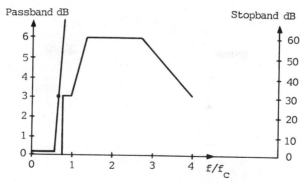

Figure 4.32 Attenuation requirements for V.32 baseband filter.

limit of the lower sideband of the sampling frequency). This is too stringent a requirement, so a sampling frequency of 14.4 kHz should be used.

Since eight samples of the carrier are calculated per period, a carrier signal that has no third or fifth harmonic can be synthesized. Therefore, the only important foldover term in (4.30) is $F(\omega + 2\omega_c)$. These components of the baseband signal around $2\omega_c$ are not reduced by any harmonic number, so they must be attenuated by the full 60 dB.

The attenuation needed around $4\omega_c$ is more debatable and is open to some compromise. Foldover from this region into the passband will cause only linear distortion, which the echo canceler in the transmitter and the equalizer in the receiver can both deal with, but there is also the problem of linear modulation of these components by the carrier fundamental. After attenuation by the output smoothing filter, they must be at a level that meets the out-of-band power specification of the appropriate PTT (see Fig. 1.22). Then after reflection from the line and further attenuation by the receiver input filter, they must be at a sufficiently low level that they could be folded into the passband by the receiver sampling (typically at 9.6 ks/s) without causing significant distortion of a low-level received signal. A typical attenuation specification for the FIR filter is shown in Fig. 4.32.

If the impulse response of this filter extends over N symbol periods, then with six samples per period, there will be $6N$ coefficients, c_i ($i = 0$ to $6N - 1$). A simplified drawing of the transmitter is shown in Fig. 4.33. The inputs to the two filters, $x(kT)$ and $y(kT)$†, are impulses at the *symbol* rate, and each is followed (at the $6f_s$ sampling rate) by five zeros. The output is

$$s\left[\frac{(6k + i)T}{6}\right] = \sum_{j=0}^{N-1} [x_{k-j} \cos \phi_{k,i} - y_{k-j} \sin \phi_{k,i}] \, c_{6j+i} \text{ for } i = 0 \text{ to } 5 \quad (4.35)$$

---

† Previously we used x_p and x_q for these, but now double subscripts would be too cumbersome, so we will, temporarily, use x and y.

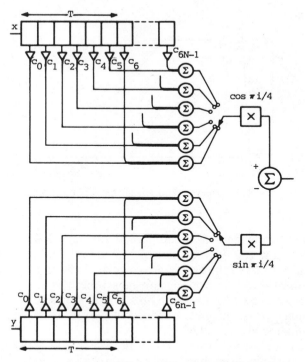

Figure 4.33 Block diagram of a basic V.32 transmitter.

where $x_k = x(kT)$, $\phi_{k,i} = (6k + i)\pi/4 + \theta$, and θ is an arbitrary phase angle that can be chosen to simplify the storage.

The transmitter can be implemented without any multiplications by storing in a ROM all possible contributions to the output at each sampling time from each input symbol and then adding the outputs of the two channels. Any point in the 32-point constellation that is used with trellis coding (see Fig.4.5a) can be defined by $x, y = \pm 1$, 3 or 5. The phase offset θ is most conveniently defined as $\pi/8$, so that the possible values of the carriers are $\pm \sin(\pi/8)$ and $\pm \sin(3\pi/8)$. If the signs of the data and the carriers are processed separately, and the signals from the two channels are added or subtracted appropriately, then only positive values need be stored, and the structure of the read-only memory (ROM) is as shown in Fig. 4.34a. As an example, the path through the ROM of an input sample of $(5,1)$, and the signs of its contributions to the output are shown in Fig. 4.34b. It is important to note that at each sampling time there are only $2N$ nonzero contributions to the output that have to be summed.

A Variation That Uses Bandpass Filters. Qureshi [Qu5] described a variation of the previous transmitter in which the signal is "modulated" first and then

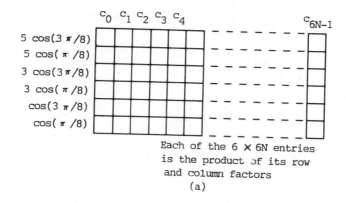

$$5 \cos(3\pi/8)$$
$$5 \cos(\pi/8)$$
$$3 \cos(3\pi/8)$$
$$3 \cos(\pi/8)$$
$$\cos(3\pi/8)$$
$$\cos(\pi/8)$$

Each of the 6 × 6N entries
is the product of its row
and column factors

(a)

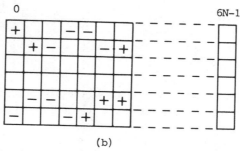

(b)

Figure 4.34 ROM contributions to output of V.32 transmitter: (a) ROM structure; (b) path through ROM of input (5,1).

passed through two bandpass filters. This is not a conventional modulation because the signal is multiplied by samples of the carrier taken at the symbol rate, which is much too slow for true sampling of the carrier. Consequently, the multiplier is a carrier that has been folded into the baseband. The product of the multiplication is in the form of impulses that have a flat, infinite spectrum, and the bandpass filters then select the appropriate part of that spectrum.

If the filters in both transmitters (the original and this variation) are realized in the conventional way (delayed input signals multiplied by stored coefficients, and then summed), the number of multiplications per sample is $(2N + 2)$ in the original low-pass implementation, and $2N$ in this bandpass variation; if the filters are realized by table look-up, the amount of storage needed is the same. The only significant advantage of this variation is that the bandpass filters do not need to be symmetrical about the center frequency and can be used to perform some preequalization of the transmitted signal, or can be designed in conjunction with the output smoothing filter for maximum efficiency.

The argument ϕ of the cosine and sine in (4.35) can be written as the sum of two terms:

$$\phi_{k,i} = [(6j + i)\pi/4 + \theta] + \left[\frac{3(k - j)\pi}{2}\right]$$

$$\triangleq \phi_{j,i} + \left[\frac{3(k - j)\pi}{2}\right]$$

and the second term associated with the x_{k-j} and y_{k-j}. Then

$$s\left[\frac{(6k + i)T}{6}\right] = \sum_{j=0}^{N-1} c_{6j+i} \cos \phi_{j,i} \, x'_{k-j} - c_{6j+i} \sin \phi_{j,i} \, y'_{k-j} \qquad (4.36)$$

where

$$x'_{k-j} = x_{k-j} \cos\left[\frac{3(k - j)\pi}{2}\right] - y_{k-j} \sin\left[\frac{3(k - j)\pi}{2}\right] \qquad (4.37a)$$

and $\qquad y'_{k-j} = x_{k-j} \sin\left[\frac{3(k - j)\pi}{2}\right] + y_{k-j} \cos\left[\frac{3(k - j)\pi}{2}\right] \qquad (4.37b)$

This represents a rotation of the vector (x,y), and since the argument of this rotation is an integer multiple of $\pi/2$, it is equivalent to a further encoding of the input data in which the output is advanced by $3\pi/2$ each symbol time; it can be incorporated into the original encoding.

The low-pass filters have been transformed to two different bandpass filters by multiplying their coefficients by samples of the carriers at the $6f_s$ rate; they can be modified if necessary. For ROM look-up implementation, there are now two $(6N \times 3)$ ROMs instead of the previous one $(6N \times 6)$ ROM.

Chapter 5

Receiver Structures and Components

This chapter will serve as an introduction to the complex subject of receiver design. It should be used like a travel guide — read before starting out, and then consulted frequently when studying individual subjects.

The main operations that may be needed in the receiver of a private-line modem, in the approximate order in which they will be performed, are as follows:

1. *Filtering.* This may be done in the passband and/or in the baseband after demodulation.
2. *Compromise Equalization.* This may be for amplitude and/or delay distortion.
3. *Automatic Gain Control (AGC).*
4. *Demodulation.* This is considered here in a much narrower sense than the "demod" that is half of a modem; here it means only frequency translation from passband to baseband. If the demodulation is to be homodyne or synchrodyne (for coherent detection), there must also be some way of recovering a carrier of the correct frequency and phase.
5. *Timing Recovery and Sampling.*
6. *Adaptive Equalization.*
7. *Signal Detection and Decoding.*

If any part of the processing in the receiver is to be done digitally, an Analog-to-Digital Converter (ADC) will also be needed. However, the design of these has been the subject of whole books and will not be discussed here. We will consider only the positioning of the ADC in the receiver and those aspects of its design that are linked with the AGC.

This chapter will deal in detail only with operations 1 – 4 and 7 (operations 5 and 6 and the ancillary operation of carrier recovery each need chapters of their own), but it will also show how interdependent all seven operations are and how important it is to consider this dependence when designing a receiver. Several other operations are necessary to make a modem suitable for operation as a "data set" on the PSTN; these are discussed in Chapter 12.

Not all the operations will be needed in every receiver, and often two or more are combined in a particular implementation so as to become almost indistinguishable. Some examples of this combination, and also of the interdependence are:

1. The compromise equalizer may be included in the bandpass filter.
2. The gain control part of the AGC may be performed in the ADC by using a multiplying DAC or a recycling ADC with a controllable reference voltage.
3. The timing and carrier recovery may be very interdependent. However, in order to shorten the time needed to train a receiver before data can be sent, considerable effort has been devoted to seeking recovery methods that can proceed independently, concurrently and at approximately the same speed.
4. The timing recovery and adaptive equalization can become intertwined if the timing is adjusted so as to minimize some parameter of the signal that is learned by the equalizer.
5. The adaptive rotator (Section 6.4.2) that is often used for carrier recovery can be considered to be part of the adaptive equalizer.

Some Typical Receiver Configurations. It must already be apparent that there are very many possible receiver configurations with different combinations of components. The capabilities and usefulness of these will become clear only after Chapters 5 – 10 have been studied, but it should be useful at this stage to show some typical receiver configurations and indicate the sections where their components are discussed.

1. A basic analog receiver that might be suitable for 4PSK or 4QAM is shown in Fig. 5.1. It uses forward-acting carrier recovery (Section 6.2) and baseband timing recovery (Section 7.4).
2. A somewhat more sophisticated analog receiver for 4QAM is shown in Fig. 5.2. It uses a forward-acting AGC with internal feedback (Section 5.3.1) to improve the operation of the carrier and timing recovery loops and to reduce the problems of d.c. offset in the basebands, and decision-directed feedback carrier recovery (Section 6.4).

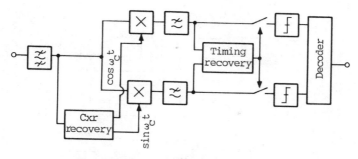

Figure 5.1 Basic analog receiver for 4QAM.

3. A nearly all-digital receiver with an analog (or switched-capacitor) front-end filter (Section 11.1) and a linear adaptive equalizer — typical of most first-generation V.22 bis modems — is shown in Fig. 5.3. It uses an AGC with feedback from the final slicers (Section 5.3.2), free-running demodulation, digital passband timing recovery with some internal feedback (Section 7.5), and a linear equalizer followed by an adaptive rotator (Sections 6.4.2 and 8.2.2).

4. An all-digital receiver with a DFE for a second-generation V.22 bis is shown in Fig. 5.4. It uses free-running, low-rate sampling and digital filtering, followed by an interpolator to generate symbol-rate samples for an AP/DFE (Section 9.1.6).

Passband or Baseband Processing? Several of the operations can be performed either in the passband or in the baseband after demodulation. The mathematics are slightly different because in the first case the signal is a real double-sideband signal and in the second case it is a complex baseband signal, but the theoretical results are exactly the same in the two cases. However, for

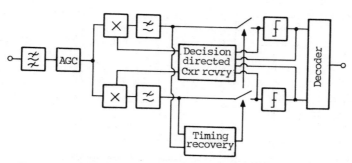

Figure 5.2 Analog receiver with decision-directed carrier recovery.

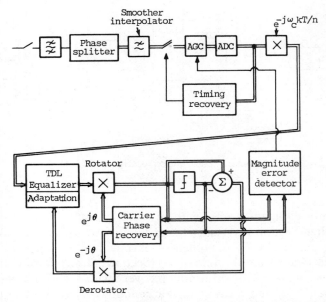

Figure 5.3 Nearly-all-digital receiver with switched-capacitor filters and linear equalizer.

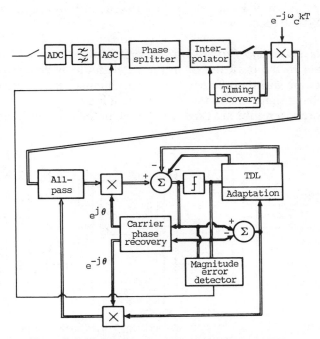

Figure 5.4 All-digital receiver with interpolator and DFE.

TABLE 5.1   Sources of Receiver "Noise"

Source	V.22		V.33
	ΔSNR	$10 \log(P_S/P_{imp})$ dB	ΔSNR
ACI from transmitter	−0.5	29.4	0
AGC jitter	−0.1	36.4	
Quantization noise from ADC			−0.2
Round-off noise in filters	−0.2	33.4	−0.1
Upper sideband of demodulation	−0.1	36.4	−0.1
Carrier phase jitter	−0.4	31.6	−0.4
Timing jitter	−0.2	31.6	−0.4
Residual ideal MSE[a] of equalizer	−0.2	33.4	−0.5
Extra MSE due to tap jitter	−0.3	31.6	−0.3
Noise enhancement by equalizer	−1.0		−2.0
Total change of SNR	−3.0 dB		−4.0 dB

[a] Mean-squared error.

most systems, practical considerations will favor one band over the other for each operation.

Some early modems performed adaptive equalization in the passband, but it seems that more recent developments of carrier recovery have made that unnecessary. We will consider only baseband adaptive equalizers.

Cumulative Effect of Imperfections in the Separate Operations. Of course, none of the operations listed above can be performed perfectly. The cumulative effect of the imperfections could be found by simulation, but this would be very laborious. Even if such simulation is to be part of the development process, preliminary estimates of the effects of the individual imperfections can be very useful in deciding design trade-offs, and a summary such as is shown in Table 5.1 should be prepared and updated often during the course of any complicated modem design project.

Most of the imperfections can be considered as sources of "receiver noise." They are not Gaussian because their maximum values are finite; nevertheless, if their effects are independent of each other and their powers are small compared to the Gaussian noise, they can be added cumulatively to the input noise on a power basis to produce an aggregate mean-squared error at the input to the decision device(s). Some of the effects of the imperfections do, of course, depend on others to some extent (e.g., carrier phase jitter will increase as the timing jitter increases), but the interdependence is usually secondary; all of them typically depend much more on the design of the receiver, the channel distortion, and the noise level.

Therefore, the easiest way to specify the separate operations is to define, for any expected received SNR (expressed in decibels), the maximum contribution to the change of SNR (ΔSNR, always negative) allowed for that imperfection. That is, if

$$10 \log \left(\frac{P_S}{P_N} \right) = \text{SNR}$$

and
$$10 \log \left(\frac{P_S}{P_N + P_{\text{imp}}} \right) = \text{SNR} + \Delta \text{SNR} \qquad (5.1)$$

then
$$10 \log \left(\frac{P_S}{P_{\text{imp}}} \right) \simeq \text{SNR} + [6.4 + 10 \log(-\Delta \text{SNR})]$$

The major sources of "noise" in a receiver and examples of their allowed contributions to a total specified ΔSNR of -3.0 dB when the input SNR is 20 dB are shown in Table 5.1 for the case of a V.22 bis high-band signal. (The receive signal is assumed to be at the lowest possible level and to have severe attenuation distortion.)

Preliminary estimates of the SNRs are also given for the much more complicated V.33 modem. [An input SNR has not been assumed — 30 dB is probably a reasonable figure to use as a base in (5.1).] Several interesting things can be learned by comparing these numbers with the V.22 bis numbers:

1. The V.33 is a half-duplex modem, and so has no ACI.
2. The effects of AGC jitter will be greater merely because there are more points in the constellation.
3. Quantization and round-off noise should be less because the range required of the ADC is less, and much less filtering is needed.
4. The effects of the same amount of carrier phase jitter would be greater, but the symbol rate is four times greater, and tracking of phase jitter is easier.
5. Timing jitter will be greater because the excess bandwidth is only 12.5% compared to 75% for a V.22.
6. Because of the smaller excess bandwidth and the greater distortion across the full band, the impulse response will be longer and have more ISI; the equalizer will not be able to do as good a job.
7. For the same reasons as in (6) the noise enhancement of a linear equalizer will be greater.

Robustness of a Receiver. Nearly all receivers depend on the transmission of a special sequence of signals at the beginning of communication (or after any prolonged period of silence from the transmitter) in order to train the various algorithms. If a receiver subsequently "loses lock" for any reason and needs to be retrained, this can be a very time consuming process and may greatly reduce the throughput of the data link. Therefore, considerable importance should be placed on the ability of a receiver to retain its training (even though making many errors) through gain and phase hits and impulse noise.

5.1 FILTERING

The bandpass filter (BPF) and the baseband low-pass filters (LPFs) should together provide the out-of-band attenuation and the in-band spectral shaping. However, the BPF by itself must reject any components of the input signal that might be folded into the signal band by demodulation by any carrier other than a pure sinewave, and only the LPF(s) can deal with any extraneous products of such demodulation.

The combined stop-band attenuation must reduce any out-of-band signals (whether they be from adjacent channels or from noise) so that when they are eventually folded into the band by sampling at the symbol rate,† they do not significantly augment the in-band noise. For half-duplex modems this is a simple task; 30 dB of attenuation would be more than enough. On the other hand, the problems of filtering in duplex modems are so special that they will be discussed separately in Section 11.1.

5.1.1 Passband Shaping of Band-Limited Signals

Initially we will consider only low-pass filters (or, by a simple frequency translation, symmetrical bandpass filters). This restriction is not, of course, essential to the development of the theory, but it makes a first explanation of it clearer and more easily grasped intuitively. At the end of this section we will discuss the generalization for asymmetrical bandpass channels.

We will be concerned with the design of only the receive filter(s). The problem of joint optimization of transmit and receive filter for a distorted channel is a very difficult one, and it is rarely solved in practice; the problem is usually avoided by using one of three simple design strategies:

1. Putting all the Nyquist shaping in the transmitter. In the past this was sometimes done for convenience of implementation (e.g., the Bell 201 and 208 modems, and some V.26 and V.27 modems), but nowadays it is more likely to be done to satisfy some other requirement such as constancy of envelope of the modulated signal (see Section 4.3).

2. Putting one half (on a log scale) of the required shaping in the transmitter with the expectation that the other half will be provided by the receiver. This is by far the most common arrangement, and will be the subject of most of this section.

3. Putting little or no shaping in the transmitter with the expectation that most of it will be provided by the channel. This will be discussed briefly in Section 11.3 on subscriber loop modems.

---

† If fractionally spaced equalizers are used, then there is a two-stage folding. The filter must deal with the noise and interference beyond f_s — assuming a $T/2$ equalizer — and the equalizer with any remaining in the band from $f_s/2$ to f_s. This is discussed in more detail in Section 8.7.

5.1.1.1 *Theoretically Optimum Matched Filters.* A basic premise of data transmission is that if a message of N bits is sent, the lowest possible error rate in the presence of Additive White Gaussian Noise (AWGN) is achieved with a receiver that correlates the noisy received signal with all possible 2^N noiseless signals and chooses the message that results in the largest correlation. The correlation operation could be implemented by passing the received signal through 2^N separate filters whose impulse responses are the time inverses of the 2^N possible signals — delayed enough so that they are causal — and then sampling each output at $t = kT + \tau$; each filter is said to be "matched" to a possible signal.

The problem with such a receiver is that for large N, the amount of computation becomes enormous and impractical.

5.1.1.2 *Suboptimum Filters for Symbol-by-Symbol Detection.* Instead of waiting until a whole message has been received and stored, the signal can be processed as it is received, and hard decisions can be made at each sampling time about which bit or symbol was sent. If the processing is constrained to be linear, then the best procedure [Fo1] is to pass the received signal through a single filter (which, in practice, may comprise a band-pass filter, a low-pass filter, and an adaptive equalizer), sample the output at the symbol-rate, and input the samples to a threshold detector.

It is usually assumed that WGN is added at the input of the receiver, as shown in Fig. 5.5a. This is the condition under which all modems are tested in laboratories because it is the only one that is easily definable and repeatable. In reality, however, the noise may be added at many points in the channel, and the aggregate noise will be at least partially filtered by the channel. This more general case can be approximated by adding the noise from a single source at one point in the channel as shown in Fig. 5.5b. Then the study of the system can be simplified by temporarily combining the transmit filter and that part of the channel before the noise source into one block with transfer function $F_1(f)$ and that part of the channel after the noise source and the receive filter into a second block $F_2(f)$ as shown in Fig. 5.5c.

We will consider the design of the receive filter for several different combinations of transmit filtering and noise sources.

Half-Shaping, No Channel Distortion, AWGN. If the end-to-end shaping is to result in no ISI, then, as discussed in Section 3.1, the ideal receive shaping is given by

$$|G_R(f)| = |G_T(f)| = [G_{\text{Nyq}}(f)]^{1/2} \qquad (5.2a)$$

and
$$\text{Arg}[G_R(f)] = -\text{Arg}[G_T(f)] \qquad (5.2b)$$

where $G_{\text{Nyq}}(f)$ is any shaping that satisfies the Nyquist condition for zero ISI:

$$G_{\text{Nyq}}(f) + G_{\text{Nyq}}(f_s - f) = T \qquad (5.3)$$

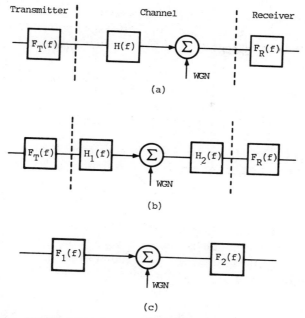

Figure 5.5 Systems with various arrangements of filtering: *(a)* laboratory setup with AWGN; *(b)* channel split before and after noise source; *(c)* equivalent filtering before and after noise source.

[Note that this is the folded form of the Nyquist requirement, and that, because $G_{\text{Nyq}}(f)$ is assumed to be real, we need not consider its complex conjugate.]

The SNR at the input to the detector for this ideal case can be calculated as follows. The transmitted signal power

$$P_T = \langle a^2 \rangle \int_{-f_s}^{f_s} |G_T(f)| \, df = \langle a^2 \rangle f_s \tag{5.4}$$

the signal power at the input to the detector is

$$P_s = \langle a^2 \rangle \int_{-f_s/2}^{f_s/2} |G_T(f) \, G_R(f) + G_T(f_s - f) \, G_R(f_s - f)|^2 \, df = \langle a^2 \rangle f_s \tag{5.5}$$

and the noise power there is

$$P_N = N_0 \int_{-f_s/2}^{f_s/2} [|G_R(f)|^2 + |G_R(f_s - f)|^2] \, df = N_0 f_s \tag{5.6}$$

The resulting SNR of $\langle a^2 \rangle / N_0$ is the best that can be achieved under these or, indeed, under any conditions. Furthermore, it can be proved [LS&W] by methods that are beyond the scope of this book that this pair of filters and a simple threshold detector constitute the best system possible for transmitting, through an undistorted channel, a PAM signal in which each level corresponds to a unique set of bits.

It is true that, as originally proved by Shannon [Sh1], much higher bit rates, with arbitrarily low error rates, are possible, but their attainment requires a much finer quantization of the transmit signal and sophisticated coding to allow the detection and correction of errors in the receiver.

Half-Shaping, Channel Distortion, Zero ISI at Detector. For the system shown in Fig. 5.5c, the requirement of zero ISI means that

$$F_1(f) F_2(f) + F_1(f_s - f) F_2(f_s - f) = T \exp(j2\pi\tau) \qquad \text{for } 0 < f < f_s \quad (5.7)$$

where the delay τ is included to make the impulse response of the receive filter causal; for the sake of simplicity only, we will assume $\tau = 0$ in the rest of this section. Then by substituting $F_2(f_s - f)$, found from (5.7), into a generalized form of (5.6), the noise power at the detector is found to be

$$P_N = 2N_0 \int_0^{f_s/2} |F_2(f)|^2 + \left[\frac{T - F_1(f) F_2(f)}{F_1(f_s - f)} \right]^2 df$$

This integral can be minimized by separately minimizing the integrand at each frequency. This is done by making $F_1(f) F_2(f)$ real; that is, by setting

$$\text{Arg}[F_2(f)] = -\text{Arg}[F_1(f)] \qquad (5.8a)$$

and

$$|F_2(f)| = T \frac{|F_1(f)|}{D(f)} \qquad (5.8b)$$

where

$$
\begin{aligned}
D(f) &= |F_1(f)|^2 + |F_1(f_s - f)|^2 \\
&= |F_T(f)H_1(f)|^2 + |F_T(f_s - f)H_1(f_s - f)|^2
\end{aligned}
\qquad (5.9)
$$

Hence the receive filter itself is defined by

$$\text{Arg}[F_R(f)] = -\text{Arg}[F_T(f) H_1(f) H_2(f)] \qquad (5.10a)$$

and

$$|F_R(f)| = T \frac{|F_T(f) H_1(f)|}{[|H_2(f)| D(f)]} \qquad (5.10b)$$

With this filter the (minimum) sampled noise power is

$$P_N = \int_0^{f_s/2} \frac{2N_0}{D(f)} df \qquad (5.11)$$

It can be seen that $1/D(f)$ is symmetrical about $f_s/2$, and therefore, as shown in Section 3.1.1, it can be considered as the transfer function of a sampled-data filter with sampling rate f_s. On the other hand, the rest of the Right-Hand Side (RHS) of (5.10 b) is a general function of f and must be realized by either a continuous filter or a sampled-data filter with a sampling rate high enough to prevent foldover. This part of the receive filter is "matched" to a single pulse of the receive signal; that is, it equalizes the delay of the signal and makes the *voltage* spectrum of the signal and the *power* spectrum of the noise equal at the input to the sampler.

Equations (5.9) and (5.10) can be considered as defining either a single receive filter or a combination as shown in Fig. 5.6.

Half-Shaping, Channel Distortion, AWGN, Minimized Noise and ISI. If the channel has significant attenuation distortion, then the equalizer part of the filter, in striving to exactly satisfy the requirement for zero ISI, might seriously enhance the noise. Therefore, it is better to minimize a combination of filtered noise and residual ISI. The ISI will usually extend over many symbol intervals, and the cumulative effect of many terms can be considered to be Gaussian; its mean-square value can therefore be added to the noise to form a composite MSE that should be minimized:

$$\text{MSE} = 2 \int_0^{f_s/2} N_0[|F_2(f)|^2 + |F_2(f_s - f)|^2]$$
$$+ \langle a^2 \rangle \, [T - F_1(f) \, F_2(f) - F_1(f_s - f) \, F_2(f_s{-}f)] \, df$$

The solution of this is as given in (5.9), except that now $D(f)$ must be defined by

$$D(f) = |F_1(f)|^2 + |F_1(f_s - f)|^2 + \frac{N_0}{\langle a^2 \rangle} \tag{5.12}$$

Intuitively, this seems a reasonable result; the matched filter part is the same as before, but now the noise is added to the signal power in the denominator in order to prevent the equalizer transfer function from becoming too large — and thereby greatly enhancing the noise — if the signal is strongly attenuated at any frequency.

It can be seen that if the channel introduces any unpredictable distortion, both the matched filter and the symbol-rate equalizer would have to be adaptive; this would be very difficult to implement. Furthermore, the amplitude

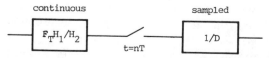

Figure 5.6 Ideal receiver as a combination of a matched filter and a symbol-rate equalizer.

response of the matched filter duplicates that of the part of the channel before the noise source and thereby makes the task of the symbol-rate equalizer much harder. Therefore, matched filters, per se, are rarely used. Nevertheless, the concept is useful for understanding other filter and equalizer structures such as compromise equalizers (Section 5.2), fractionally spaced equalizers (Section 8.7), and prefilters for decision-feedback equalizers (Section 9.1).

To help in this understanding, the separate functions that the ideal combination shown in Fig. 5.6 performs should be emphasized:

1. It filters out the noise beyond the band of the signal (i.e., for $|f| >$ $(1 + \alpha)f_s/2$); this is the only function that does not have to be adaptive.
2. It matches the attenuation of the channel and thereby minimizes the effect of the noise, in the band $(1 - \alpha)f_s/2 < |f| < (1 + \alpha)f_s/2$. Note that if we are considering the filter and equalizer as a combination, we can disregard the matching of the attenuation in the band $|f| < (1 - \alpha)f_s/2$ because that is counteracted by the equalizer.
3. It equalizes the delay of the channel in that same band and thereby minimizes the loss of power due to foldover when the signal is sampled at the symbol rate.
4. It delays the signal so that it can be sampled at particular times. We shall see later that the performance of this ideal combination of matched filter and symbol-rate equalizer is very insensitive to the sampling time. The time is therefore fixed in some simple way (see Section 7.6) and the filter adjusted to match it.
5. It equalizes both the folded attenuation and delay.

The receiver configuration that is often used — a fixed filter, compromise equalizer, and a symbol-rate adaptive equalizer — is suboptimum because it does not perform the second and third functions. It also chooses the sampling time to fit the signal rather than fixing the time and adjusting the equalizer. Nevertheless, it is much simpler to implement, and is often quite adequate.

Special Case of a Palindromic†　Channel.   It can be seen that if the approximation is made that the noise is added in the "middle" of the channel, then

$$H_1(f) = H_2(f)$$

and
$$F_R(f) = \frac{F_T(f)}{D(f)}$$

Thus the optimum combination is the same filter matched to the transmit filter that we found in the undistorted case, plus a delay equalizer and a symbol-

† This word may seem inappropriate here, but it perfectly describes a channel that looks the same in both directions.

rate equalizer. In this case matching the attenuation of the channel becomes unnecessary, and the simpler configuration of fixed filter, compromise equalizer, and symbol-rate equalizer is suboptimum only in that it does not equalize the delay in the excess band.

Although this may seem to be a rather special case, it is a good model for data transfer on the PSTN, where a typical connection comprises a distorting two-wire loop, a noisy four-wire path, and another distorting two-wire loop.

Full Shaping in the Transmitter, No Channel Distortion. We will consider here the particular case of an SQAM or TFM transmitter for which

$$F_T(f) = \frac{1 + \cos(\pi f / f_s)}{2} \quad \text{for } |f| < f_s$$

which, for brevity, we will write as

$$F_T(u) = \frac{1 + \cos u}{2} \quad \text{for } |u| < \pi$$

where $u = \pi f / f_s$.

The simplest way of designing the receive filter would be to ignore all suggestions for matching and make $F_R(u) = 1$ for $|u| < \pi$. The noise bandwidth of the filter would be double that of the optimum given by (5.3), which has led to the frequently-quoted figure of 3 dB deterioration in SNR at the input to the detector compared to the optimum. However, this ignores the fact that because of the extra filtering in the transmitter, the average pulse energy $\langle a^2 \rangle$ could be increased by a factor of $\frac{4}{3}$ while still maintaining the same transmitted power. That is,

$$
\begin{aligned}
P_S &= \frac{4\langle a^2 \rangle}{3} \int_{-f_s}^{f_s} \left[\frac{1 + \cos(\pi f / f_s)}{2} \right]^2 df \\
&= \frac{4\langle a^2 \rangle f_s}{3\pi} \int_{-\pi}^{\pi} \left[\frac{1 + \cos u}{2} \right]^2 du \\
&= \langle a^2 \rangle f_s
\end{aligned}
$$

Therefore, the resultant loss of SNR at the detector is

$$\Delta \text{SNR} = 10 \log(2 \times \tfrac{3}{4}) = 1.76 \text{ dB}$$

Most of this loss can be regained by using a receive filter defined by (5.8); that is,

$$F_R(u) = \frac{\dfrac{1 + \cos u}{2}}{\left(\dfrac{1 + \cos u}{2}\right)^2 + \left(\dfrac{1 + \cos(\pi - u)}{2}\right)^2}$$

$$= \frac{1 + \cos u}{1 + \cos^2 u} \tag{5.13}$$

This is plotted in Fig. 5.7; it is very similar to the optimum filter shown in Fig. 2 of [J&D] as the optimum filter for TFM. With this filter the noise power at the input to the detector is, from (5.11),

$$P_N = \frac{4N_0 f_s}{\pi} \int_0^{\pi/2} \frac{1}{1 + \cos u}\, du$$

which, from [Dw], is equal to $\sqrt{2}N_0 f_s$. The average signal energy is $4\langle a^2\rangle/3$, so the SNR is $(4\langle a^2\rangle/3\sqrt{2}N_0)$, which, compared to the optimum of $\langle a^2\rangle/N_0$, represents a loss of only 0.26 dB.

It is interesting to note that although the spectrum of the output of the receive filter,

$$F_T(u)\, F_R(u) = \frac{(1 + \cos u)^2}{2(1 + \cos^2 u)} \tag{5.14}$$

satisfies the requirement for zero ISI, it is not a raised cosine shaping.

Generalization of Formulas for Asymmetrical Bandpass Channels. In this section so far we have considered only real low-pass channels; these channels can also serve as the low-pass equivalents of symmetrical bandpass channels, but for asymmetrical channels, we need to generalize the formulas. Equation (5.10) for the optimum receive filter is still valid, but $D(f)$ must now be defined by

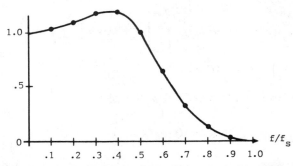

Figure 5.7 Amplitude response of ideal receive filter for a fully shaped, 100% excess bandwidth signal.

$$D(f) = \begin{cases} |F_1(f)|^2 + |F_1(f-f_s)|^2 & \text{for } 0 < f < f_s \\ |F_1(f)|^2 + |F_1(f+f_s)|^2 & \text{for } -f_s < f < 0 \end{cases} \qquad (5.15)$$

5.1.2 Passband Shaping for Time-Limited Signals

Theoretically a signal that is limited in time has infinite bandwidth, so there is no "stopband"; the passband shaping must be performed across an infinite bandwidth. In practice the signal is usually mildly band-limited, but nevertheless the received signal is often processed as though it had retained its original time-limited form.

The matched filters described in the last section could be implemented in either the frequency or the time domain, but for a band-limited signal, the former is more convenient; conversely, for time-limited signals processing in the time domain is more convenient.

Integrate and Dump: A Matched Filter for Time-Limited Pulses. For an ASK system, the single transmit pulse is a rectangle:

$$p(t) = \begin{cases} 1/T & \text{for } 0 < t < T \\ 0 & \text{otherwise} \end{cases}$$

and a noisy received signal can be correlated with this by integrating it from 0 to T. The result of the correlation (the output of the integrator) is then input to a threshold detector just as with a frequency-domain matched filter. In order to deal with a succession of modulated pulses, the output of the integrator is sampled at $t = nT$, and then immediately the integrator is reset ("dumped"); this is shown in Fig. 5.8a.

Weighted Integrate and Dump. Sometimes the time-limited pulse is shaped; the most important example of this, MSK, which is discussed in Section 4.3.2, has a basic pulse

$$p(t) = \begin{cases} \sin\left(\dfrac{\pi t}{T}\right) & \text{for } 0 < t < T \\ 0 & \text{otherwise} \end{cases}$$

In this case the correlation is performed by premultiplying the received signal by $\sin(\pi t/T)$ before integrating. A single demodulated baseband channel for an MSK receiver is shown in Fig. 5.8b.

This premultiplication may be difficult to perform with some types of hardware, and it may be better to use a frequency-domain filtering approach. The receive filter can be optimized according to the methods of the last section for any particular set of band-limiting filters in the channel (considered as distor-

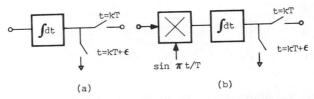

Figure 5.8 Integrate-and-dump circuits: *(a)* for BPSK; *(b)* weighted for MSK—only one dimension shown.

tion), but there are no simple closed-form expressions for the resultant sampled noise-power; simulation would seem to be the only way to evaluate the performance.

5.2 COMPROMISE EQUALIZERS

If a channel is significantly distorted, but an adaptive equalizer is not used, the distortion of a signal delivered to the threshold detector can be reduced by using a fixed equalizer that would counteract the distortion of some average channel. The attenuation and delay responses of this channel can be defined as the simple averages of the extremes (back-to-back and worst-case) or as weighted averages that consider the probabilities of each amount of distortion, as illustrated, for example, in Figs. 1.9–1.11.

Even if an adaptive equalizer is to be used, there are several reasons for using a compromise equalizer; these will be discussed in Section 5.2.1.

Parts of the compromise equalizer could be positioned after the sampler, but for economy of implementation it is usually best to place it all before and incorporate it into the receive filter. The optimum transfer function $F_{RE}(f)$ of this receive filter/equalizer combination follows easily from the previous discussion.

If the average channel has a low-pass (but now not necessarily symmetrical) transfer function $H_A(f)$, then from (5.8)

$$|F_{RE}(f)| = \frac{|F_T(f)\,H_A(f)|}{D(f)} \tag{5.16a}$$

and

$$\text{Arg}[F_{RE}(f)] = -\text{Arg}[F_T(f)\,H_A(f)] \tag{5.16b}$$

where

$$D(f) = |F_T(f)\,H_A(f)|^2 + |F_T(f \pm f_s)\,H_A(f \pm f_s)|^2 \quad \text{for } |f| < f_s/2 \tag{5.17}$$

and the + or − applies to negative or positive values of f.

If a large excess band is used ($\alpha \simeq 1$), then the amplitude response of this combination may be very different from that of the combination of a conventional receive filter with a simple equalizer for the average channel (i.e., F_T/H_A); an example of this may be helpful. Consider a V.22 high-band signal ($f_c = 2.4$ kHz, $f_s = 0.6$ kHz and $\alpha = 0.75$). From Fig. 1.9 we can see that an average (50 percentile) telephone channel would have a linear variation of attenuation across the band of about 5 dB; therefore, H_A would be as shown in Fig. 5.9. This figure also shows the optimum F_{RE} and the conventional F_T/H_A; it can be seen that they are significantly different. Numerical integration of $|F_{RE}|^2$ and $|F_T/H_A|^2$ shows that the noise power at the detector is 0.3 dB lower for the optimum filter than for the conventional one. This may not have much practical significance, but it is interesting nevertheless; the optimum filter attenuates the signal more in the upper sideband where it is already attenuated—and therefore has a lower SNR—and emphasizes it more in the lower sideband.

The end-to-end impulse response with the optimum receive filter is shown in Fig. 5.10; the impulse response with the straightforward equalizer is just that of a 75% raised-cosine shaping. Both, of course, result in zero ISI when sampled at $t = kT$, but because of the asymmetry of the optimum shaping about the carrier frequency, the imaginary part of its impulse response is not zero at all times.

5.2.1 Why a Compromise Equalizer *and* an Adaptive Equalizer?

Many of the reasons for using both types of equalizer will not be fully understandable till we have studied adaptive equalizers in Chapters 8 and 9, but

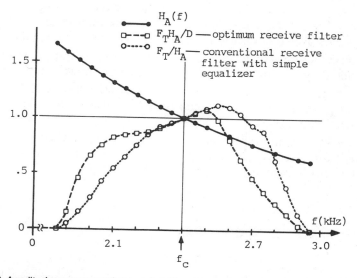

Figure 5.9 Amplitude responses of channel and two types of receive filter for average distortion of a V.22 high-band signal.

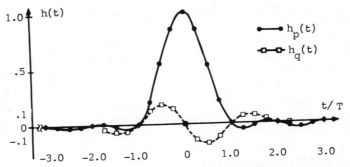

Figure 5.10 Impulse response resulting from optimum filter shown in Fig. 5.9.

nevertheless we should introduce them here—together with one reason for sometimes not using a compromise equalizer.

1. The simplest reason is that if the peak distortion of the input to the adaptive equalizer is approximately halved, the convergence time of the equalizer will be reduced. If the initial error rate without a compromise equalizer is low (i.e., the noise-free eye is open or almost open), the convergence of the equalizer will be exponential, and this speeding up will be inconsequential. However, if the initial error rate is very high, an adaptive equalizer may "stutter" or may not converge at all, and a compromise equalizer may be essential.

2. As we shall see in Chapter 7, most timing recovery methods operate "open-loop" before the sampling and the adaptive equalizer. In these cases the jitter of the recovered clock will be reduced by compromise equalization.

3. If a symbol-rate equalizer is to be used, the loss of sampled signal energy that is caused by aliasing will be reduced by a compromise delay equalizer.

4. A compromise amplitude equalizer will increase the rate of convergence of most adaptive equalizers by reducing the spread of eigenvalues of the signal's autocorrelation matrix.

5. However, a compromise amplitude equalizer should *not* be used with a decision-feedback equalizer because, like any linear equalizer, it would enhance the noise—precisely what the DFE is intended to avoid.

5.3 AUTOMATIC GAIN CONTROL

There are two levels of reasons for using an AGC, and two corresponding degrees of precision needed.

5.3.1 AGCs for Signals Without Amplitude Modulation

These signals can be detected by using zero-level thresholds or, in the case of multiphase signals, by comparing the ratio of two signals to a nonzero threshold. In either case the amplitude of the signal is not important. Nevertheless, a crude AGC is usually desirable because:

1. The bandwidth and transient response of the carrier recovery loop (see Chapter 6) will depend on the amplitude of the input signal. Conservative design of the loop can usually ensure that a variation in level of about ±2 dB can be tolerated; the AGC need adjust the signal only within this degree of precision.

2. A d.c. offset in the baseband may seriously impair the detection of low-level signals; an AGC in the passband before demodulation avoids this.

If an adaptive equalizer is used, it must either adjust the taps so that they refine the gain and thereby match the signal to a fixed level, or learn the level of the crudely controlled signal and adjust its expected signal level accordingly.

5.3.1.1 Feedforward Gain Control.
A generic feedforward AGC circuit is shown in Fig. 5.11; the received signal is effectively divided by its long-term average amplitude. An analog implementation of this has the serious problem that the variable-gain amplifier must include some nonlinear device; this, without feedback, can seldom control the output level closely enough. On the other hand, a digital implementation has the problem that the block that generates the control signal must perform a division—a tedious operation in most processors. For these reasons, feedforward control is almost never used.

5.3.1.2 Feedback Gain Control.
A generic circuit is shown in Fig. 5.12; its operation is nonlinear, and a general mathematical analysis would be very complicated. Analysis by simulation is more promising but would be useful for

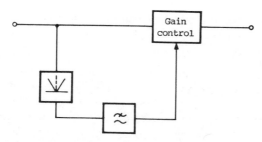

Figure 5.11 Generic forward-acting AGC control.

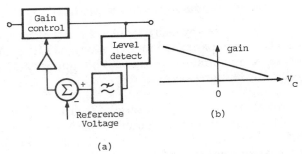

Figure 5.12 Feedback AGC: *(a)* circuit; *(b)* characteristic of gain-control device.

only a particular set of specifications and type of hardware. However, even without these, a few general comments may be helpful:

1. The gain control device should have a characteristic like that in Fig. 5.12*b*; this can be approximated quite well by a Junction Field-Effect Transistor (JFET) or a variable-transconductance amplifier.

2. The level detector can be either a full-wave rectifier or some form of peak detector with a relaxation circuit. The rectifier could be hard (an absolute value circuit), soft (a squarer), or anything in between.

3. The bandwidth of the low-pass is determined by the reaction time (from full gain before signal arrives) specified for the AGC. The wanted d.c. component of the output of the level detector is contaminated by low-frequency data-related components generated by the nonlinearity, and if fed back to the gain control, these will cause intermodulation distortion of the signal. However, this is amplitude modulation that affects the signal and the noise equally, and if the filters and phase splitters between the AGC and the threshold detector have constant amplitude and linear phase across the band, the modulation will not reduce the SNR at the slicers (even though it will reduce the opening of a conventionally generated eye pattern).

5.3.2 AGCs for Signals With Amplitude Modulation

For a multilevel QAM signal, the AGC must perform two functions: (1) it must crudely control the level of the input signals to the timing-recovery, carrier-recovery, and tap-adjustment loops so that their transient responses are acceptable and their stabilities are ensured; and (2) it must control the relationship between the signal and the threshold levels so that the error in the gain contributes an acceptably small amount to the total MSE, as suggested in Table 5.1. It can do the first only by using a coarse variable-gain device fairly early in the receiver, but the second can be done by refining this gain or by combining the coarse gain with either a precise adjustment of the equalizer taps to match some fixed thresholds or a precise learning of the signal level and appropriate setting of the thresholds and expected signal levels.

Before considering different types of AGC, we must learn how to calculate their permissible jitter and required resolution and must also examine the basic algorithms for adjusting them.

5.3.2.1 *Required Resolution.*

Suppose that the nominal gain of the AGC is unity, and that it is adjusted very carefully (i.e., after a lot of averaging)† using a step size of δ, so that the gain is randomly distributed over the range $1 \pm \delta/2$. The ratio of the mean-squared error of the amplified signal to the power in that signal will be equal to the mean-squared jitter of the AGC gain, which is $\delta^2/12$.

Therefore, if, for example, as suggested in Table 5.1, the AGC is allowed just 0.1 dB contribution to the reduction of an input SNR of 20 dB, then

$$\frac{\delta^2}{12} = 10^{-(36.4/10)} = 0.00023$$

so that
$$\delta \simeq 0.05.$$

If now, the input signal level has a range from unity to $1/R$ (i.e., $A_{\max} = 20 \log R$ dB), then the AGC must provide a gain from unity to R in steps no bigger than δ. If the steps are linear approximately R/δ steps will be needed; if the steps are logarithmic, approximately $A_{\max}/8.7\delta$ will be needed.

5.3.2.2 *Generation of the Feedback Gain-Control Signal.*

There are two methods of generating the signal that is used to control the variable-gain device. The first is simple and straightforward but rather limited in its application; it is often described as "blind" because it is an adaptive process that does not need information about the receive data for its adaptation. The second, "decision-aided," method is more powerful but also more complicated; there are a few unanswered questions about its operation.

Blind Averaging of the Amplitude of the Passband Signal. In this simpler method the output of the variable-gain device is rectified (hard or soft), reduced by an amount corresponding to the nominal level, and then low-pass filtered and amplified to produce the gain-control signal. We shall see that hard rectification is preferable in practice to soft but is more difficult to analyze in general. Therefore, we will calculate the amount of filtering for only the particular case of the 16-point constellation shown in Fig. 4.7.

The expectation (average) of the amplitude of the sampled signal, $|s(kT)|$ (abbreviated to $|s|$ for convenience), is

---

† We will consider the case of continual and somewhat random adjustment—requiring a much smaller step size—later.

$$E[|s|] = \frac{\sqrt{2} + 2\sqrt{10} + \sqrt{18}}{4} \simeq 3.0$$

and its variance is

$$\mathrm{Var}[|s|] = \frac{(\sqrt{2} - 3)^2 + 2(\sqrt{10} - 3) + (\sqrt{18} - 3)^2}{4} \simeq 1.03$$

Therefore, the relative variance is

$$\frac{\mathrm{Var}[|s|]}{E^2[|s|]} \simeq 0.11$$

and, using the argument in Section 5.3.2.1, this must be reduced by some form of filtering to $\sigma^2 = 0.00023$.

If the amplitude of the AGC error signal were simply averaged over a block of N symbols, the resultant variance would be the input variance divided by N; therefore, N must be determined from $0.11/N = 0.00023$, leading to $N \simeq 500$. In practice, block averages and moving averages are tedious to perform; approximately the same noise bandwidth and transient response can be achieved with a simple first-order low-pass filter with a 3 dB point at $f_s/3N$.

Such a narrow filter would result in a convergence time for the AGC that would be unacceptably long for some modems (e.g., a V.22 bis modem with a symbol rate of only 600 s/s). However, it is common to use a simple signal with only a few discrete frequency components (see Section 4.2.5) for initialization, and the d.c. component could be retrieved from such a signal after rectification by using a relatively wide-band low-pass filter; therefore, a wide filter could be used for initialization and its bandwidth reduced after the AGC had converged and before the arrival of random data.

This resolves the conflict of the initialization and steady-state jitter requirements, but if the gain of a channel can change during the transfer of data by more than a few tenths of a decibel in a thousand symbol periods, then such a low-jitter, narrow-band loop would be unable to follow the gain, and serious reduction of the tolerance to noise would result. Since fairly rapid changes of gain ("hits") of as much as ± 6 dB are possible on the PSTN, this method of simple averaging is recommended only for more stable media.

The variance of a soft-rectified (squared) signal is even greater than that of a hard-rectified one, and such a signal would require about one and a half times as long an averaging period; therefore, squaring should never be used in combination with blind averaging.

For larger constellations (of 32, 64, and 128 points), the variance of the unfiltered rectified signal is greater than for 16 points *and* the limits on the jitter of the AGC become tighter; consequently blind averaging or filtering is not suitable.

Decision-Aided Adjustment. The variance (jitter) of the control signal can be greatly reduced if the data-related variation of the amplitude is first removed. This can be done by comparing the signal at any point in the receiver to that that would be produced at that point if the transmitted signal were passed through a channel with nominal gain and then by filtering and amplifying the difference to produce the gain control signal.

For simplicity of implementation, this comparison is best done in the baseband at the inputs to the slicers as shown, for one dimension, in Fig. 5.13 — ignoring the dotted lines for the moment. The amplitudes of the finely quantized signals are subtracted from the coarsely quantized outputs of the slicers, and the differences added and filtered. If the signal has been equalized, the only error in these differences is noise plus accumulated imperfections. If these latter are assumed to be small compared to the noise, the bandwidth of the AGC loop filter can be calculated as follows.

The suggestion in Table 5.1 is that for an input SNR of 20 dB, the "signal-to-jitter" ratio should be 36.4 dB. Since the addition of the error signals from the two basebands doubles the error component but increases the noise by only 3 dB, this means that the filter must reduce the noise power by 13.4 dB; the noise can be assumed to be white across the Nyquist band ($|f| < f_s/2$), so the bandwidth of the filter must be approximately $f_s/44$.

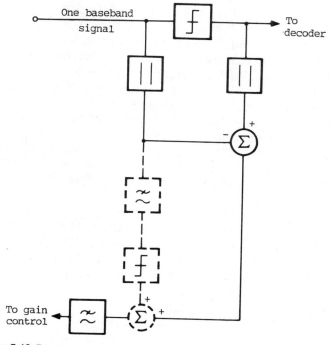

Figure 5.13 Decision-aided AGC with (dashed line) addition to avoid false locks.

This bandwidth is much greater than can be allowed when using blind averaging, and the AGC loop will be able to follow most smooth changes of attenuation in the channel. However, if a large gain hit occurs, serious problems can arise.

The Problem of Gain Hits. Nearly all decision-aided feedback loops (for carrier or timing recovery, or for the AGC) have the problem that if the error somehow becomes large enough that, even without noise, the decision about which symbol was transmitted is wrong, then the error will appear to be of the wrong sign and the loop may converge to a false stable point. (We will see another example of this phenomenon in Section 6.4.3. on carrier recovery).

As a simple example of this effect, consider a 16-point constellation. Since the errors are calculated separately in each baseband, we need consider only one dimension, in which the nominal points are ± 1 and ± 3. Then if the gain of the channel is G, the signal input to the slicer, which has a threshold of 2.0, is $\pm G$ or $\pm 3G$. The estimate of the gain error is

$$E(G) = \frac{(G - [G]) + (3G - [3G])}{2} \tag{5.18}$$

where $[x]$ is the coarsely quantized value of x; that is

$$[x] = 1 \text{ if } x < 2 \quad \text{and} = 3 \text{ if } x > 2$$

The value of $E(G)$ is plotted in Fig. 5.14a; it can be seen that if G rises above 2 or falls below $\frac{2}{3}$ the slicer errs, and a discontinuity in $E(G)$ occurs. For $G > 2$ the sign of the error $E(G)$, which controls the direction in which the AGC loop moves, remains positive, and the loop will reconverge to $G = 1$. On the other hand, for $G < \frac{2}{3}$, the sign of the error changes, and the loop will converge to a stable false zero at $G = \frac{1}{2}$. Thus if the signal level decreases by more than 3.52 dB, the AGC will lock up incorrectly.

For larger constellations there are many more stable but false zeros; the $E(G)$ curve for 64 points (eight in each dimension) is shown in Fig. 5.14b; it can be seen that if the signal level decreases by more than 1.94 dB, the loop will lock incorrectly.

An Improved Measure of Gain Error. The measure of the gain in (5.19) compares the signal amplitudes with the nominal amplitudes. The problem of false lock can be reduced somewhat by instead comparing the gain with the nominal gain of unity†; that is, for the L points in each dimension of a square constellation,

$$E_1(G) = \frac{2}{L} \sum_{k=1}^{L/2} \frac{(2k - 1)G}{[(2k - 1)G]} - 1 \tag{5.19}$$

---

† We will use a similar strategy in dealing with false lock of a carrier recovery algorithm in Section 6.4.3.

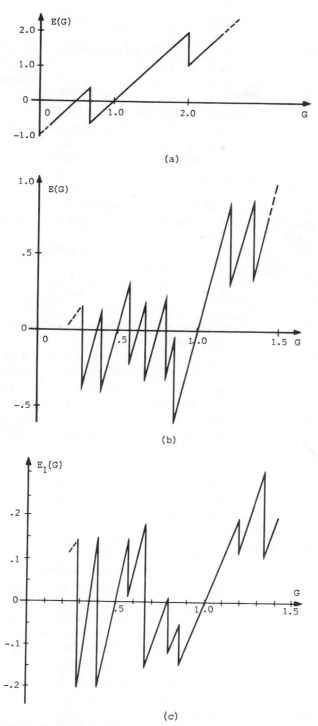

Figure 5.14 Average error signal of decision-aided AGC: *(a)* 16QAM; *(b)* 64QAM; *(c)* 64QAM with modified measure of gain.

{Note that the division can be performed as a multiplication by one of a small number $(L/2)$ of stored values, $1/[(2k-1)G]$}.

For 16 points, the shape of $E_1(G)$ is the same as that of $E(G)$; however, for 64 points (see Fig. 5.14c), it is quite different. The false zero at $G = 0.79$ is only just stable and would not be if noise and AGC jitter were added. Therefore, the AGC loop could tolerate a level drop of 3.52 dB before converging to the false stable zero at $G = 0.5$; this could be a significant advantage in deciding the bandwidth and transient response of the extra filter (indicated by dotted line in Fig. 5.13) used in the "mixed-strategy" method.

Avoidance of False Lock by Monitoring the Detected Data. False lock occurs only if the input level drops, and then the first thing that happens is that the outer points seem to have vanished.

One suggestion that has been made for getting out of false lock is to monitor the decisions in each channel, and if no outer point occurs in either channel in N symbols, then to increase the gain by something more than 3.5 dB (> 1.9 dB for 64-point, etc.). The probability of no outer point being transmitted in N symbols is $(\frac{1}{4})^N [(\frac{9}{16})^N$ for 64-point], so if there were no noise, a value of N of about 15 (36 for 64-point) would give a very reliable indication (with only a 10^{-9} probability of a false indication) of an attenuation hit.

However, after a hit of more than 3.5 dB, the decision-aided part of the control would converge to a very jittery overall gain of $\frac{1}{2}$. What were originally outer points would now have an average value of $\frac{3}{2}$, and their noise-induced error rate would be quite high. Whether it would be low enough for a reliable determination of the absence of outer points in a reasonable time is a very complicated question, and this method cannot be recommended until it has been answered.

Avoidance of False Lock by Using a Mixed Strategy. The decision-aided method fails if the error is large, whereas blind averaging cannot make the error very small without using a very long averaging time. It would seem wise to use the best parts of each method.

A one-dimensional version of a promising mixed method is shown as a dashed addition to Fig. 5.13. The decision-aided part proceeds as before, and the rectified baseband signal is also input to a fairly wideband low-pass filter. If the output of this filter, which is too jittery for direct gain control but is nevertheless a reliable indicator of an attenuation hit, falls below a threshold of $\frac{4}{5}$ (the average value of a signal that has experienced a critical 3.5 dB drop in level), a single pulse is added to the input of the main loop filter. This pulse should have the effect of temporarily increasing the gain by 3.5 dB and of thereby restoring the signal to the region where the decision-aided algorithm can operate correctly. A timer should also be started when the pulse is injected, because if the attenuation hit is greater than 3.5 dB, further pulses will have to be injected later after the first has passed through the low-pass filter and increased the gain.

Readers should be aware that decision-aided AGC circuits are controversial, and the secrets of the few that have been used successfully are carefully guarded. The one described above is the best that I know about, but it has not been tested.

5.3.2.3 *Gain Control Configurations.* In order to achieve the precision necessary for a multipoint constellation, it is essential that the AGC control signal be generated digitally, but it may then be used in several different ways: it may operate on the signal either before or after the A/D conversion, or it may even be converted back to an analog signal.

Simple Digital Multiplication. The simplest control method is shown in Fig. 5.15*a*; the A/D-converted signal is simply multiplied by the control word. This is most suitable for signals without a lot of ACI and without a wide variation of receive levels; for other signals, the number of bits required from the ADC will have to be great enough to deal with the combination of the largest receive signal and the ACI, and also with the smallest receive signal. For an FDX modem for use on the PSTN (e.g., a V.22 bis), as many as six extra bits may be needed to deal with the wide range of signals. Nevertheless, this is the only arrangement possible if the filtering of the ACI is to be done digitally.

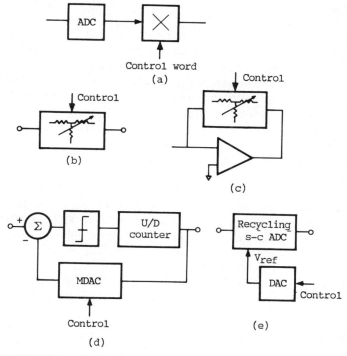

Figure 5.15 Various implementations of gain control: *(a)* purely digital; *(b)* digitally controlled analog attenuator; *(c)* digitally controlled analog amplifier; *(d)* embedded in successive-approximation ADC; *(e)* providing reference voltage for a switched-capacitor (s-c) recycling ADC.

Gain Control Before the ADC. To reduce the demands on the ADC a digitally controlled analog attenuator (as shown in Fig. 5.15*b*) or amplifier (as shown in Fig. 5.15*c*) can be used. However, such an attenuator with a full 12-bit control would be an expensive component. A cost-effective compromise is to use only enough bits for the gain control (typically about six) as are necessary to reduce the variation of the signal level to that that allows predictable operation of the recovery and adjustment loops and leads to mostly error-free slicing; the thresholds are then adjusted to match the learned value of the signal, which may now vary over a much narrower range.

The delay through the equalizer would now be between the gain control and the level learning, and there would be small transient perturbations to the signal when the more significant bits, which control the gain, were changed. The long-term average effect of these perturbations would depend in a very complicated way on the number and magnitude of the leading taps of the equalizer; no analysis has been published.

Gain Control Within the ADC. The digitally controlled attenuator in the feedback loop in Fig. 5.15*c* can be incorporated into a successive approximation ADC if a Multiplying D/A Converter (MDAC) is used as shown in Fig. 5.15*d*. The same limitations on the size of the control word would apply as in the previous implementation.

Another way of using the control word inside the ADC loop, which is particularly applicable for switched-capacitor implementation, is shown in Fig. 5.15*e*. The control word is converted to a d.c. voltage, which is then used as the reference voltage for a recycling ADC [Le5]. Satisfactory eight-bit conversion over a 30 dB input range has been achieved in the laboratory, but I have seen no reports of results with integrated circuits.

5.4   DEMODULATION

Demodulation, the process of shifting the frequency band of the received signal back down to baseband, may, for the sake of convenience, be performed in several stages. However, we need not be concerned here with any intermediate stages; it is the overall demodulation that is important. This can be one of three types: (1) coherent — preferably described as either homodyne or synchrodyne — demodulation; (2) noncoherent, or free-running demodulation; or (3) differential demodulation. We will briefly discuss all three here, but the emphasis in Chapters 6, 7, and 8 will be on synchrodyne demodulation.

5.4.1   Coherent, Homodyne, or Synchrodyne Demodulation

The term "coherent ("running together") demodulation" is often used, but it is rather misleading; the overall process of "coherent detection," in which the phase of the signal is deduced from its relation to a locally generated steady carrier, may use either homodyne, synchrodyne, or free-running demodulation followed by adaptive rotation. A nice distinction between the first two was proposed by Tucker [Tu]; "homodyne" ("same movement") means the use of the same carrier (transmitted as a pilot along with the signal), and "synchrodyne" means the use of a locally generated carrier that is somehow synchronized with that implied by the signal. These are typically used for SSB and QAM, respectively. Methods of synchronizing carriers are discussed in Chapter 6.

Basic homodyne and synchrodyne demodulators are shown in Fig. 5.16a for a one-dimensional signal (SBB) and in Fig. 5.16b for a two-dimensional (QM) signal. The low-pass filter(s) must reject the upper sideband product of the demodulation that begins at $f_c - (1 + \alpha)f_s/2$; even for wide-band systems ($f_s/f_c > \frac{1}{2}$) such as a V.33, the 46 dB attenuation needed to prevent an input SNR of 30 dB from being reduced by the upper sideband "noise" by more than 0.1 dB is simple to achieve.

The multipliers are often much easier to implement if the carriers are square waves. This does not impose any greater burden on the low-pass filters because the extra unwanted product of modulation — the lower sideband of the third harmonic of the carrier — is superimposed (9.5 dB down) on the upper sideband of the fundamental.

5.4.1.1 *Phase Shift Method of Demodulating.*

The upper sideband can also be eliminated by using the Hartley method shown in Fig. 5.17. The phase shifters should be designed together with the bandpass filter as explained in Appendix I; the phase and amplitude matches needed for the 46 dB rejection of the USB suggested in the previous section are easy to achieve digitally, not difficult when using switched-capacitor circuitry, and quite difficult in an active-RC implementation.

If the carriers used were truly sinusoidal then the low-pass filters would no longer be needed; in the very useful method of passband sampling and sampled

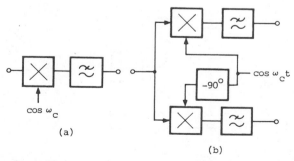

(a)

(b)

Figure 5.16 Demodulation (free-running or synchrodyne): (a) one-dimensional; (b) two-dimensional.

demodulation that is described in Section 5.5, just a few samples of these carriers are used.

5.4.2 Noncoherent, or Free-Running, Demodulation

This can be implemented as shown in Fig. 5.16b or 5.17, except that now the oscillator is free-running at the nominal carrier frequency; this is therefore a special case of heterodyne demodulation in which the nominal intermediate frequency is zero. This process is usually followed by adaptive equalization and rotation as described in Section 6.4.2; the combination is equivalent to synchrodyne demodulation.

5.4.3 Differential Demodulation

Just as either synchrodyne or free-running demodulation is a necessary preliminary of coherent detection, differential demodulation is a preliminary of *comparison detection*. The most important use of this method is for QPM signals, so we will confine our discussion to these.

A two-dimensional differential demodulator is conventionally drawn as shown in Fig. 5.18, but this requires some careful examination. For example, it is often said that the carrier frequency f_c must be an integer multiple of the symbol rate f_s, but this is neither necessary nor sufficient. The phase shift of the upper, "in-phase," band-pass delay network is best written as

$$\phi_p(\omega) = (\omega_c - \omega)T + \phi_c$$

so that

$$\frac{d\phi_p}{d\omega} = -T \qquad \text{and} \qquad \phi_p(\omega_c) = \phi_c$$

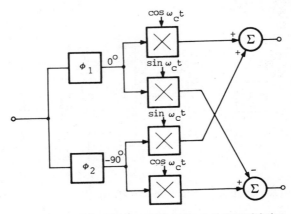

Figure 5.17 Hartley (phase-shift) method of demodulating.

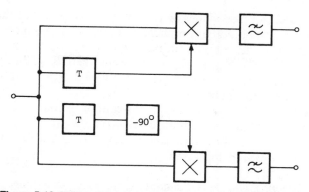

Figure 5.18 Differential demodulation of a two-dimensional signal.

It would appear that only one delay of T should be needed, but the circuit is drawn this way only to emphasize the separate operations; the $-90°$ phase shifter is not realizable by itself, and it must be combined with the quadrature delay to form a single network whose phase shift is

$$\phi_q(\omega) = (\omega_c - \omega)T + \phi_c - \frac{\pi}{2}$$

In order to understand this circuit, it is easiest to consider a received signal derived from one of the four repeated patterns ($\Delta\phi = 0, 90°, 180°,$ or $270°$) described in Section 4.2.5. The received signal is

$$s(t) = A_m \cos 2\pi \left(\frac{f_c + mf_s}{4} \right) t + (1 - A_m) \cos 2\pi \left(\frac{f_c - f_s + mf_s}{4} \right) t$$

where $m = 0, 1, 2,$ or 3. The outputs of the two low-pass filters at $t = kT$ are

$$x_p(kT) = \cos \left(\phi_c - \frac{k\pi}{2} \right)$$

$$x_q(kT) = \cos \left(\phi_c + \frac{(1 - k)\pi}{2} \right)$$

If the delay networks are designed so that $\phi_c = 0$, then

$$x_p(kT) = \cos \left(\frac{k\pi}{2} \right)$$

$$x_q(kT) = \sin \left(\frac{k\pi}{2} \right)$$

from which the original value of k could be recovered.

However, these sampled values of x_p and x_q can be ± 1 or 0, and slicing for least noise sensitivity would be complicated. In practice, it is more convenient to set ϕ_c equal to any odd integer multiple of $\pi/4$. Then at the sampling times, x_p and x_q can have only the values $\pm 1/\sqrt{2}$, and simple zero-level thresholds can be used for the slicers; the decoding logic to derive k can be adjusted to fit the particular multiple of $\pi/4$ chosen.

If the phase shift at the carrier frequency cannot be maintained because of either frequency shift in the medium or element variations in the delay networks, then one or both of the baseband signals will have a d.c. offset, and the sensitivity to noise will be increased.

Similarity to Synchrodyne Demodulation. Figure 5.19 shows a generic quadrature demodulator with two possible ways of generating the in-phase and quadrature channel multiplicands. It can be seen that the only difference between differential and synchrodyne demodulation is the source of the "carriers."

Limiting Before Multiplying. It would seem that one of each pair of multiplicands could be limited first so as to make implementation of the multipliers (now binary times continuous instead of continuous times continuous) easier; this has almost certainly been done in practical modems, but unfortunately, all the textbooks that discuss differential demodulation assume both multiplicands to be continuous, and I have seen no analysis of the effects of limiting.

5.4.4   Relative Merits of Coherent and Comparison Detection

Comparison detection has the following advantages:

1. For simple receivers that do not require adaptive equalization, it is less expensive to implement.

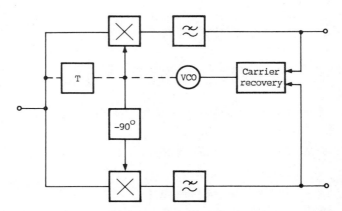

Figure 5.19 Generic QAM demodulator showing two sources of "carriers."

2. For modems that require "very fast" initialization, the fact that no time is needed for a local carrier oscillator to acquire lock may be a significant advantage. For frequency-hopped systems that are able to maintain the same data timing and AGC setting from hop to hop, the instantaneous "carrier-lock" feature of comparison detection may be an overwhelming advantage.

On the other hand, for Time-Division Multiple-Access (TDMA) systems that require only "fast" acquisition and cannot maintain data timing from burst to burst, it would seem that the importance of advantage 2 will diminish as better timing and carrier recovery methods are developed. For example, if carrier and timing acquisition can be made independent and concurrent (methods of almost achieving this are discussed in Chapters 6 and 7), then carrier acquisition may not add very much to the total initialization time.

Comparison detection has the following disadvantages:

1. It is inferior in performance; because two noisy signals are used in the detection instead of one noisy and one steady, the effective SNR is reduced. For a large number of phases the asymptotic deterioration is 3 dB; for the most common case of four phases it is 2.3 dB.
2. It is much more difficult to combine with adaptive equalization.

Partly because I think that the disadvantages usually outweigh the advantages, and will do so even more as the science of receiver design progresses, and partly because the problems of comparison detection are nearly all particular to data speeds and types of hardware of which I have no experience, we will not consider comparison detection any further.

5.4.5 Sampled Demodulation

If sinusoidal carriers are used, so that no low-pass filtering is needed, the demodulator shown in Fig. 5.17, which can be either synchrodyne or free-running, can be followed directly by a sampler at the symbol rate or some small integer multiple of that rate, as shown in Fig. 5.20a. Then, since the multipliers are memoryless devices, the samplers can be moved through them as shown in Fig. 5.20b. The signal and its Hilbert transform must now be multiplied by samples (at the *symbol* rate or some multiple thereof) of the local carrier—a task that usually requires DSP and a look-up table containing all needed values of the sinusoidal carriers.

In many modems (particularly those for the telephone network) the transmit carrier frequency is a simple rational multiple of the symbol rate; that is,

$$f_c = \frac{M f_s}{N}$$

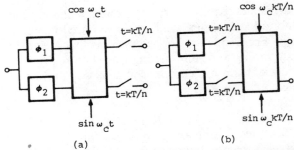

Figure 5.20 Sampling: *(a)* after demodulation; *(b)* before demodulation.

where M and N are small integers. If this relationship could be preserved at the receiver, the look-up table would need at most $2N$ entries, and the method would be practically feasible. However, if the transmitter and/or the receiver use more than one stage of modulation, or if there is a frequency shift in the transmission channel, the rational relationship will be destroyed; the sampling times will precess through a carrier period and a very large, finely structured look-up table will be needed.

In these very common cases a "free-running" demodulation should be performed by a carrier that is locked to M/N times the locally derived symbol clock. The look-up table shrinks back to $2N$ entries, and the resulting slow rotation of the constellation can be counteracted by an adaptive rotator as described in Section 6.4.2 and 8.2.2.

For most telephone modems, this leads to a very simple demodulating algorithm. For example, a V.33 modem with $f_s = 2.4$ kHz and $f_c = 1.8$ kHz may use sampled carrier values of only 0 or ± 1.

5.5 SAMPLING

In an elementary receiver an analog signal is processed (filtered, etc.), demodulated down to baseband, and input to a threshold detector. The continuous output of the detector is then sampled digitally at the learned symbol rate, and the result is passed to a decoder. This is the basic sampling process.

If an adaptive equalizer that uses samples taken at the symbol rate is used, the same basic sampling is performed, but now, as we shall see in Section 7.5, the best phase of the sampling clock must be defined differently, and different clock recovery methods must be used.

5.5.1 Higher Sampling Rates

In many receivers it is desirable to use sampled-data processing from as early a stage as possible. The sampling rate for this processing must be high enough to fully characterize the passband signal—that is, it must be greater than

$[2f_c + (1 + \alpha)f_s]$ and therefore often much greater than the symbol rate. Since the signal must eventually be sampled at the *transmit* symbol rate, which is recovered adaptively, there are five alternatives for this early sampling:

1. It can be free-running at a frequency (L times the *nominal* f_s) that is high enough that when the symbol rate sampling selects one of L samples,† the quantizing error in the sampling time is acceptable. This approach may be feasible for switched-capacitor filtering where hardware is usually not timeshared and high sampling rates in each section are acceptable—even, according to some designers, desirable—but it is too prodigal for DSP where, in order to economize in hardware, it is necessary to keep the early sampling rate as low as possible.

2. It can be free-running at a medium frequency; the output of the preliminary filter is smoothed with a simple continuous LPF.

3. It can be free-running at a low frequency; the samples at the recovered symbol rate can then be calculated by interpolation. This is discussed in Section 7.5.2.3.

4. It can be free-running at a low frequency; the delay in a fractionally spaced equalizer (see Section 8.7) is then adjusted so that one of each L (now a small number) samples is acceptable. Timing recovery for fractionally spaced equalizers is discussed in Section 7.6.

5. It can be at a low frequency that is locked to (i.e., is an integer multiple of) the recovered symbol clock. This is the most commonly used method, but it has the disadvantage (which may be serious) that the delay through the sampled prefilters is inside the timing recovery loop.

5.6 DETECTION AND DECODING

Unless convolutional coding (see Chapter 10) is used, detection and decoding are simple processes. The baseband signals are quantized to their nearest nominal levels by using (1) threshold slicing, (2) table look-up, (3) Viterbi decoding, or (4) error detection and correction, and then decoded to yield the original input data.

For the decoding, there are three possible combinations that depend on whether differential phase encoding and/or differential demodulation are used.‡ If differential encoding and synchrodyne demodulation are used, the two detected bits that represent the signal's quadrant must be decoded according to the inverse of Table 4.1.

---

† This process of selecting one of L samples is often called *decimation*, but this is a complete and aggravating departure from the original meaning of the word (reduction by one-tenth); a new word is needed.

‡ The combination of absolute phase encoding and differential demodulation is never used.

Effects of Single-Symbol Errors. If the dibits must be decoded in this way, a single error in either the p or q channel will advance or retard the "present quadrant" by one, and then, at the next symbol time, similarly change the "previous quadrant." This will result in double errors, which would emerge from the decoder as one of the four patterns:

xeex,   xexe,   exex,   or   exxe

for a 4QAM modem such as a V.22 or

xexxexxx,   xexxxexx,   exxxexxx,   or   exxxxexx

for a 16QAM modem such as V.22 bis.

The relation between encoding, demodulation, and the plurality of errors can be summarized by

	Synchrodyne Demodulation	Differential Demodulation
Absolute phase encoding	Single	Not used
Differential phase encoding	Double	Single

Chapter 6

Carrier Recovery for Synchrodyne Demodulation

As we have seen in the previous chapter, there are three methods of demodulation that achieve the same result: (1) conventional synchrodyne demodulation in which the received signal is shifted directly from its passband to d.c. by a carrier locked to that implied in the signal, (2) a two-stage (or even multistage) process in which the carrier for the last stage is locked to the remaining "carrier" of the signal, and (3) a two-stage process in which the first carrier is free-running at the nominal carrier frequency and the second is an adaptive rotator (always implemented digitally) that multiplies the complex baseband signal by $\exp(j\phi)$, where it is assumed that ϕ is only slowly varying.

The first two methods to be considered may use either a bandpass filter (BPF) or a phase-locked loop (PLL) to refine a harmonic of the required carrier; the third method can use only a PLL. For timing recovery, there is a similar flexibility; either a BPF or a PLL can be used for most of the methods described in Chapter 7. Therefore, in Section 6.1 we will consider those aspects of BPFs and PLLs that are particularly applicable to carrier and timing recovery; for a more comprehensive discussion, the reader is referred to one of the standard works on the subject, [Ga4] or [Li1].

Interaction of Carrier Phase and Sampling Time. We shall find that most methods of carrier recovery operate on the sampled signal, and the phase chosen depends upon the sampling time. Conversely, in Chapter 7 we shall find that many methods of timing recovery operate on the baseband signal(s), and the timing phase chosen depends upon the phase of the demodulating carrier. Joint acquisition of carrier and timing has been considered in [Ko1], [Me1], and [MS&T], but we will take the more traditional approach; we will study methods of performing each task, and then consider ways of reducing the dependence of both the convergence rate and the steady-state result on the other parameter.

6.1 BANDPASS FILTERS AND PHASE-LOCKED LOOPS

Basically a PLL is a circuit in which a composite low-pass filter, consisting of the actual loop filter and the integrator implicit in the Voltage-Controlled Oscillator (VCO) is transformed into a BPF by the feedback through the VCO and the phase detector. As far as jitter of the output signal is concerned, the performance of a PLL will be the same as that of a BPF with the same (transformed) response. However, other aspects of the performance will be different and may determine the choice of one or the other.

6.1.1 Static Phase Error

The phase shifts through both BPF and PLL are approximately proportional to the difference between the input frequency and the center frequency, but in a PLL the constant of proportionality is divided by the d.c. gain of the loop. This means that the sensitivity to detuning (of the input and/or of the device itself †) can be greatly reduced in a PLL. In fact, in many PLLs implemented by analog circuitry or by Digital Signal Processing (DSP), the d.c. gain is made infinite so that the loop can track frequency offset with no phase error. This is more difficult to do in a truly digital PLL, but an approximate solution was described by Bradley in a private communication. The method can be summarized as follows.

In a conventional first-order lag/lead digital PLL (Fig. 1 of [C&L], Fig. 16 of [L&Cl]) the phase detector decides whether the input lags or leads the reference and passes the binary decision to an Up/Down (U/D) counter (also called a *sequential* or *random-walk filter*); when the counter overflows (underflows), it inserts (deletes) one pulse into (from) a pulse train, which is then input to a frequency divider. In order to increase the d.c. gain, the output of the first U/D counter is input to a second; when the first counter overflows *or* underflows, the state of the second counter, which may be either positive or negative, defines how many pulses are inserted into or deleted from the pulse train. The second counter would need to be infinitely long in order to provide the equivalent of infinite gain at d.c.; in practice, typical lengths are from four to eight bits.

The phase shift through both BPF and PLL also varies with the bandwidth. In a BPF it is inversely proportional, but in a PLL the relationship is more complicated; sometimes, as the bandwidth is reduced, the gain must also be

---

† The technology of VCOs has lagged behind that of filters, but for most means of implementation (microwave, *LC*, crystal, switched-capacitor, active-*RC*), the deviation of the center frequency should not be more than twice that of a filter.

reduced in order to keep the loop stable, and the phase shift for a given frequency deviation increases accordingly.

6.1.2 Well-Behaved Aquisition

If the initial phase of the VCO in a PLL is within $\pm \pi/2$ of the final phase† the phase error will decay to an acceptably low value in a few time constants. In contrast to this, Gardner [Ga2] has shown that the phase error of an initially quenched BPF decays in about one-tenth of a time constant.

If the filter is not quenched and has been previously excited by a signal of a different phase from a different transmitter (e.g., in the timing recovery for a TDMA system or a four-wire multidrop polled system), the average acquisition time becomes comparable to that of a PLL without hang-up.

6.1.3 Hang-Up

If the initial phase of the output of a PLL is exactly 180° away from the desired final phase, most phase detectors will generate a zero error voltage‡; this is an unstable equilibrium or top dead center, to borrow terms from mathematics and automotive engineering. Noise will eventually push the loop slightly in one direction, but it will start very slowly. There was some discussion in [Ga3], [M&P], and [Ga6] as to whether the loop actually changes direction (equivocates or — more precisely — vacillates); the consensus seemed to be that it does not. Nonetheless, the phenomenon persists and may result in acquisition times that are intolerably long.

The response of the phase detector at the 180° point can sometimes be made discontinuous by using decision-aided techniques (see Section 6.4), and this greatly reduces — but does not completely eliminate — the chance of hang-up; however, it is difficult to use this approach in timing recovery, and in Chapter 7 we will consider other methods of avoiding hang-up.

6.2 FORWARD-ACTING METHODS

In a QPSK or 4QAM system the phase of the carrier is changed by an integer multiple of $\pi/2$ each symbol time. Therefore, if the frequency of the modulated signal could be quadrupled by some means, the phase of this signal would change only by multiples of 2π, and the signal would contain a discrete component at $4f_c$; the modulation would have been "stripped off." This is a rather simplistic explanation of the process — indeed, there are some data patterns for

---

† Depending on the type of phase detector used, the desired output may be either in phase, or 90° out of phase, with the input.

‡ This statement may seem wrong for QAM carrier recovery, since, as we shall see later, the unwanted zeros are at odd integer multiples of $\pi/4$! However, the output of the BPF or the VCO of a PLL is usually at $4f_c$, and the final carrier is generated by a divide-by-4 counter.

which quadrupling does not produce any component at $4f_c$—but nevertheless, it does convey the essence of the method.

The simplest circuit for doing this is shown in Fig. 6.1a; a hard rectifier (absolute value device) can be used in place of the fourth-power device, or two soft rectifiers (squarers) can be used in tandem with a $2f_c$ filter in between, as shown in Fig. 6.1b. The main problems with this circuit are:

1. Difficulties of building the limiters and frequency divider at four times the, often already very high, carrier frequency.
2. Jitter of the output caused by the input data pattern. This is called *pattern jitter* or *self-noise* and was analyzed in detail by Gardner [Ga5].
3. Variations of the carrier frequency and center frequency of the filter. This could be solved by using a PLL, but one of the main uses for such simple modulation–demodulation schemes is in burst-mode transmission (TDMA) via satellite, where very rapid acquisition becomes essential; it is sometimes considered that the potential hang-up problems of PLLs eliminate them from consideration. Another solution is to use an Automatic Frequency Control (AFC) to track frequency variations, but an AFC may be complicated and expensive.

An alternative solution, suggested in [K . . .] and [LL&B], is to use a pair of identical BPFs that form an auto-tracking filter as shown in Fig. 6.2. Any phase shift through the first filter is repeated by the second filter but doubled by the squarer, so that the resultant phase shifts cancel when the multiplier generates the difference frequency of its two inputs. This circuit has some of the properties of a PLL, and in order to make its initial phase convergence fast enough, the bandwidth of the BPFs has to be greater than that of a single BPF; the resultant SNR of the recovered carrier is about 8 dB worse than that with the system

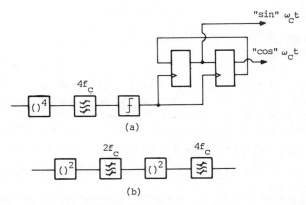

Figure 6.1 Frequency quadrupling and dividing to generate $4f_c$ and f_c: *(a)* using a fourth-power circuit; *(b)* using two squarers.

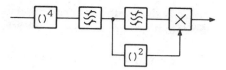

Figure 6.2 Auto-tracking filter for frequency quadrupling.

shown in Fig. 6.1. It is not clear whether such a limitation is fundamental to all such self-tracking methods.

Extension to Larger Constellations. This simple method can be used for constellations with 16 points and more,† but it has been generally assumed that the pattern jitter would be intolerable. However, it can be shown that, at least for 16QAM, the outer points dominate, and the fourth-power signal has a component at $4f_c$ that is usable if a very narrow band PLL is used to refine it. Whether the independence from data decisions and equalizer convergence that this forward-acting method offers outweighs the problems of such a narrow-band PLL remains to be seen.

6.3 FEEDBACK METHODS

The feed-forward methods described in the previous section operate on the received signal to generate a signal at $4f_c$; they may use a BPF or a PLL to refine this signal, but that is rather incidental. Only one of the feedback methods to be considered, *reverse modulation,* allows this choice; the rest (*demodulation– remodulation,* Costas loops, and decision-feedback methods) generate a very low frequency control signal that can only be input to a VCO. Much has been written on the theory of these loops [L&S1 – L&S2], [Si2 – Si5], [W&A]; we can attempt only the briefest of summaries here.

Let us assume that a transmit baseband signal $[x_p(t) + jx_q(t)]$, where $x_p(kT)$, $x_q(kT) = \pm 1$, is received undistorted, but (as a result of modulation, propagation, and demodulation) rotated by θ_e. The received baseband signal is

$$x'_p + jx'_q = (x_p \cos \theta_e - x_q \sin \theta_e) + j(x_q \cos \theta_e + x_p \sin \theta_e) \quad (6.1)$$

The best error signal to use to drive the local carrier VCO is some function of the *maximum a posteriori* (MAP) estimate of the error in the phase of the local carrier (i.e., the difference of its phase from that of the received but suppressed carrier). The MAP estimate of θ_e is

$$g(\theta_e) = x'_p \tanh(x'_q) - x'_q \tanh(x'_p) \quad (6.2)$$

† It will certainly not work for eight-phase because the fourth-power signal could have two exactly opposite phases; that is, it has no component at $4f_c$.

The methods differ mainly in how they approximate the hyperbolic tangent functions; the best approximations are as follows. At high SNR:

$$\tanh x \simeq \mathrm{sgn}\, x \qquad \text{"hard" limiting} \tag{6.3}$$

so that

$$g(\theta_e) = x_p' \,\mathrm{sgn}(x_q') - x_q' \,\mathrm{sgn}(x_p') \tag{6.4}$$

and at low SNR:

$$\tanh x \simeq x - \frac{x^3}{3} \qquad \text{"soft" limiting} \tag{6.5}$$

so that

$$
\begin{aligned}
g(\theta_e) &= x_p' \left(x_q' - \frac{x_q'^3}{3} \right) - x_q' \left(x_p' - \frac{x_q'^3}{3} \right) \\
&= x_p' x_q' \frac{(x_p'^2 - x_q'^2)}{3}
\end{aligned}
\tag{6.6}
$$

6.3.1 Conventional Costas Loops

Two versions of this classical loop that are suitable for QPSK and 4QAM are shown in Figs. 6.3 and 6.4. It was shown in [L&S1] and [Si4] that both are stochastically equivalent to the frequency-quadrupling method discussed in Section 6.2. This means that the trade-offs between acquisition time and noise

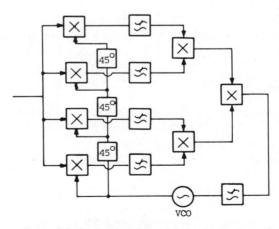

Figure 6.3 Conventional quadriphase Costas loop.

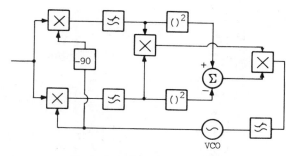

Figure 6.4 MAP estimation loop.

bandwidth (and consequent jitter of the recovered carrier) are the same in the three cases. All three methods are optimum in conditions of *low SNR*.

The second, simpler circuit can be more directly related to the theory because it can be seen that its phase error signal is given by (6.6), which can be combined with (6.1) to give (after a little algebra)

$$g(\theta_e) = \frac{2}{3} \sin 4\theta_e \qquad (6.7)$$

The Costas loop circuits look more complicated than the frequency-quadrupling one, but their implementation may be simpler, because now no operations (limiting, flip-flop toggling, etc.) need to be performed at $4f_c$; all are either in the baseband or at f_c.

6.3.2 Costas Loop with Hard-Limiting (Polarity Type)

If the phase error approximation given in (6.3) is used, the circuit shown in Fig. 6.5 results. In conditions of *high SNR* this will be 3 dB better than those in the previous three methods. Several variations of this that hard-limit both inputs to the multiplier — and are therefore appropriate for very high data rates — have been described [L&V].

Since the hard-limited signals become, when sampled, the decisions about the transmitted signal, this method has sometimes been called *decision-directed*. However, it seems best to reserve that name for those methods that also use timing information (Section 6.3.5).

6.3.3 Demodulation – Remodulation

Several variations of the circuit in Fig. 6.5 have been proposed; one was described in detail by Weber and Alem [W&A] and is shown in Fig. 6.6. In this the limited baseband signals *remodulate* the locally generated carrier, and the resultant passband signal is phase-compared to the received signal so as to gener-

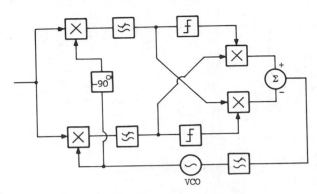

Figure 6.5 Polarity-type Costas loop.

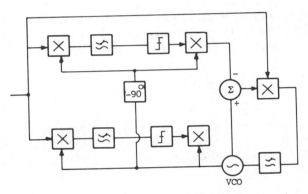

Figure 6.6 Demodulation–remodulation to generate phase-error signal.

ate a control signal for the VCO. The method is stochastically equivalent to the hard-limited Costas loop; it uses one more phase detector than the Costas loop, and it may require careful matching of the delays through the demodulation–remodulation paths (see Section 6.3.4). Its advantages (if any) over the Costas loop probably lie only in subtle matters of implementation.

6.3.4 Reverse Modulation

Another variation was described in [DI&K] and [YH&W] and analyzed in [Ga2]; it is drawn in Fig. 6.7 so as to emphasize the similarities to the other methods. The limited baseband signals are now used to *reverse-modulate*† the *received signal* and thereby generate a component at the carrier frequency. This

---

† This method has also been called *remodulation,* but that obscures the differences from the other methods.

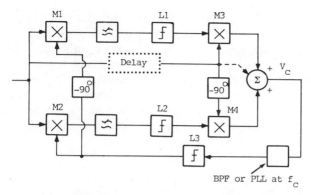

Figure 6.7 Reverse modulation to generate signal at f_c.

may be refined with a PLL, as originally suggested, but the main attraction of the method would seem to be that, as suggested by Gardner [Ga2], it can avoid the hang-up worries of PLLs by using a BPF — now centered at f_c instead of $4f_c$ as is needed for frequency quadrupling.

Theoretical analysis of this circuit for all values of the phase error θ_e would be very complicated, but around $\theta_e = 90n°$ ($n = 0,1,2,3$) the circuit behaves like a conventional PLL and, just like the other methods, drives θ_e to the nearest multiple of $90°$. It was shown in [Ga2] that the method is stochastically equivalent to the other hard-limiting methods; that is, the phase jitter resulting from random noise is the same and is 3 dB less than for the methods that use soft-limiting.

It has also been shown [DI&K] that the pattern jitter resulting from random data and any pulse shape other than rectangular is much less than with the frequency-quadrupling method; however, it is not clear whether this is also true for the Costas loops and the demodulation–remodulation method.†

Delay Compensation. The circuit in Fig. 6.7 shows demodulators, realized by multipliers $M1$ and $M2$ (usually called *balanced mixers* at the high frequencies for which these circuits are used) and LPFs. Since the operation of the circuit depends on the phase relationship between the pairs of inputs to the remodulator multipliers, $M3$ and $M4$, compensation for the delay through the LPFs would have to be included in the received signal line to the reverse modulator, as shown in [YH&W] and the dotted box in Fig. 6.7.

Start-Up. The advantage of Gardner's version, which uses a BPF, is that it avoids one form of hang-up — the one that is inherent in all PLLs; the disadvantage is that it suffers from another form of hang-up — it may not start from a

---

† The amount of pattern jitter depends nonlinearly on the bandwidth of the filters and the ISI and is very difficult to calculate. Most studies of these demodulation methods have assumed rectangular pulses. As bandwidth becomes more valuable, other pulse shapes must be studied.

quiet state. If the BPF output has decayed or been quenched after the previous signal has finished, limiter $L3$ will probably latch in one output state, the other two limiters will follow some phase-shifted versions of the received signal, and V_c will have components around d.c. and $2f_c$ only, which will not pass through the BPF. Ways of "jump-starting" the circuit depend on the form of the training signal that is used at the beginning of each new transmission; if the signal has a discrete component at f_c (e.g., alternating $+90°$ and $-90°$ phase changes†), the received signal can be temporarily routed directly to the input of the BPF, as shown by the dashed line in Fig. 6.7.

6.4 DECISION-DIRECTED METHODS

The main disadvantage of all the previous methods is that they estimate the phase error continuously, and for band-limited signals with nonrectangular baseband pulse shapes, which necessarily extend over several symbol periods, this means that the estimate at any time depends on the past, present, and future data. Consequently, even with no noise added to the signal, the control signal delivered to the VCO will be corrupted by *pattern noise*. For signals that use less than 100% of excess bandwidth and operate in conditions of high SNR, pattern noise is greater than input Gaussian noise, and it and the required speed of carrier acquisition are the determining factors for the bandwidth of the loop filters. Pattern noise is discussed in detail in Section 7.4.2 in the context of timing recovery, where it is more serious because it is unavoidable.

For QPSK and 4QAM, the slicing that is needed to generate the receive data is equivalent to hard-limiting, so the estimate of the phase error is given by (6.4).

6.4.1 Without an Adaptive Equalizer

If the received signal is band-limited but undistorted and timing can be recovered first, pattern noise can be avoided by estimating the phase error only at the times of maximum eye opening; the form of the estimate given by (6.4) eliminates any dependence on the present data, and since there is no ISI, the estimate will also be independent of past and future data. In an analog implementation using hard-limiting to generate the data decisions, the baseband signals are put into Sample-and-Hold (S&H) circuits as shown in Fig. 6.8, and a steady estimate of the sampled phase error is input to the loop LPF throughout each symbol period.

If $|\theta_e| < \pi/4$, then $\mathrm{sgn}(x_p') = x_p$ and $\mathrm{sgn}(x_q') = x_q$, so that combining (6.1) and (6.4) gives

$$g(\theta_e) = x_p' x_q - x_q' x_p = -(x_p'^2 + x_q'^2) \sin \theta_e = -2 \sin \theta_e \qquad (6.8)$$

† We shall see that this signal is also useful for rapid clock recovery; it is used in many voice-band modems.

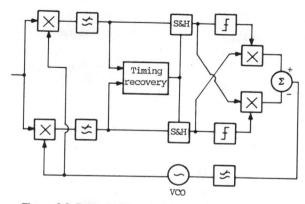

Figure 6.8 Basic decision-directed carrier recovery loop.

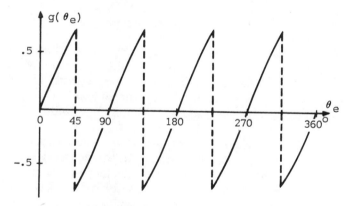

Figure 6.9 $\mathscr{S}$ curve for 4QAM.

If $|\theta_e| > \pi/4$, then $(x_p' + jx_q')$ will be rotated into another quadrant, and the phase error will be seen as the difference from the nominal phase, $(2N + 1)\pi/4$, in that quadrant; $g(\theta_e)$ will therefore repeat at intervals of $\pi/2$, as shown in Fig. 6.9; this is called the $\mathscr{S}$ curve for the PLL. It can be seen that without noise it is an almost ideal sawtooth phase-detector characteristic; the zeros at $(2N + 1)\pi/4$ are very unstable because any slight change of phase either way generates a maximum error signal that rapidly drives the loop to one of the stable zeros.†

---

† This is a consequence of the hard-limiting that is used in the decision-making. The non-decision-directed carrier recovery methods have conventional sinusoidal $\mathscr{S}$ curves.

6.4.1.1   *Design of the Loop Filter.* The loop filter should be designed according to standard PLL theory [Ga4] or [Li1], which can be summarized very briefly as follows:

A model of a PLL is shown in Fig. 6.10. The output phase is given by

$$\theta_o = \left(\frac{1}{p}\right) F(p)\, K_v K_d\, (\theta_i - \theta_o)$$

so that the closed-loop transfer function is

$$\frac{\theta_o}{\theta_i} = \frac{K_v K_d\, F(p)}{p + K_v K_d\, F(p)} \tag{6.9}$$

There is seldom any advantage in making $F(p)$ any more than first order; that is,

$$F(p) = \frac{A_0(1 + p/p_1)}{(1 + p/p_2)} \tag{6.10}$$

The open-loop d.c. gain (excluding the integrator) $A_0 K_v K_d$ should be chosen to render acceptable the phase error resulting from the maximum frequency offset of channel *plus* VCO. The poles of the closed-loop transfer function should usually be critically damped and their magnitude chosen to achieve the best compromise between transient response and noise bandwidth.

6.4.1.2   *Delay Within the Loop.* The low-pass filters and the sample and hold both introduce delay into the loop and make the design a very imprecise operation. Typically, the transient response of the loop will be more oscillatory than predicted by simple theory, and if the product of the delay and the loop bandwidth is too great, the loop may even become unstable.

In some systems it is very tempting to perform the band-limiting and adjacent channel rejection in the baseband after demodulation, but if fairly fast

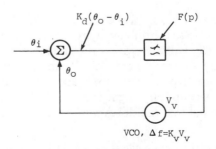

Figure 6.10 A basic PLL.

carrier acquisition is required, the resultant delay in the LPFs may make this impractical.

6.4.1.3 Rapid Carrier Acquisition.

Decision-directed methods have been preferred for all voice-band modems because of their superior noise/jitter performance, but they have been eschewed for systems requiring very fast acquisition (e.g., TDMA) because of worries about having to acquire receiver timing before the sampled carrier recovery can start. However, it was shown in [Bi1] that is is possible, with the right training signal, to acquire carrier and timing simultaneously and independently. Perhaps decision-directed methods should be reevaluated for these systems.

It will be shown in Section 7.4.5 that timing recovery can be performed in the passband before demodulation and therefore can be totally independent of the phase of the demodulating carrier. Even if recovery is performed in the baseband (generally preferable because of the lower frequencies involved, and because there is then no need to compensate for the delay through the demodulating LPFs), it is still possible, with the right sort of nonlinear operation, to make the recovery independent of the carrier phase.

If the training signal is repeated 180° phase changes, the baseband signals are sinewaves at $f_s/2$. Samples taken at any time $(nT + \tau)$† result in

$$g(\theta_e) = 2\left(\cos \frac{\pi\tau}{T}\right)(\sin \theta_e) \qquad (6.11)$$

so that for any $\tau \neq \pm T/2$, the carrier loop can get started immediately.

If measures to prevent hang-up of the timing PLL, such as are discussed in Section 7.1.1 are used, the sampling time can be quickly kicked to within $\pm T/4$ of the correct time, and the carrier loop receives an error signal that is more than 0.707 of the maximum.

6.4.2 With an Adaptive Equalizer

If a linear baseband equalizer is to be used, its presence inside the carrier loop would place very severe constraints upon the loop filter; the bandwidth of the filter would have to be very small and the acquisition time of the loop very long; furthermore, there would be no possibility of tracking any phase jitter.

To solve this problem, we must enlarge on the idea introduced in Section 5.4.2: demodulation by a free-running carrier followed by adaptive rotation. Two possible configurations for this are shown in Figs. 6.11 and 6.12. The phase detector is designed as before, and the only difference is that the loop filter must now incorporate the integrator that was previously implicit in the VCO.

Since the carrier loop now operates on sampled signals, the transfer function of the loop LPF could be a direct transformation of $F(p)$ in (6.10). However,

† The time scale is defined so that $\tau = 0$ is the best sampling time.

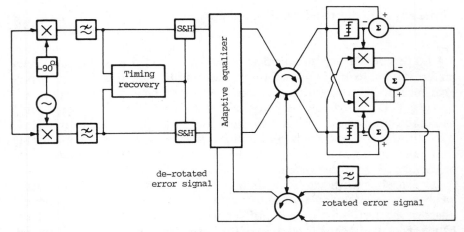

Figure 6.11 Free-running demodulation using baseband timing recovery and adaptive rotator.

there is now no significant advantage in using a pole that is not at d.c.,† so a simpler form for the LPF/integrator combination is preferable:

$$F(z) = \frac{z - z_1}{(z - 1)^2} \tag{6.12}$$

Then the closed-loop transfer function is

$$\frac{\theta_o}{\theta_i} = \frac{K_v K_d (z - z_1)}{(z - 1)^2 + K_v K_d (z - z_1)} \tag{6.13}$$

and $K_v K_d$ and z_1 can be calculated so as to provide the appropriate bandwidth and critical damping.

6.4.2.1 Tracking Phase Jitter. As we discussed in Section 1.4.2.8, phase jitter (extraneous phase modulation) in the telephone system typically has components at very low frequencies (<10 Hz), at 20 Hz (the ringing frequency), and the fundamental and harmonics of the power frequency. If any of these components can be expected to have magnitudes greater than about 2° peak, the carrier loop should be designed so that the demodulating carrier can track them and thereby counteract the phase modulation; this is done by designing the filter to have high-gain, narrow passbands at each frequency [Bi2].

† For an active-*RC* implementation of a *p*-domain filter, the pole is usually put at a nonzero frequency to prevent latch-up of the operational amplifier due to bias current. The problem does not arise in a digital implementation.

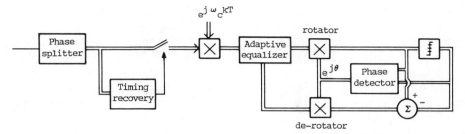

Figure 6.12 Free-running demodulation using passband timing recovery.

Conventional filters of order greater than two are usually designed as the tandem connection of biquadratic sections; that is,

$$F(z) = \prod_{i=0}^{N} F_i(z)$$

However, to provide multiple passbands, it is much simpler to use a parallel connection:

$$F(z) = \sum_{i=0}^{N} F_i(z)$$

Then, if each section is narrow, it has high attenuation in the passbands of the others, and all sections can be designed separately.

The low pass should have a transfer function as given in (6.12), and, ideally, each of the bandpasses should be a conventional bilinear BPF with, for the sake of simplicity, infinite gain at its center frequency. However, such a BPF has no delay—part of its input appears immediately at its output—and so would result in an unrealizable delay-free loop. Therefore, an extra z^{-1} factor must be included in each of the $F_i(z)$ that is,

$$F_i(z) = \frac{A_i(z^2 - 1)}{z^2 - 2 \cos(2\pi f_i T)z + 1} z^{-1} \qquad (6.14)$$

The products $A_i K_v K_d$—defined as K_i for brevity—control the bandwidth of each portion of the closed-loop response. If the symbol rate (the sampling rate for the loop) is high compared to each f_i, and if each passband is narrow (i.e., $K_i \ll 1$), the denominator of the closed-loop transfer function has approximate factors

$$D_i(z) \simeq (1 + K_i)z^2 - 2 \cos (2\pi f_i T)z + (1 - K_i) \qquad (6.15)$$

so that the radius of the pole is approximately $(1 - K_i)$, and the Q is $\pi f_i/(K_i f_s)$. The closed-loop bandwidths around d.c. and all the f_is can be added to give an approximation of the overall noise bandwidth and an estimation of the SNR penalty incurred in tracking the phase jitter. In general, since the peak jitter at f_i, $i \neq 0$, is very much less than the initial static phase error (typically $<5°$, compared to $<45°$), fast convergence of the tracking parts of the loop is not as important as fast d.c. convergence. Therefore, in order to reduce the overall noise bandwidth, the BPF bandwidths should be made much less than the low-pass bandwidth.

Filter Configuration. Two configurations for a second-order digital filter section described by Jackson [Ja] are shown in Fig. 6.13. The 1D configuration in Fig. 6.13*a*—the most common—has the disadvantage that the values that are stored depend only on the denominator, so that if, as in all our cases (LPF and BPFs), both poles have magnitudes very close to unity, the stored values must be drastically scaled to prevent overflow—with a resulting loss of dynamic range. The 2D configuration in Fig. 6.13*b* avoids this problem without requiring any extra storage elements.

The bandpass sections of (6.14) are best realized with a separate delay, so the low-pass transfer function of (6.12) should be rewritten in a similar form:

$$F_0(z) = \frac{A_0(z - z_1)}{(z - 1)} \frac{z}{(z - 1)} z^{-1} \tag{6.16}$$

Then the total filter/integrator/delay combination is as shown in Fig. 6.14.

An Example. A loop was designed for a V.22 bis modem ($f_s = 600$ Hz) to track jitter at 60 and 120 Hz. With a low-pass bandwidth of 20 Hz and BPF Q values

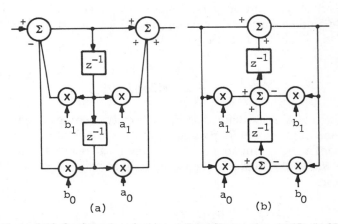

Figure 6.13 Configurations for biquad digital filter section: *(a)* 1D. *(b)* 2D.

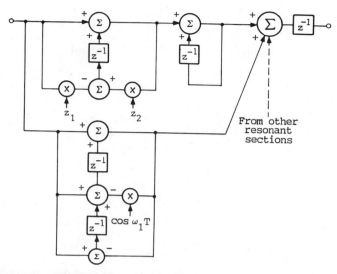

Figure 6.14 Complete carrier loop filter and integrator to track phase jitter.

of 3 and 6, respectively (much too low; 10 and 20 would have been more appropriate!), the SNR penalty incurred by the complete carrier loop was 0.7 dB (too high; 0.4 dB could have been achieved).

6.4.3 Multilevel Signals and the Problem of Phase Hits

As originally described in [Si5], a QAM signal has a discontinuity in its phase-detector output whenever, because of rotation, any signal point crosses a threshold, and is therefore interpreted as a different point. For 4QAM, the discontinuities occur only at $(2N + 1)\pi/4$, but for 16QAM, as can be seen from Fig. 6.15, in the first quadrant they occur at $\theta_e = \pm 16.9°$, $18.4°$, $20.8°$, and $45°$. Equation (6.4) must now be modified to

$$g(\theta_e) = x'_p \hat{x}_q - x'_q \hat{x}_p \tag{6.17}$$

where $\hat{x}_p$ and $\hat{x}_q$ (also $= \pm 1, \pm 3$) are the sliced estimates of x'_p and x'_q, and the $\mathcal{S}$ curve becomes as shown in Fig. 6.16. It can be seen that with no noise whatsoever there are three false but stable zeros at $20.7°$, $27.9°$, and $35.5°$. Noise has the effect of rounding (damping) this curve, and even a small amount (SNR < 30 dB) causes the first and last of these zeros to vanish. However, the zero at $27.9°$ persists, even with an SNR of 15 dB; consequently, a phase hit of greater than about $20°$ would cause the loop to converge to the false zero.

Phase hits of greater than $20°$ are very unlikely on the telephone system (see

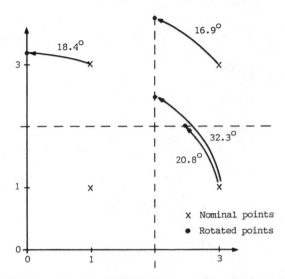

Figure 6.15 One quadrant of a 16QAM showing rotation of points to thresholds.

Fig. 1.21), but if one should occur, a receiver might not recover† until another two- or four-point training signal was sent. For larger constellations (32CR, 64SQ, and 128CR), there are many more discontinuities in the $\mathscr{S}$ curve, and much smaller phase hits would be sufficient to cause convergence to a false zero. It is clear that some other form of $g(\theta_e)$ is needed.

Each estimate of $\sin \theta_e$ in (6.17) is weighted by the energy in the sample. Since the outer points cross thresholds first, their effects should be decreased by estimating $\sin \theta_e$ directly without the weighting. This would involve a very tedious calculation, but $\tan \theta_e$ can be estimated by

$$\tan \theta_e = \frac{x'_p \hat{x}_q - x'_q \hat{x}_p}{x'_p \hat{x}_p + x'_q \hat{x}_q} \tag{6.18}$$

This would still require a division—impractical in an analog implementation, and a time-consuming operation in most digital processors—so the approximation should be made that the denominator $\approx (\hat{x}_p^2 + \hat{x}_q^2)$, the energy in the sliced signal. Since there are only a few nominal signal points in each quadrant, the deweighting can be performed by a switchable analog gain or by a DSP multiplication:

$$g(\theta_e) = (x'_p \hat{x}_q - x'_q \hat{x}_p) \left(\frac{1}{\hat{x}_p^2 + \hat{x}_q^2} \right) \tag{6.19}$$

† At $\theta_e = 27.9°$ the average phase error is zero, but its variance is quite large; reception of a preponderance of inner points ($\pm 1, \pm 1$), which still generate phase errors of the correct polarity, *might* push the phase back into the segment around $\theta_e = 0$.

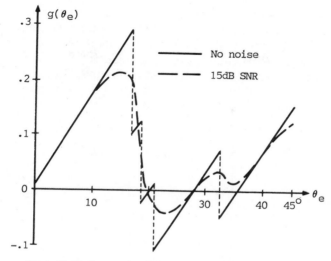

Figure 6.16 $\mathscr{S}$ curve for 16QAM using $g(\theta_e) = x'_p \hat{x}_q - x'_q \hat{x}_p$.

A plot of this $g(\theta_e)$ for 16QAM at several SNRs is shown in Fig. 6.17; it is very likely that the weak false zero shown for SNR > 20 dB would be overcome by random data and that the loop would always converge back to $\theta_e = 0$. However, one cannot be so confident about larger constellations, and more work is needed to devise ways to protect these against phase hits.

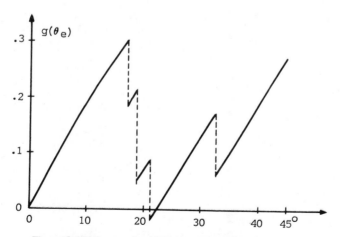

Figure 6.17 $\mathscr{S}$ curve for 16QAM using estimate of $\tan \theta_e$.

A Caveat. The level of noise added to all points — inner and outer — is the same, so the SNR of the latter is obviously greater; as long as θ_e stays in the main segment around zero, the outer points are better for estimating the phase error, and deweighting them will have the effect of reducing the average SNR of the loop — by as much as 3 dB for the larger constellations. For any given medium, the probability of phase hits must be weighed against the increased phase jitter that would result from guarding against them.

Chapter 7

Timing Recovery

The most important timing recovery task in the receiver is to generate a signal at the symbol rate f_s that can be used first to sample the partially processed received signal† and then to synchronize the decoding of this signal into the best estimate of the transmitted data. Transmission of a discrete tone at f_s (or any multiple thereof) — whether separately or as a result of the data code chosen — would be wasteful of power, because such a tone can convey no data information. Therefore this chapter will deal only with the cases where the received signal does not have an explicit spectral component at f_s or any fraction or multiple thereof, and some nonlinear operation must be performed to generate such a component. Depending on the type of system, the wanted tone may then be merely contaminated by the other products of the nonlinearity or almost completely buried by them. Just as for carrier recovery, there are are two basic methods of refining such a signal — a BPF or a PLL. Most of the factors to be considered in the choice of one or the other are the same as for carrier recovery (considered in Section 6.1), but there are a few that are special to timing recovery; these are discussed in Section 7.1.

In order to understand the conditions under which timing recovery methods must operate and the criteria by which they are judged, we will consider, in Section 7.2, some typical baseband systems and calculate the best phases of the sampling clock for each. There are two "best phases" in each case because, although the best phase is always, rather obviously, that that minimizes the

---

† This "partially processed signal" is the received signal that has been passed through any combination of filter, AGC, and demodulator; in the discussion of timing recovery it will be referred to simply as the data signal.

error rate of the slicer and decoder, a more precise definition of it depends on whether an adaptive equalizer is used.

Before discussing actual systems for timing recovery, we will discuss in Section 7.3 general characteristics of the systems: passband or baseband, forward-acting or feedback, operation with scrambled or unscrambled start-up signals, and so on.

In Sections 7.4 – 7.6 we will discuss timing recovery methods used in the cases of no equalizer, symbol-rate equalizers, and fractionally spaced equalizers respectively and compare the sampling phases that each method generates with the appropriate optimum phase. Finally, the special problem of timing recovery for partial response systems will be dealt with in Section 7.7.

7.1   PHASE-LOCKED LOOPS FOR TIMING RECOVERY

7.1.1   Avoiding Hang-Up

Hang-up may be a more serious problem for timing than for carrier recovery for two reasons: (1) many receivers are configured so that timing must be recovered before carrier recovery can start (even though, as discussed in Section 6.4.1, that is neither desirable nor necessary), and (2) most phase detectors used for timing recovery have a sinusoidal $\mathscr{S}$ curve (in contrast to the sawtooth curve of the decision-aided carrier recovery loops), so that if a loop is nearly 180° away from the correct phase, it will start very slowly. Fortunately it is easier to detect and correct hang-up in timing recovery loops.

The correction is done by "kicking" the VCO† away from the 180° point by either injecting a pulse so as to change its phase by 180° or by toggling the first stage of a frequency divider.

The detection process is more controversial. Some authors, [TT&M], have suggested using a complementary quadrature arm (phase detector and loop filter), but there is a simpler method. This was shown but not adequately described in [Bi1]; the following should be an improvement on both the method and the description.

The solid lines in Fig. 7.1a show a conventional PLL in which the parameters of the phase detector, low-pass filter, and VCO are such that in the stable state the output lags behind the input by 90°.‡ With this phase relationship the output should be low whenever a positive-going (p-g) transition of the input occurs; if it is high, there is a possibility of hang-up. Therefore, the dotted circuitry shown in Fig. 7.1a detects p-g transitions and uses them, if necessary, to toggle the output-select flip-flop. If it is feasible to run the oscillator at $2f_s$, the simplification shown in Fig. 7.1b is possible.

---

† Often, as we shall see, the best arrangement is a VCO followed by a frequency divider; in this chapter "VCO" means any such combination.

‡ The modification that is needed when the desired output is in phase with the input is described in Section 7.4.1.

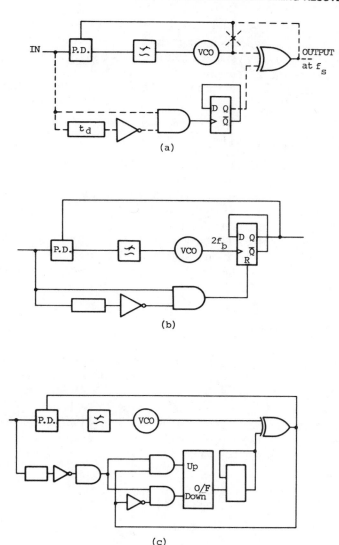

Figure 7.1 PLL with modifications to prevent hang-up: *(a)* basic; *(b)* a simplification; *(c)* modification to prevent vacillation around 90° points.

If the input is moderately jittery, this simple circuit may cause occasional vacillation around the ±90° points as jittery p-g transitions successively set and reset the output-select flip-flop. The effect will be less severe than the original 180° hang-up because the error voltage generated at the ±90° points is maximum, and the loop will move away much faster; nevertheless, it cannot be ignored. A solution is to insert an up/down counter in the set line as shown in

Fig. 7.1c; the counter is incremented by a "raw" toggle pulse and decremented by its complement (i.e., whenever a p-g transition occurs and the output is low); the flip-flop is toggled only when the counter overflows. For protection against vacillation around the $\pm 90°$ points a maximum count of four (two stages) is adequate; for extremely jittery input (see Section 7.4.1), more stages may be needed.

7.1.2 Band-Limiting the Input of a PLL

In most PLLs the signal from the VCO that is input to the phase detector is a square wave. It therefore has frequency components at $(2n + 1)f_s$, and since all the nonlinear operations (except pure squaring) that are used to generate the component at f_s also generate extraneous components over a very wide band, it is necessary, in order to avoid aliasing, to band-limit the input to the PLL to $2f_s$.† The choice, therefore, is not between a BPF and a PLL, but between a narrow BPF and the combination of a medium BPF and a PLL. The bandwidth requirements of the medium BPF will be calculated in each section where one is recommended.

7.2 TYPICAL BASEBAND SYSTEMS

As an example of the problems presented by the different cases, and to provide a basis for comparing the different methods, we will develop further the example that was discussed by Lucky, Salz, and Weldon [LS&W]. This is a typical full-response system in which a signal with an excess bandwidth factor α is passed through a channel with attenuation and delay that are linear and parabolic functions of frequency, respectively. The scaling parameters, A and β, are the attenuation (Atten) in decibels and the normalized group delay at the band edge, so that

$$\text{Atten} = A \left(\frac{2f}{f_s} \right) \text{dB}$$

and

$$\text{Delay} = \beta T \left(\frac{2f}{f_s} \right)^2 \tag{7.1}$$

where, as always,

$$f_s T = 1$$

Attenuation distortion alone results in an impulse response and eye pattern that are symmetrical about some center time. The best timing phase is the same by all criteria, and most recovery methods should find it with no trouble. There

---

† It can be seen that such a PLL is comparable to a sampled-data filter with a sampling rate of $4f_s$.

should be no static phase error (assuming no frequency offset), and the only effect of the distortion should be to increase the jitter of the recovered clock.

Delay distortion, on the other hand, causes the impulse response and the eye pattern to be asymmetrical, and this may be very pronounced if a large excess bandwidth is used. We will therefore first consider the problems of timing recovery as only β, the normalized delay at the band-edge frequency, is varied. Later we will consider the effects of adding attenuation distortion.

Figures 7.2a, b, and c show the variation of peak distortion with sampling phase for three excess bandwidths (large, medium, and small) and several values of β. RMS distortion is often just as interesting as peak distortion, but, although it is much less than the latter (e.g., for $\alpha = 0.5$ and $\beta = 2.0$ the minimum peak and RMS distortions are 0.5 and 0.25, respectively), its variation with sampling phase is very similar, and, most importantly, its minimum occurs at very nearly the same phase; we need not plot it separately.

7.3 GENERAL CHARACTERISTICS OF TIMING RECOVERY SYSTEMS

7.3.1 Timing Recovery from the Passband or the Baseband Signal?

For modulated signals, the most straightforward method of processing in the receiver is to demodulate first and then recover the timing from the baseband signal(s). If conventional quadrature modulation is used, both baseband channels can and should contribute to the timing recovery. If they are staggered, as in SQAM or SQPSK, then the two channels contain opposite timing information, and the timing signal from one (whether it be a zero-crossing pulse or the output of a rectifier) must be inverted before being added to that from the other.

Sometimes it is desirable to recover the timing directly from the modulated passband signal; this is discussed in Sections 7.3.5 and 7.4.3, but it must be emphasized that this can be done only with conventional quadrature modulation. For staggered quadrature systems and SSB systems there is little or no timing information in the passband signal,† and the timing phase derived from the baseband signal is totally dependent on the phase of the demodulating carrier.

7.3.2 Forward-Acting or Feedback Timing Recovery

The basic characteristics of these two methods are:

1. *Forward-Acting.* The continuous data signal is processed to produce a clock signal that is used to sample this same data signal. Subsequent processing is done on the sampled signal.

---

† After all, timing information in a passband signal must be contained in the envelope (to make it independent of the carrier phase), and the staggered modulation schemes were devised to minimize the variation of the envelope.

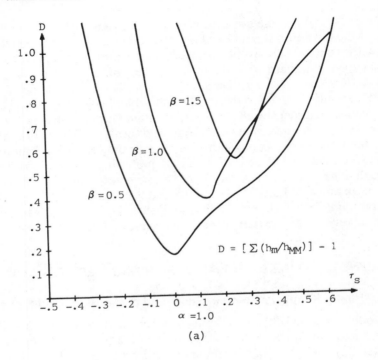

$$D = [\Sigma (h_m/h_{MM})] - 1$$

$\alpha = 1.0$

(a)

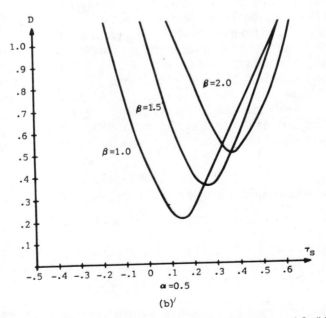

$\alpha = 0.5$

(b)

Figure 7.2 Variation of peak eye distortion with sampling time: *(a)* $\alpha = 1.0$; *(b)* $\alpha = 0.5$; *(c)* $\alpha = 0.25$.

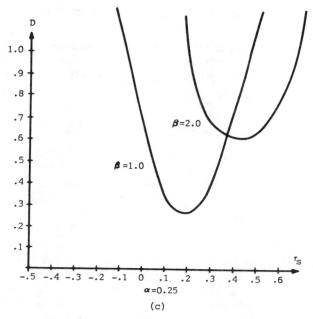

Figure 7.2 Continued

2. *Feedback.* The data signal is sampled at as low a rate as possible and, at some point later in the mainstream processing, an estimate is made of the accuracy of the sampling phase, and this is fed back to the original sampler.

Most practical methods have some of the characteristics of both methods, and nothing would be gained by discussing whether any particular method is more one than the other. However, a few general comments may be helpful in understanding the methods and the organization of this chapter:

1. Receivers without adaptive equalizers nearly always use a forward-acting method, because the amount of processing of the *sampled* signal needed to estimate the error in the sampling phase is usually comparable to that needed for a simple equalizer. We will consider one feedback scheme that uses a very simple measure of the error but will find that it is useful for only certain types of distortion.

2. For receivers with adaptive equalizers, the ideal approach is to jointly optimize the timing phase, the demodulating carrier phase, and the equalizer coefficients [Ko1]; however, for full-response systems wth symbol-rate equalizers, this requires a *lot* of processing. For these systems, the forward-acting band-edge component maximization (BECM) method (Section 7.4.1) pro-

duces a timing phase that is much better than that produced by any suboptimum feedback method of comparable complexity.

3. On the other hand, for full-response systems with fractionally spaced equalizers, and also for partial-response systems, it is quite easy to develop an adequate measure of timing phase error, and the feedback method is competitive with the forward-acting method.

7.3.3   Scrambled or Unscrambled Start-Up Signals

Most forward-acting timing recovery methods will acquire lock from any state with random data, but some feedback methods will fail if their initial phase error is near 180°; this is because they are decision-directed and the initial error rate is very high. This is much more serious than hang-up in a PLL, because the wrong timing phase may be stable. Therefore, in order to give the designers of receivers the greatest choice of methods, many data communication protocols require the transmission of a simple repeated pattern that is rich in timing information whenever the transmitter and receiver may have drifted apart (after quiet periods, or when a new transmitter in a multidrop system starts up).

For full-response systems this pattern is always one that contains a strong component at the band-edge frequency(ies). In the baseband the reversal pattern is most common. In the passband a signal that allows concurrent and almost independent acquisition of carrier and timing, as described in [Bi1], is desirable; for QAM there are two possibilities—repeated 180° phase changes or alternating $+90°$, $-90°$ phase changes as discussed in Section 4.2.5.† For partial-response systems there is less consensus; the best candidates are signals that contain strong components at $f_s/4$ in the baseband or at $(f_c \pm f_s/4)$ in the passband [Qu4].

7.3.4   Higher Multiples of the Symbol Clock

Subsidiary timing recovery problems arise if the data is encoded in M bits per symbol and/or the preprocessing of the receive signal or the timing recovery operation itself is done using sampled-data techniques (switched-capacitor or digital). In the first case the output of the data to the receiving terminal in serial form necessitates the generation of a clock signal at the frequency Mf_s; this can be done very easily by a digital PLL such as is described in Section 7.4.1. In the second case aliasing at the interface between the preprocessing and the symbol-rate sampling can be most easily avoided if the preprocessing is done at an integer multiple Lf_s of the recovered symbol rate. This higher frequency is easy

---

† Occasionally, in modems that usually do not use adaptive equalizers (e.g., a V.26 modem), these patterns can cause problems. The timing phase acquired initially is that of the band-edge components, and as we shall see, this is sometimes a long way away from the phase that gives the maximum eye opening. When random data starts, the error rate will be high until the timing moves to a better phase.

to generate, but it must then be fed back to the original sampling; this is an example of a mixed feedback/forward-acting method that will be discussed in Section 7.4.3.

7.4 TIMING RECOVERY FOR NO ADAPTIVE EQUALIZER

Calculation of the best sampling phase or delay (that that minimizes the error rate for some SNR) is extremely complicated, but it is certain that the best phase is always somewhere between the two phases that minimize the peak distortion and the RMS distortion of the sampled signal. As we have seen, the difference between these bounding phases is small and either would be perfectly satisfactory in practice. Unfortunately, although it is easy to calculate them mathematically, it is usually very difficult to generate them practically.

Therefore, we will approach the problem pragmatically. What has been done, and how good is it? We will describe various methods that have been developed† and consider the conditions (line distortion, system requirements, hardware constraints) under which each is suitable. The primary criterion for judging the methods is how well they approximate the best sampling phase, so we will examine the results of the methods before describing their implementation.

Since we are concerned here with systems without an adaptive equalizer, we will consider first the distortionless case and then amounts of channel distortion that close the eye to 50% ($D = 0.5$); this would degrade the SNR by about 5 dB. Most of the methods were evaluated for $\alpha = .25, .5$ and 1.0, and appropriate amounts of pure delay distortion and a mixture of attenuation and delay distortion as defined in (7.1) by the parameters A and β. Two methods that impose constraints on the sampled impulse response (see Section 7.4.4) were also evaluated for pure attenuation distortion. The normalized steady-state sampling delay τ/T and the ratio (in decibels) of the maximum eye opening to the eye opening at that delay are shown in Table 7.1.

In forward-acting methods, because the data signal contains no discrete spectral component at the symbol rate, some nonlinear operation must be performed to generate one. The methods differ in whether (1) this is done on the passband or the demodulated baseband signal, (2) the input to the nonlinear operator is band limited or has infinite band-width, and (3) a BPF or a PLL or a PBF/PLL combination is used to refine the tone at the symbol frequency.

7.4.1 Threshold-Crossing Detector and a Digital PLL

This is the simplest method of all in both concept and implementation. A square-wave clock is adjusted in frequency and phase so that one edge is aligned with the average of all the threshold crossings of the data signal; the signal is then

---

† More detailed discussions of some of the methods can be found in the references cited in [Fr3].

TABLE 7.1 Eight Unequalized Systems[a]

$\alpha/A/\beta$	Best	Median	Mean	Absolute Value	BECM	Square	$h_{-1}=h_1$	$h_1=0$
1.0/0/1.4	0.25	0.29–0.4 (2.1–13.6)	0.29 (2.1)	0.45 (∞)	0.47 (∞)	0.57 (∞)	0.14 (3.3)	0.25 (0)
1.0/6.2/0.62	0.20	0.14–0.19 (0.2–0)	0.16 (0.1)	0.19 (0)	0.21 (0)	0.28 (1.0)		
1.0/7.0/0	0	0 (0)	0 (0)	0 (0)	0 (0)	0 (0)	0 (0)	b (∞)
0.5/0/2.0	0.35	0.38–0.44 (0.3–1.1)	0.39 (0.4)	0.52 (6.5)	0.67 (∞)	0.73 (∞)		
0.5/7.2/0.72	0.17	0.15 (0)	0.15 (0)	0.16 (0)	0.24 (0)	0.26 (.4)		
0.25/0/1.7	0.31	0.31–0.35 (0–0.1)	0.32 (0)	0.37 (0.1)	0.57 (∞)	0.58 (∞)	0.26 (0.03)	0.20 (3.7)
0.25/7.8/0.78	0.20	0.14 (0.7)	0.14 (0.7)	0.15 (0.6)	0.26 (0.4)	0.26 (0.4)		
0.25/8.0/0	0	0 (0)	0 (0)	0 (0)	0 (0)	0 (0)	0 (0)	0.30 (9.8)

[a] Showing the best normalized sampling time, the times found by various methods, and the reduction (in decibels) of eye opening incurred by sampling at those times compared with sampling at the best time.
[b] Method fails completely. See Section 7.4.4.

sampled on the other edge. Theoretically, the method could be used for multi-level signals, but this would mean that the timing recovery would be dependent on the amplitude of the signal (and therefore on the state of the AGC); in practice, the method is used only for binary signals and the only threshold used is zero.

For reasons that will become apparent, this method is used only for low- and medium-speed data signals. These signals are usually band-limited, so that the first operation must be a zero-crossing detector (hard limiter). In principle, however, the method could be used, without the limiter, for infinite-bandwidth non-return-to-zero (NRZ) pulses.

A basic circuit that implements this method is shown in Fig. 7.3a. A clock with frequency Nf_s is used to sample the sign of the data signal. When two successive samples are of opposite sign, the present phase of the derived symbol clock is tested. Since the intention is to put the positive edge of this clock near the center of the horizontal eye opening, if the clock is high (low), a pulse is inserted into (deleted from) the nf_s clock train in order to advance (retard) the divide-by-N counter by one count.

Steady-State Phase. Since the circuit advances or retards the clock by the same amount regardless of how much the clock is late or early, the effect is to set the negative edge at the *median* of the zero crossings. The median is actually a

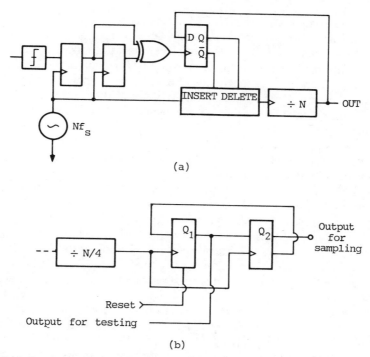

(a)

(b)

Figure 7.3 Median-finding circuits: *(a)* basic; *(b)* modification to frequency divider to allow resetting.

range, with equal numbers of zero crossings before and after, and none within the range or "null zone." The recovered clock will drift to one end or the other of the range depending on the relative frequencies of the transmit and receive oscillators.

If the impulse response is short and distorted, the distribution of zero crossings is discontinuous, and the range may become wide. In the 1.0/0/1.4 example the impulse response has only three significant lobes, and the zero crossings are distributed around three local "means"; the median is anywhere from 0.79 to 0.90. It can be seen from the results in Table 7.1 that a drift to the late end of this range would be catastrophic.

However, for smaller excess bandwidths or for a mixture of amplitude and delay distortion, the penalties in SNR would not be serious.

Steady-State Jitter. This simple circuit is a PLL with no filter, and the jitter is determined only by the step size. This, in turn, is determined by the requirement to track frequency offset between the clock sources in transmitter and receiver. If crystal oscillators are used, this should be less than 0.02% (200 ppm), so that, since there is only a 50% probability of a transition in each symbol period, N must be less than 2500.

A method of calculating the RMS jitter for such a system was given in [F&B]. For a voice-band modem with $f = 600$ s/s (V.22) or 1200 s/s (V.26), it would be about 10% — a just acceptable amount. The jitter can reduced by including an up/down counter as described in Section 7.1.1; if the count and the frequency divisor are N_c and N_d, respectively, the same frequency-tracking ability is maintained by setting $N = N_c N_d$.† The value of N_c that minimizes the jitter is very difficult to calculate; the choice is usually determined by the need for compatibility with the circuit that controls the acquisition of the loop.

Acquisition. The phase error when the loop first receives a signal has a rectangular probability distribution over the range $-\pi$ to $+\pi$; the magnitude of the average is $\pi/2$. The probability of a zero crossing occurring in a symbol period is $\frac{1}{2}$, and each insertion or deletion causes a phase change of $2\pi/N_d$. Therefore, the average number of zero crossings required for convergence is $N_c N_d/2$, and if the data were truly random, the maximum would be $N_c N_d$. However, there may be considerably fewer than $N/2$ transitions in N symbol periods, and in order to reduce to an acceptably low value the probability of there not being enough transitions for full convergence, it may be necessary to allow considerably more than N periods.‡

It is frequently the case that the values of N required for acquisition and steady-state jitter are very different. The best solution is to turn off the up/down counter (i.e., set $N_c = 1$) during acquisition; then N_c and N can be chosen separately to satisfy the acquisition and jitter requirements, respectively.

The discussion so far has assumed that the information provided by each zero crossing is correct, and that the loop can always start off immediately in the right direction. This, unfortunately, is not the case; the problem of hang-up in PLLs, discussed in Sections 6.1.3 and 7.1.1, may be particularly serious here. The basic solution proposed there is still appropriate, but some modifications are needed.

First, because the phase detector is of the early/late type and the negative-going (n-g) transitions of the desired output are aligned with the average of the input transition pulses, a 90° shifted version of the output must be generated in order to define the acceptable and unacceptable quadrants of the output. This is most easily done by implementing the last two stages of the frequency divider as a two-bit Johnson counter as shown in Fig. 7.3b; when a transition of the input occurs, the first stage of the counter should be reset.

Second, if the spectral shaping and/or channel distortion are such as to reduce the horizontal eye opening to less than 50%, the up/down counter that was previously introduced to protect against occasional vacillations about ±90° must be enlarged to prevent kicking the loop out of lock. For example, Fig. 7.4 shows the binary (i.e., after the limiter) eye that results from the system

---

† The circuit shown in Fig. 7.3a is a special case with $N_c = 1$ of this more general circuit.
‡ For example, if an acceptable failure rate is 1 in 10^5, and $N = 30$, then $2N$ symbol periods will be needed.

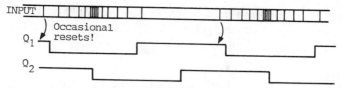

Figure 7.4 Binary eye with less than 50% opening, showing occasional false resets.

with $\alpha = 0.5$ and $\beta = 2.0$ (0.5/0/2.0), together with a supposedly locked clock signal. It can be seen that, because the horizontal opening is only 40%, if there were no up/down counter, extreme transitions of the eye would occasionally reset the "converged" clock!

The maximum count of the up/down counter should be determined by the expected worst-case horizontal eye opening; no tests or simulations have been performed yet, but it is estimated that a three-stage counter should always be adequate.

Systems for Which the Method is Suitable. All timing recovery methods that use a controlled divide down from some high frequency clock have one fundamental limitation — the need for a minimum resolution in the derived clock of $\pm \delta T$ means that some digital logic must be performed at the frequency $f_s/2\delta$ — typically $50f_s$. This puts an upper limit on the data rate at which these methods can be used.

This relatively crude method has two disadvantages: drift of the steady-state phase to one end of the null zone when the impulse response is short and distorted, and excessive data-related jitter when the excess bandwidth is small. The method is only suitable as an inexpensive solution in cases of large excess bandwidths and low delay distortion.

7.4.2 Zero-Crossing Detector in the Baseband plus a Symbol-Rate Filter

The null-zone effect of the median method can be eliminated by making the size of each correction proportional to the phase error. This means that, in the steady state, the transitions of the VCO or VCO/frequency-divider combination are at the *mean* of the zero crossings of the signal. As can be seen from Table 7.1, the mean is later than the best phase if there is only delay distortion, and earlier if there is a mixture of amplitude and delay distortion. Nevertheless, it is the only approximation that is acceptable under all conditions.

As in the previous method, the recovered timing is a function of only the phase information in the input data signal. If the signal is band-limited, the first component must be a hard limiter to convert the zero crossings into transitions of NRZ pulses.

A circuit for finding the mean was originally described by Saltzberg [Sa1], and there have been many variations since. A much later version by Le-Ngoc and Feher [L&F], uses a BPF to refine the component at f_s; it is shown in Fig. 7.5. Each transition generates a pulse of length ΔT which is input to a BPF tuned to f_s. The filter transfer function is

$$F(p) = \frac{2\sigma p}{p^2 + 2\sigma p + (2\pi f_s)^2}$$

it has unity gain at its center frequency f_s and a Q of $2\pi f_s/\sigma$.

The performance of this circuit with random data input can be best understood by first considering the special case of a system with $\alpha = 1.0$ and no channel distortion; the transitions occur randomly at $t = nT$ with a probability of one-half. For such a system Le-Ngoc and Feher considered the spectrum of the sequence of Return-to-Zero (RZ) pulses and concluded that, in order to reduce the jitter to acceptable levels, the filter must reject most of the power in the continuous spectrum around f_s. This is unreasonably pessimistic because most of that power results from amplitude modulation of the component at f_s and does not affect the zero crossings of the filter output. This was clearly explained by Gardner [Ga5] and others (see references in [Ga5]) who considered separately the in-phase and quadrature components of the self-noise (also known as *pattern noise* or *systematic jitter*).

The RMS jitter of the zero crossings of the filter output is caused only by the quadrature components and is approximately proportional to $1/Q$ (not $1/\sqrt{Q}$ as would be calculated by considering all extraneous components as "noise"); the constant of proportionality is, fortunately, small but, unfortunately, very difficult to calculate. Pending a more rigorous mathematical analysis, some estimate of it can be obtained as follows.

The response to random data or, more precisely, to a sequence of RZ pulses $a_k p(t - kT)$, where

$$p(t) = 1 \text{ for } 0 < t < \Delta T \text{ and } = 0 \text{ otherwise}$$

and
$$a_k = 0,1 \text{ with probability } \tfrac{1}{2}$$

must lie somewhere between the response to a single isolated pulse ($a_k = \delta_k$, the Dirac impulse function, generated by a single change of the input data; that is, by the sequence . . 00001111 . .) and the response to a continuous sequence of pulses ($a_k = 1$; generated by . . 010101 . .). These responses can be calculated, using simple Laplace transform theory, for the two extremes of pulse length ($\Delta \ll 1$ and $\Delta = \tfrac{1}{2}$), and their zero crossings can be compared to those of a perfect system ($Q = \infty$).

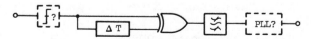

Figure 7.5 Zero-crossing detector plus BPF.

1. *Single Pulse.* The filter rings at a frequency,

$$f_r = \left(1 - \frac{1}{8Q^2}\right)f_s$$

and the nth zero crossing lags by

$$\frac{1}{2\pi Q} + \frac{n}{8Q^2} \qquad \text{for } \Delta \ll 1$$

and by only $\qquad \dfrac{n}{8Q^2} \qquad \text{for } \Delta = \tfrac{1}{2}$

2. *Repeated Pulses.* The zero crossings lead by

$$\frac{(1 - \pi/4)}{2\pi Q} \qquad \text{for } \Delta \ll 1$$

and by $\qquad \dfrac{(\pi^2 - 8)}{16\pi Q} \qquad \text{for } \Delta = \tfrac{1}{2}$

If the impulse response of the system is short, these extreme data patterns play a significant part in the determination of the output jitter, so that, by a very rough approximation

$$\text{RMS jitter} \approx \frac{0.6}{2\pi Q}$$

$$\approx \frac{0.1}{Q} \qquad \text{for } \Delta \ll 1$$

and

$$\approx \frac{0.25}{2\pi Q}$$

$$\approx \frac{0.04}{Q} \qquad \text{for } \Delta = \frac{1}{2}$$

One can see, therefore, that the "self-jitter" that is introduced by the filter in the process of restoring the missing zero-crossings is vey small. In practice, it will usually be insignificant compared to the jitter resulting from distortion of the zero crossings of the input. This can be explained informally thus: self-jitter is due to the absence (because of the randomness of the data pattern) of zero crossings that would otherwise be in the right place, whereas the greater jitter resulting from distortion is due to zero crossings that are present but in the wrong place!

The additions to the spectrum of the input that are caused by distortion are mostly in the form of phase modulation, and the power of this that passes to the output is inversely proportional to the Q of the filter. The ratio of input to output RMS jitter is therefore proportional to Q; this provides a simple way of calculating the necessary Q.

Often the answer is very disquieting; a Q of 100 might be required but would be impractical in many implementations. The solution is to augment the BPF with a PLL that has the advantage that, as discussed in Section 6.1.1, the sensitivity to the center frequency is divided by the d.c. gain of the loop. The BPF should now remove only those parts of the spectrum of the input that would cause aliasing in the phase detector; a Q of 10 is typical.

Optimum Amount of Delay T. Several authors [Sa1] and [Sp] have stated that $\Delta = \frac{1}{2}$ is optimum because it maximizes the power of the component at f_s; also, we have shown that this choice minimizes the self-jitter. On the other hand, we have shown that the self-jitter is often insignificant compared to that caused by distortion of the transitions of the input (which the choice of Δ cannot affect). There is, therefore, no hard-and-fast rule for the choice of Δ; several factors must be considered. If it is important to have a large amplitude of the f_s component (because of difficulties of providing gain in the BPF or problems of d.c. offset in the BPF or the limiter), $\frac{1}{2}$ should be used; if delay is expensive, then $\Delta \ll 1$ should be used; if the input zero crossings have very little distortion and it is important to keep the Q of the BPF as low as possible, $\frac{1}{2}$ should be used.

7.4.2.1   A Modification for SQAM and SQPSK.

If the baseband signals of a quadrature modulation system are staggered but the two transition pulses are added as in a conventional system, the generated components at f_s will be 180° out of phase and will cancel each other; a simple remedy is to invert one of the signals as shown in Fig. 7.6.

7.4.3   Other Nonlinear Operations in the Baseband plus a BPF

Another way of generating the component at f_s before filtering is to perform a nonlinear operation continuously (instead of only when a zero crossing occurs). The most commonly used operations are rectification and delay-and-multiply.

Delay-and-multiply circuits were developed to deal with infinite-bandwidth NRZ pulses. For these, they are equivalent to the transition pulse generator of the previous section (the output of the multiplier is zero unless there has been a

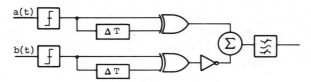

Figure 7.6 Modification to zero-crossing detector for SQAM.

data transition). They have also been used for band-limited signals, with the "proof" of the optimality of a delay of $T/2$ carried over. However, it seems that this is unjustifiable, and that a delay of zero is just as good. Since a delay-and-multiply with zero delay is a squarer, we need only consider rectification (of which squaring is a special case) for base-band operations from now on.

Rectification may be "hard" (an absolute value circuit), "soft" (a squarer) or somewhere in between (most practical circuits that use diodes or amplifiers) and is useful only on band-limited signals.

When they are used in the systems for which they were originally conceived, there is little difference between the theoretical performance of these methods, and the choice is usually made on the basis of ease of implementation using a particular type of hardware.

Steady-State Phase. It can be seen from Table 7.1 that for large excess bandwidths ($\alpha \simeq 1.0$), as delay distortion of the channel increases, the sampling phase derived from an absolute value operation can become very poor, and that derived from squaring can become completely useless. The reason for the large delay generated by the squarer can be understood by realizing that the component at f_s is created when pairs of components of a random data signal at frequencies ($f_s/2 - \Delta f$) and ($f_s/2 + \Delta f$) are mixed by the squaring. If the phases of the components are $\theta-$ and $\theta+$, the phase of the resultant will be ($\theta_- + \theta_+$). If the data sequence is a repeated 010101, then the data signal is a sinewave at $f_s/2$ and $\Delta f = 0$; then the phase of the timing signal is $2\theta(f_s/2)$. However, the delay distortion of channels is nearly always such that ($\theta_- + \theta_+$) is a minimum when $\Delta f = 0$, so that for all other data sequences, the contribution to the phase is greater than this. The steady-state phase of the timing signal is some sort of average (we do not need to be any more precise than that at this stage) of all of these contributions, and is therefore greater than the band-edge phase. As the excess bandwidth is decreased or the attenuation (always increasing with frequency) is increased, the role of the band-edge component in determining the timing phase increases, and the squarer's phase approaches the band-edge phase from above; however, as confirmed by Table 7.1, it can never be less. It is well known that band-edge timing is very poor for some distorted but unequalized channels [Ly1]; "squarer" timing is worse!

Hard rectification produces higher-order even powers of the signal in addition to the square, so that lower frequency components, which have been delayed less, can also contribute to the timing signal; consequently, the delay of the timing signal is less, and generally better than, the band-edge delay. But even so, it can still sometimes be too great.

Steady-State Jitter. As in the previous method, if the zero crossings of the data signal are not distorted, much of the power in the input to the filter is contained in amplitude modulation of the component at f_s (see Fig. 2 of [F&B]) and does not affect the "self-jitter" of the system. The necessary Q of the filter, and the resultant transient response and steady-state jitter can be calculated as in Section 7.3.2. It is generally agreed [Ga2] and [D&M] that of all the different types of nonlinearity, hard rectification results in the least jitter.

Systems for Which the Method is Suitable. If the distortion in the channel is mainly attenuation (e.g., in high-speed binary transmission on twisted-pair cable), this method may work as well as the zero-crossing pulse generator and may be slightly easier to implement.

On the other hand, if there is a significant chance of encountering some channels with mainly delay distortion (e.g., a connection on the PSTN via loaded subscriber loops and high-quality† carrier systems), then, at the risk of raising some hackles or lowering some spirits, it must be stated that the method is useless!

An Alternative to a BPF: The Wave Difference Method. Suzuki and colleagues [STOT] and Agazzi and colleagues [ATMH] have described a sampled-data version of this rectify-and-filter method. The sampling rate is quite low (typically $4f_s$), and the BPF is realized as a commutating filter. The result is that there is a lot of aliasing, which increases the jitter of the recovered clock; however, the jitter is apparently acceptable for high-speed (up to 160 kbit) transmission on twisted-pair subscriber loops.

It should be noted that the name "wave difference method" is misleading for full-response systems because the wave that is "differenced" is not the data signal, but a squared or rectified version of same. As we shall see in Section 7.6.2, the name makes more sense when the method is used for partial-response signals.

7.4.4 Feedback Methods Using Symbol-Rate Sampling

If a one-dimensional signal is sampled at the symbol rate, it is easy to learn selected terms of the impulse response and then adjust the sampling phase so as to drive some function of these terms to zero. Lucky and colleagues [LS&W] suggested that since the time derivative of the impulse response is typically greatest one symbol period after the maximum of the response, it would be a good idea to drive the first trailing echo, h_1,‡ to zero. Mueller and Muller [M&M] investigated this further and alternatively suggested that some symmetry of the response be imposed by forcing $h_{-1} = h_1$. Either of these can be effected by a simple application of the principles of decision-feedback equalization (see Chapter 9). However, before we consider the implementation, we must consider the possible results.

---

† High quality for voice transmission that is—flat amplitude response but large delay distortion.
‡ In the rest of this book we have recognized that the impulse response is causal and have numbered its terms from h_1 to h_M. When considering timing recovery, however, we are not interested in the number of leading terms, and it is much easier to abandon temporarily the "causal" convention, and consider the main term to have subscript zero.

Steady-State Phase. Table 7.1 agrees with [M&M] on two points: (1) if there is mainly delay distortion, the phase derived from $h_1 = 0$ is almost perfect; and (2) if there is mainly amplitude distortion, that derived from $h_{-1} = h_1$ is likewise almost perfect.†

However, for large excess bandwidths and amplitude distortion, $h_1 = 0$ fails completely (the impulse response has no zero crossings!), and even for $\alpha = 0.25$, it incurs a penalty of almost 10 dB because of a greatly reduced eye opening. *Furthermore,* for mainly delay distortion, $h_{-1} = h_1$ causes too early sampling for all excess bandwidths and incurs an SNR penalty of more than 3 dB.

The conclusion from this has to be that deriving timing by imposing either of these simple constraints on the sampled impulse response is useful only if the types of channel distortion can be accurately predicted. If they can, and the appropriate constraint is chosen, then the method is attractive for baseband systems where symbol-rate DSP is required. The reader is referred to [M&M] for more details.

7.4.5   Timing Recovery from the Passband Signal

Possible Advantages of Passband Operation. If the modulated signal has any envelope modulation, it is possible to recover the timing signal before demodulating. This has been advocated for systems that require fast acquisition because timing recovery can get started without having to wait for the carrier to converge. Let us examine this argument more closely.

If a passband signal

$$s(t) = x_p(t) \cos \omega_c t - x_q(t) \sin \omega_c t \tag{7.2a}$$

and its Hilbert transform,

$$\hat{s}(t) = x_p(t) \sin \omega_c t + x_q(t) \cos \omega_c t \tag{7.2b}$$

are squared and added, the result

$$s_2(t) = [s(t)]^2 + [\hat{s}(t)]^2 = [x_p(t)]^2 + [x_q(t)]^2 \tag{7.3}$$

is exactly what would be obtained by operations in the baseband. So, squaring is as good in the passband as it is in the baseband. But since a passband operation is usually more expensive to perform, we must ask whether it is necessary.

If $s(t)$ is demodulated by carriers $\cos(\omega_c t + \theta_c)$ and $\sin(\omega_c t + \theta_c)$, the baseband signals are as given by (6.1):

$$\begin{aligned} x_p' &= x_p \cos \theta_c - x_q \sin \theta_c \\ x_q' &= x_q \cos \theta_c + x_p \sin \theta_c \end{aligned} \tag{7.4}$$

† The impulse response is almost symmetrical about $t = 0$, so $h_{-1} = h_1$ is the obvious choice.

so that, when squared and added

$$(x'_p)^2 + (x'_q)^2 = x_p^2 + x_q^2$$

and the result is independent of the phase of the demodulating carrier; timing recovery should be able to start immediately.

As Gardner [Ga2] has pointed out, this equivalence of passband and baseband operations holds only for soft rectification, which is difficult to perform in practice. Therefore, let us examine the performance of a timing recovery circuit with the other extreme of hard rectification; then, having established a method and developed some insight, we can decide how to estimate the performance when using a particular nonlinear device — by interpolation between the results for hard and soft rectification or by more precise simulation?

First consider an undistorted system with $\alpha = 1.0$; a stylized representation of a baseband eye pattern is shown in Fig. 7.7a. It can be seen that when this is rectified the heavily drawn traces will reinforce each other and provide a strong component at f_s. If the error in the carrier phase is any odd multiple of $\pi/4$, the cross-coupling between the channels is maximum, and each eye looks like Fig. 7.7b. Now there are proportionally fewer traces (again drawn heavily) that will provide clock information when rectified, and there are some (drawn dashed) that will provide a clock component of the opposite phase. The magnitude of the clock component $|s(f_s, \theta_c)|$ was calculated (using an exact impulse response, not just the stylized one) for several carrier phase errors, and the results are plotted in Fig. 7.8.

The effect of this reduction of the clock component on convergence of the timing recovery in a burst or polled mode depends on the type of recovery system. If only an unquenched narrow BPF is used, the rate at which the new clock component can overcome the old component that is stored in the filter is proportional to the magnitude of the new component, which increases as the carrier converges. Let us make the following simplifying assumptions:

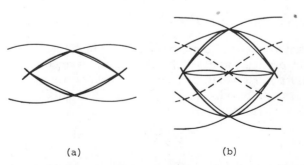

(a)   (b)

Figure 7.7 Binary eyes in QAM system: *(a)* zero carrier phase error; *(b)* $\pi/4$ carrier phase error.

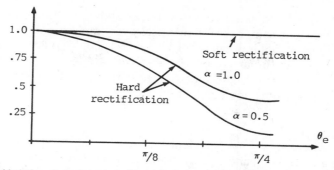

Figure 7.8 Variation of clock component in rectified baseband signal with carrier phase error.

1. The curve in Fig. 7.8 for $\alpha = 1.0$ can be approximated by

$$|S(f_s,\theta_c)| = |S(f_s,0)|\,(1 - 0.7\theta_c)$$

2. The carrier and clock loops are designed, assuming no interaction, to converge exponentially with the same time constant t_c.

Now with interaction, the error in the sampling delay τ_e will decay according to

$$\frac{d\tau_e}{dt} = \left(1 - 0.7\theta_c \exp\left(-\frac{t}{t_c}\right)\right)\tau_e$$

The solution of this differential equation is a sum of exponentials:

$$\frac{\tau_e}{\tau_0} = \exp(0.7\theta_0)\,\exp(-t/t_c)$$

$$\times [1 - 0.7\theta_0 \exp(-t/t_c) + 0.245\theta_0 \exp(-2t/t_c)\, \ldots\,] \quad (7.5)$$

where θ_0 and τ_0 are the initial errors of the carrier phase and timing. If θ_0 is a maximum of $\pi/4$, then in one time constant the timing error will decrease to 0.52 of the initial error, compared to 0.37 for no interaction; subsequently, the error will be about 1.7 times what it would be with no interaction. Or, putting it another way, carrier convergence will add a maximum of half of a time constant to the convergence time of the clock.

This method of analysis can be used for other amounts of excess bandwidth (the effects of carrier phase error on the clock component increase as α is reduced) and other models of the nonlinearity (the effects decrease for softer rectification), but it would become very complicated for higher order loops.

On the other hand, if a medium BPF followed by a limiter and a PLL are used, the old clock component should decay very rapidly, and the 37% of new clock component that is present even at $\theta_c = \pi/4$ should suffice to establish the zero crossings of the input to the PLL (albeit with a little more jitter) and get the PLL started at full speed. The unconverged state of the carrier should not slow down the clock acquisition at all.

Nonlinear Operations in the Passband. To extract timing information from a QPSK signal (the passband equivalent of baseband NRZ pulses) a delay-and-multiply must be used. Feher and Takhar [F&T] showed that the optimum delay is an integer multiple of half the carrier period; this ensures that there is no cross-coupling between the two channels. If the data signal is band-limited QAM, then as in the baseband, the integer multiplier can be reduced to zero, and the circuit becomes a squarer.

7.5 TIMING RECOVERY FOR SYMBOL-RATE ADAPTIVE EQUALIZERS

As we have seen, there are many different timing recovery methods for systems without adaptive equalizers. However, when an adaptive equalizer is used, there is much less variation; the theoretical goal of the timing recovery depends on the type of adaptive equalizer, but the practical implementation is often the same for all types. This is because for the high-speed, multilevel data signals for which adaptive equalizers are typically used, timing recovery is much more critical, and one basic technique, known as Band-Edge Component Maximization (BECM), becomes paramount.

The average timing phase generated by BECM is very nearly the best available for symbol-rate sampling and equalization. Even if sampling at a higher rate is necessary (see Sections 7.6 and 8.7), BECM may still be advantageous because it reduces the jitter on the timing signal. To help in understanding the relationship between acquisition and jitter of the timing signal, and adaptation of the equalizer, an analogy may be useful:

Consider two parallel moving walkways that are moving at slightly different speeds (the transmit and receive free-running clocks). In order to learn most precisely about events on the first walkway, an observer on the second should ideally walk at the difference speed. If, however, by analogy with most binary first-order DPLLs, the observer has only two possible speeds, full (which clearly must be greater than the maximum possible difference speed) and zero, then the observer must start and stop. If this observer occasionally gets the wrong information and walks in the wrong direction, it will be necessary to observe the first walkway while passing it at more than twice the difference speed; the observer may not be able to adapt quickly enough.

7.5.1 Timing Recovery for Baseband (One-Dimensional) Signals

Consider a baseband data signal, $s(t)$, with a Fourier transform

$$S(f) = |S(f)| \exp(j\theta(f))$$

The transform of the signal sampled at $t = kT + \tau$ that was given in (3.12c) can be expanded to take account of the phase of $S(f)$:

$$
\begin{aligned}
S(f,\tau) &= |S(f)| \exp j[2\pi f\tau + \theta(f)] \\
&\quad + |S(f_s - f)| \exp - j[2\pi(f_s - f)\tau + \theta(f_s - f)] \\
&= |S(f)| \cos[(2\pi f\tau + \theta(f)] \\
&\quad + |S(f_s - f) \cos[2\pi(f_s - f)\tau + \theta(f_s - f)] \\
&\quad + j\{|S(f)| \sin[2\pi f\tau + \theta(f)] \\
&\quad + |S(f_s - f)| \sin[2\pi(f_s - f)\tau + \theta(f_s - f)]\}
\end{aligned}
\tag{7.6}
$$

Therefore, the spectrum of the sampled power is

$$
\begin{aligned}
|S(f,\tau)|^2 &= |S(f)|^2 + |S(f_s - f)|^2 \\
&\quad + 2 \cdot |S(f)| \cdot |S(f_s - f)| \cdot \cos[2\pi f_s\tau + \theta(f) + \theta(f_s - f)]
\end{aligned}
\tag{7.7}
$$

Optimum Timing for Infinite Equalizers. Mazo [Ma1] has shown that in order to minimize the MSE of a linear equalizer in the presence of white Gaussian noise with variance σ^2, τ should satisfy

$$\frac{d}{d\tau} \int_0^{f_s/2} \frac{1}{|S(f,\tau)|^2 + \sigma^2} \, df = 0 \tag{7.8}$$

Solution of this equation to find what will be called τ_{MMSE} from now on is very complicated and probably not practically feasible. It will be shown later that instead of minimizing the MSE, it is adequate in nearly all cases to maximize the total sampled power; the reduction of SNR at the output of the equalizer will rarely exceed 0.1 dB.

If a nonlinear equalizer [Decision-Feedback Equalizer (DFE) or Maximum-Likelihood Sequence Estimator (MLSE)] is to be used, the aim is to maximize the SNR of the sampled signal; since the power of the sampled noise does not depend on the sampling time, the SNR is again maximized by maximizing the power of the sampled signal.

Maximizing the Sampled Power. This is achieved by fixing τ so that

$$\frac{d}{d\tau} \int_0^{f_s/2} |S(f,\tau)|^2 \, df = 0 \tag{7.9}$$

By combining (7.7) and (7.9), we find that the value of τ that maximizes the sampled power τ_{MSP} is given by

$$\frac{d}{d\tau} \int_0^{f_s/2} |S(f)\, S(f_s-f)|\cos[2\pi f\tau + \theta(f) + \theta(f_s-f)]\, df = 0$$

which leads to

$$\tan[2\pi f_s \tau_{\mathrm{MSP}}] = -\frac{\displaystyle\int_0^{f_s/2} |S(f)\, S(f_s-f)|\sin[\theta(f) + \theta(f_s-f)]\, df}{\displaystyle\int_0^{f_s/2} |S(f)\, S(f_s-f)|\cos[\theta(f) + \theta(f_s-f)]\, df} \tag{7.10}$$

In Section 7.3.3 we considered, and discarded, the strategy of squaring $s(t)$ and selecting the component at f_s; now we must look at it again. The spectrum of $[s(t)]^2$, which we will denote by $S2(f)$, is calculated by convolving $S(f)$ with itself; the component at f_s that is selected by the filter is therefore given by

$$\begin{aligned} S2(f) &= \int_{-\infty}^{\infty} |S(f)|\exp j\theta(f)\,|S(f_s-f)|\exp j\theta(f_s-f)\, df \\ &= 2\int_0^{f_s/2} |S(f)\,S(f_s-f)|\exp j[(\theta(f) + \theta(f_s-f)]\, df \end{aligned} \tag{7.11}$$

and the phase angle of this is equal to $-2\pi f_s\tau_{\mathrm{MSP}}$ as given by (7.10). We thus have the interesting result that the phase angle derived from squaring and filtering exactly maximizes the sampled power.

However, this result is not very useful by itself. We have assumed completely "random" data and a clock recovery filter with infinite Q, so that the clock phase is the steady average. In practice, the clock will jitter about the average and, if a particular input pattern lasts long enough, the clock will migrate to the phase determined by that pattern. We will show how, in an extreme case, complete spectral "nulling" may occur, and then discuss how to prevent this.

The Phenomenon of Spectral Nulling. It can be seen from (7.7) that complete cancellation of the terms of the sampled spectrum will occur at a frequency f if

$$|S(f)| = |S(f_s-f)|$$

and

$$2\pi f_s\tau + \theta(f) + \theta(f_s-f) = \pm\pi.$$

Since, for the great majority of baseband systems, $|S(f)| > |S(f_s-f)|$, nulling can occur only at $f_s/2$, and then τ must be given by

$$\pi f_s\tau = -\theta(f_s/2) \pm \pi/2$$

With such a disastrous choice of timing phase, the response to the repeated data sequence 0101 would be zero, and the channel could not be perfectly equalized even with an infinitely long linear equalizer; in trying to do so, the equalizer would make the gain at the band edge very large, and the consequent noise enhancement would be unacceptable.

An example of this effect can be seen in a system with $\alpha = 1.0$ and $\beta = 2.0$. The sampling delay that maximizes the eye opening (31%) is 0.39 T, and that that maximizes the sampled energy is 0.78 T, but the delay that would be called for by the repeated data sequence 0011 is only 0.168 T. Figure 7.9 shows the sampled spectrum for both delays; it can be seen that the timing extreme produced by the half-speed reversals would be disastrous.

Reduction of Jitter by Prefiltering. Wide, slow jitter (and the resultant reduction of sampled power) can be reduced by filtering the signal before rectifying, and thereby limiting the range of frequencies that can contribute to the timing signal. Several authors [F&B] and [Ly2], have shown that the quadrature component of the self-noise, and hence the jitter, can be eliminated if the total transfer function applied to the signal before the nonlinearity (i.e., transmit filter, channel, receive filter, and timing-recovery prefilter) is symmetrical about $f_s/2$. There are, however, two practical problems with this approach:

1. In order to make efficient use of the available bandwidth, high-speed data systems typically use small excess bandwidths and sharp cut-off shaping filters; this means that even with considerable detuning of the prefilter, symmetry is very difficult to achieve.
2. Unpredictable distortion in the channel makes it impossible to ensure symmetry under all conditions.

A solution of these problems is to make the prefilter sufficiently narrow to reject those frequencies that would lead to serious errors in the sampling phase, but unfortunately there has been nothing published on how narrow that is.

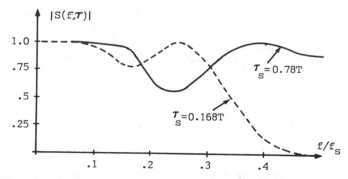

Figure 7.9 Sampled spectrum for optimum sampling phase and for phase produced by repeated 0011 pattern.

There certainly can be a trade-off between the bandwidths of the prefilter and the BPF/PLL following the nonlinearity. For example, a very narrow prefilter needs a postfilter only to reject any extraneous products of the "nonsoft" rectification; conversely, a very narrow BPF/PLL could use any rectified data signal directly because the probability of any adverse sequences having enough time to drive the sampling phase away from its mean would become vanishingly small.

A pragmatic compromise that is often adopted is to (1) make the Q of the prefilter as high as can be easily implemented in the chosen technology without resulting in any significant static phase error over all possible variations in component values and (2) design the medium-Q BPF only to reject d.c. and to attenuate the odd harmonics of f_s (which would cause aliasing in the phase detector) by about 20 dB; a Q of 4.0 is usually adequate.

The sampling delay that results from this compromise is obviously between that for no prefilter τ_{MSP} and that for a very narrow filter τ_{BE}.

Prefiltering for Band-Edge Component Maximization. Making the prefilter very narrow is equivalent to evaluating the integrals in (7.10) at only $f_s/2$, where $|S(f)\,S(f_s - f)|$ is usually a maximum. That is,

$$\tan[2\pi f_s \tau_{\mathrm{BE}}] = -\tan\,[2\theta(f_s/2)]$$

or
$$\tau_{\mathrm{BE}} = -\frac{\theta(f_s/2)}{\pi f_s} \qquad (7.12)$$

This is the antithesis of spectral nulling; the optimum delay is approximated by the phase delay at the band edge, and the band-edge component of the sampled signal is said to be maximized.†

However, this is a more severe approximation than the one that approximates τ_{MMSE} by τ_{MSP}; its accuracy may depend quite strongly on the amounts of excess bandwidth and attenuation and delay distortions. To test this, the three cases of $\alpha = 0.25$, 0.5, and 1.0 with now enough distortion (pure delay and a mixture) to just close the eye were analyzed. In each case the maximum sampled power that would be available if there were no delay distortion was calculated, and the magnitude of the signal scaled so that this number became unity; then the powers (in decibels) obtained by sampling at both τ_{MSP} and τ_{BE} were calculated. Similarly, the quantity

$$\Delta\mathrm{SNR} = 10 \log \int_0^{f_s/2} \frac{1}{|S(f,\tau)|^2}\, df \qquad (7.13)$$

---

† Strictly speaking, there is no discrete component at the band edge or at any other frequency, but the amplitude, subject to a phase ambiguity of $\pm\pi$, is maximized.

which is the reduction in SNR caused by a linear equalizer at high SNR, was evaluated for both band-edge and optimum delays. The results are shown in Table 7.2.

It can be seen that only in the extreme case of $\alpha = 1.0$ and $\beta = 2.6$ is the power significantly reduced (by 0.66 dB) by sampling at the symbol rate. It appears, therefore, that symbol-rate sampling is adequate in most baseband systems.

Furthermore, only in that extreme case is there any significant difference (0.13 dB) between the band-edge sampled power and the maximum power, or between the reduction of SNR with band-edge sampling and the minimum possible reduction (0.31 dB); in all other cases the difference is less than 0.1 dB. So, at least in the baseband, band-edge timing is a very good approximation to the optimum for either linear or nonlinear equalizers.

Desirability of a PLL for BECM. The combination of a prefilter and a postfilter is usually insufficient to make the jitter either small enough that it can be ignored or slow enough that the equalizer can follow it, and they must be augmented by a PLL. The structure of this should be as described in Section 6.1.1, and one example of its design is given in Section 7.5.2.

7.5.2 Timing Recovery for a Two-Dimensional Signal

The timing for a two-dimensional signal (QAM or QPSK or any variation thereof) can be recovered from the passband signal or from the baseband signal after demodulation, using the techniques of this section or Section 7.3.1, respectively. If the signal has been transmitted through a bandpass channel whose response is symmetrical about the carrier frequency f_c, the equivalent baseband impulse response is real, and all the discussion and results of the previous section apply just as well to the passband.

If, however, the response is asymmetrical, then, as Qureshi and Forney [Q&F] and Ungerboeck [U3] have shown, the opportunities for cancellation in

TABLE 7.2 Three Sampling Times and Some Sampled Powers and Reductions in SNR for More Severely Distorted Systems

Channel			Sampling Times			Sampled Power (dB)		ΔSNR (dB)	
α	A	β	τ_{MSP}	τ_{MMSE}	τ_{BE}	Maximum	at τ_{BE}	Minimum	at τ_{BE}
1.0	0	2.6	0.99	1.02	0.87	−0.66	−0.79	−0.87	−1.18
1.0	11.0	1.1	0.47	0.42	0.37	−0.09	−0.12	−2.8	−2.85
0.5	0	3.0	1.09	1.09	1.0	−0.11	−0.17	−0.10	−0.17
0.5	13.0	1.3	0.45	0.45	0.44	0	0	−3.45	−3.47
0.25	0	2.8	0.95	0.95	0.93	0	−0.01	0	−0.01

the sampled spectrum (aliasing) are greater, and in extreme cases the viability of symbol-rate sampling becomes questionable†. This can be shown as follows.

When a pass-band signal, $\mathrm{Re}[s(t) \exp j\omega_c t]$, (band-limited to $f_c - f_s < f < f_c + f_s$) is sampled at $t = kT + \tau$, and then demodulated down to baseband by multiplying each sample by $\exp j\omega_c t$,‡ the more general transform of the sampled and demodulated signal $S(f,\tau)$ that was given in (3.12b) can be expanded to include the phase of $S(f)$:

$$S(f,\tau) = |S(f - f_s)| \exp j[2\pi(f - f_s)\tau + \theta(f - f_s)] + |S(f)| \exp j[2\pi f\tau + \theta(f)]$$
$$+ |S(f + f_s)| \exp j[2\pi(f + f_s) + \theta(f + f_s)] \quad (7.14)$$

where $S(f)$ and $\theta(f)$ need no longer have any symmetries about zero frequency.

The spectrum of the sampled power is given by

$$|S(f,\tau)|^2 = |S(f)|^2 + |S(f - f_s)|^2$$
$$+ 2|S(f) \, S(f - f_s)| \cos[2\pi f_s\tau + \theta(f) - \theta(f - f_s)] \quad (7.15a)$$

for $f > 0$ and by

$$|S(f,\tau)|^2 = |S(f)|^2 + |S(f + f_s)|^2$$
$$+ 2|S(f) \, S(f + f_s)| \cos[2\pi f_s\tau + \theta(f + f_s) - \theta(f)] \quad (7.15b)$$

for $f < 0$.

If the sampling time could be chosen separately for each frequency component, or, equivalently, if the delay of the channel were equalized before sampling, the argument of each of the cosine functions would be zero. Therefore, the component of the power that is *lost* by sampling is

$$\Delta P(f,\tau) = 2\,|S(f) \, S(f - f_s)| \{1 - \cos[2\pi f_s\tau + \theta(f) - \theta(f - f_s)]\}$$
$$\text{for } f > 0 \quad (7.16a)$$

and $2\,|S(f) \, S(f + f_s)| \{1 - \cos[2\pi f_s\tau + \theta(f + f_s) - \theta(f)]\}$
$$\text{for } f < 0 \quad (7.16b)$$

The multiplying term in ΔP, $|S(f) \, S(f \pm f_s)|$, is usually greatest at the band edges $\pm f_s/2$, so it is useful to expand $\theta(f)$ as a Taylor series about those frequencies. Then the argument of the cosine in (7.16), defined as ϕ, becomes

---

† It must be emphasized that this is a consequence of the asymmetry of the channel and cannot be ameliorated by the method of timing recovery; symbol-rate sampling is no more or less effective whether the timing is recovered from the passband or the baseband.

‡ We will be concerned only with the sampled power, so we do not have to consider the phase of the demodulating carrier.

$$\phi = 2\pi f_s \tau + \theta(f_s/2) - \theta(-f_s/2) + (d\theta_+ - d\theta_-)(f \mp f_s/2)$$

$$+ (d^2\theta_+ - d^2\theta_-)\frac{(f \mp f_s/2)^2}{2} \quad (7.17)$$

where $d\theta_+$, $d\theta_-$, $d^2\theta_+$, and $d^2\theta_-$ are the first and second derivatives of $\theta(f)$ evaluated at the upper and lower band edges, respectively, and the upper and lower signs apply for $f > 0$ and < 0, respectively.

The sum of the first three terms can be made zero by setting

$$\tau = \tau_{BE} = \frac{\theta(-f_s/2) - \theta(f_s/2)}{2\pi f_s} \quad (7.18)$$

which is the complex baseband equivalent of (7.12). If the channel is symmetrical, then $d\theta_+ = d\theta_-$, and the terms involving the first derivatives of the phase cancel, and, just as in the one-dimensional case, the power loss depends only on the second derivatives. However, if $d\theta_+ \neq d\theta_-$, then there is a contribution to the phase mismatch that is proportional to the delay difference between the two sidebands; this may be much more severe.

If the phase of the channel is plotted as a function of frequency, the "band-edge" sampling delay is the slope of the chord drawn between $\theta(-f_s/2)$ and $\theta(f_s/2)$. This is shown in Fig. 7.10 for the case of a parabolic delay with a minimum at 2.0 kHz (cubic phase). At some frequency just below the upper band edge $(f_s/2 - \Delta f)$, it can be seen that the argument of the cosine in (7.16a) is

$$2\pi f_s \tau + \theta(f) - \theta(f - f_s) \simeq \Delta f(d\theta_+ - d\theta_-)$$

and a spectral dip occurs if this argument is equal to π. The difference between the delays at the two ends determines the distance of the spectral dips from the bandedges, and it can be seen from (7.15) that the relative levels of the signals at the two frequencies, separated by f_s, determine the depth of the dip (equal levels cause a null).

The worst possible situation arises when this type of phase response (resulting from greater delay at the low end) is combined with high attenuation at the upper end of the band. This could occur on the PSTN if the connection comprised long unloaded local loops and several carrier links. Using the linear and parabolic approximations to attenuation and delay suggested in Chapter 1, one could characterize a 99 percentile channel (90 percentile each of attenuation and delay distortions, since these two are seldom correlated) by

$$A = \begin{cases} 0 & \text{for } f < 1.1 \text{ kHz} \\ 8.0(f - 1.1) \text{ dB} & \text{for } f > 1.1 \text{ kHz} \end{cases} \quad (7.19)$$

and

$$\frac{d\theta}{d\omega} = 1.5(f - 2.0)^2 \text{ ms}$$

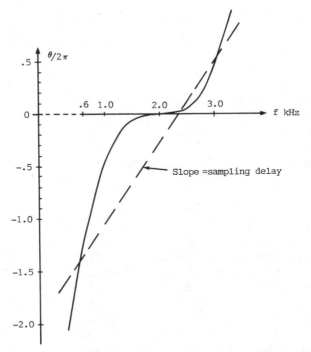

Figure 7.10 Cubic phase curve (delay minimum at 2.0 kHz) with BECM timing delay.

Figures 7.11*a* and 7.11*b* show the sampled spectra (using band-edge timing) for two modems recommended by the CCITT: V.27, a 1600 baud QAM system with f_c = 1800 Hz and 50% excess bandwidth and V.32, a 2400 baud QAM system with f_c = 1800 Hz and 12.5% excess bandwidth. Superimposed on each figure, for reference, is the sampled spectrum that would be obtained if the delay were equalized before sampling.

The magnitudes of the signals were then scaled so that the total available sampled powers (with no delay distortion) were unity. Table 7.3 shows the

TABLE 7.3   Sampled Power and SNR Loss for a Worst-Case Channel with Low-End Delay Distortion and High-End Attenuation Distortion

	Sampled Power (dB)		Δ SNR (dB)	
	At τ_{MSP}	At τ_{BE}	At τ_{MMSE}	At τ_{BE}
V.27	−0.14	−0.15	−1.60	−2.20
V.32	−0.01	−0.01	−4.3	−4.40

powers (in decibels) obtained by sampling the normalized signals at τ_{BE} and τ_{MSP} and the Δ SNR (defined by (7.13)) incurred by sampling at τ_{BE} and τ_{MMSE}.

It can be seen that the differences in the powers and Δ SNRs between the three choices of timing are very small for the V.27 signal and negligible for the V.32; this is in agreement with [Go2]. If symbol-rate sampling is to be used, then τ_{BE} (which is much easier to find than the other two) is perfectly acceptable.

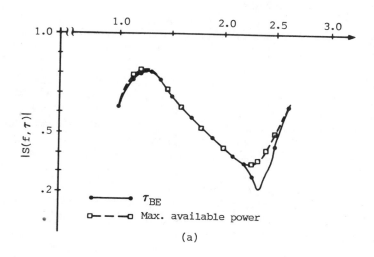

(a)

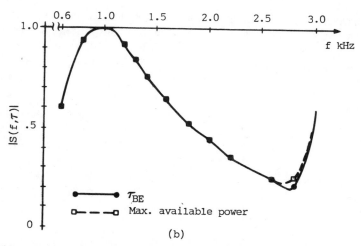

(b)

Figure 7.11 Maximum available power and sampled power: *(a)* for a distorted V.27; *(b)* for a distorted V.32.

For the system with 50% excess bandwidth, the losses incurred in symbol-rate sampling are significant but not serious, and for the sharply cut-off system, they are not even significant. This statement seems to contradict the conclusions of [Q&F] and [Un3], but I would not be so foolish! The results are presented this way in order to make the point that symbol-rate timing recovery per se works quite well even in cases of extreme distortion. The problem is that, whereas the performance of an infinite equalizer depends only on the total sampled power or SNR (for nonlinear and linear equalizers, respectively), the performance of a *finite* equalizer also depends strongly on the shape of the sampled spectrum and is particularly degraded by the localized dips that result from aliasing. This subject will be dealt with in more detail in Section 8.7.

7.5.2.1 *Implementation of BECM in the Passband.* A simple extension of the BECM technique to the passband is shown in Fig. 7.12. The two prefilters should, ideally, each have the same bandwidth as the single filter at $f_s/2$ in the baseband implementation, but this might make the Q of the upper filter impractical. The calculation must therefore proceed in the reverse direction and the baseband system be designed assuming a prefilter bandwidth of $(f_c + f_s/2)/Q_{max}$.

The necessary bandwidth of the filter centered at $f_s/2$ depends on (1) how the multiplication (nonlinearity) is performed, (2) the ratio of the carrier and clock frequencies, and (3) whether the input to the phase detector is hard-limited. The calculations are usually very complicated, but, fortunately, they are rarely necessary. We will see that a filter is nearly always needed, so its Q should be made as high as practical; this will always be sufficient to prevent aliasing.

In active-RC and switched-capacitor technology the best way to perform the multiplication is to hard-limit the output of the upper filter, and then use this binary signal to select either the output of the lower filter or its negative [Bi3]; this is the passband equivalent of hard rectification in the baseband. If each filter output has unit amplitude, the product has components at $[2(n - 1)f_c + nf_s]$ and $[2nf_c + (n - 1)f_s]$ with amplitude $1/[2(2n - 1)]$ for all integer values of n. A filter is certainly needed to reject these components regardless of the ratio of f_c to f_s.

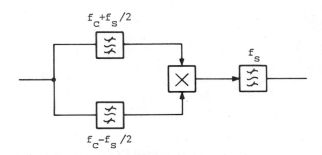

Figure 7.12 Basic circuit for BECM in the passband.

Other multiplication techniques using diodes and so forth essentially soft-limit one of the multiplicands, and still generate all of those components. The component at $2f_c$ is at the same level as the wanted f_s, and if $2f_c \sim (2m + 1)f_s$, then filtering is needed. The components for other values of n are at lower levels, and careful simulation would be needed to determine whether a filter is necessary to reject them.

In fact, the only combination of conditions under which filtering is not needed is when (1) the multiplication is perfect—producing only components at f_s and $2f_c$, (2) $2f_c$ is very different from $(2m + 1)f_s$—so that the low-pass filter in the PLL can reject the component there, and (3) the product signal is input to a linear phase detector (one in which the output is proportional to both the phase error and the amplitude of the input).

These conditions are sometimes satisfied in a digital implementation.

7.5.2.2 Digital Implementation of BECM in the Passband.
A straightforward digitalization of Fig. 7.12 would perform the band-edge filtering, multiplication, and f_s filtering at a sampling rate of Lf_s, and the desired sampling times would then be at the zero crossings of the output of the f_s filter. There are only L samples per symbol period; the timing could be established from these in four possible ways:

1. Select the one sample out of each L that is closest to the zero crossing. Unfortunately, L can seldom be large enough that the errors incurred in this are acceptable.

2. Change the phase of the original sampling so as to drive to zero one sample out of L. If any filtering for band-limiting or adjacent channel rejection has been performed on the sampled signal, the delay through this filter will be inside the timing control loop; this will certainly make the design of the loop difficult, and may make the achievement of a satisfactory transient response impossible. A block diagram is shown in Fig. 7.13a.

3. Use the value of the smallest sample to control an interpolation between two of the original L samples, as shown in Fig. 7.13b. This would be simple to implement because there would be no feedback, and interpolated samples would need to be generated only at the symbol rate (or perhaps at $2f_s$ if a $T/2$ FSE were used). The interpolation phase, however, would depend on the amplitude of the timing signal, and would consequently be very jittery.

4. Place an interpolator before the timing recovery circuit, as shown in Fig. 7.13c, and drive to zero one sample of the signal at f_s by adjusting the interpolation phase.

It can be seen that methods 2 and 4—the only practical ones in most cases—both have some of the characteristics of the previous forward-acting methods,

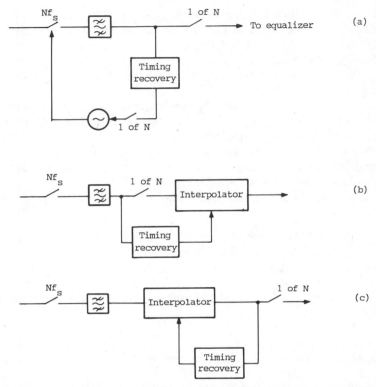

Figure 7.13 Relationship between timing recovery and input sampling: *(a)* controlling phase of original sampling; *(b)* feedforward interpolation; *(c)* feedback interpolation.

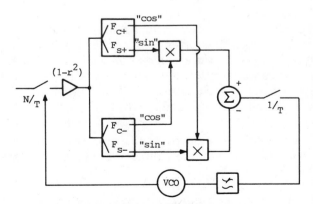

Figure 7.14 Block diagram of BECM circuit.

but also need feedback and a rather special PLL; we will consider the design of those parts of the loop that are common to both methods in the rest of this section and then discuss the interpolator in Section 7.5.2.3.

A preliminary version of the loop derived from Fig. 7.12 would have four filters (two band edge, f_s, and loop low-pass), and would be difficult to design; we must try to simplify it.

The only unwanted product of the digital multiplication is at $2f_c$; Godard [Go1] showed that this can be canceled without filtering by performing a complex multiplication (demodulation) that generates only the lower sideband. A simplification of Godard's circuit is shown in Fig. 7.14; each of the two arm filters (tuned to $f_c \pm f_s/2$) now has two orthogonal outputs. The four transfer functions are

$$F_{c+}(z) = \frac{z}{D_+},$$

$$F_{s+}(z) = \frac{z^2 - 1}{D_+},$$

$$F_{c-}(z) = \frac{az}{D_-},$$

and
$$F_{s-}(z) = \frac{z^2 - 1}{D_-},$$

where
$$D_+ = z^2 - (1 + r^2) \cos \phi_+ \, z + r^2$$
$$D_- = z^2 - (1 + r^2) \cos \phi_- \, z + r^2$$
$$\phi_+ = \frac{2\pi(f_c + f_s/2)}{L}$$

and
$$\phi_- = \frac{2\pi(f_c - f_s/2)}{L}$$

and the multiplying factor on F_{c-} to ensure that the upper sideband cancels is

$$a = \frac{\sin \phi_-}{\sin \phi_+}$$

It can be seen that all four filters have a gain at their center frequency that is proportional to $1/(1 - r^2)$; the initial scaling factor of $(1 - r^2)$ is necessary to prevent saturation.

When the clean signal at f_s has been generated, the function of phase detection is performed by selecting one sample out of L; this sample is input to a low-pass filter whose output drives a VCO (preferably crystal-controlled). The last part of the VCO in the PLL must be a frequency divide-by-L, with its input

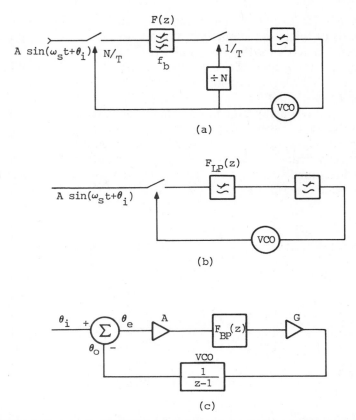

Figure 7.15 Equivalent circuits of BECM loop: *(a)* in passband; *(b)* in baseband; *(c)* as a conventional PLL.

and output controlling the original sampling and the symbol-rate sampling, respectively.

The following method of calculating the parameters of the loop by reducing it to an equivalent conventional baseband loop was described by Sorbara, a colleague, in a private communication.

Since the arm filters are each quite narrow band, we can postulate that the input to the system is $A\,[\sin(\omega_- t + \theta_-) + \sin(\omega_+ t + \theta_+)]$,† where $\omega_- = 2\pi(f_c - f_s/2)$, $\omega_+ = 2\pi(f_c + f_s/2)$, and θ_- and θ_+ are the phase shifts of the channel at the lower and upper band edges, respectively. The output of the

---

† The value assigned to A should depend on the start-up data signal used. If the full power (normalized to unity) is transmitted as band-edge components (e.g., in a V.27), then $A = 1/\sqrt{2}$; if only half the power is there (e.g., in a V.32), then $A = 1/2$. If a special synchronizing signal is not used, and the data is always random, the magnitude of the band-edge components is proportional to $1/\sqrt{Q}$; this is discussed later in this section.

adder will be samples at $t = kT/L$ of $A \sin(2\pi f_s t + \theta_i)/\sin \phi_+$, where θ_i, the input phase, equals $(\theta_+ - \theta_-)$. Therefore, we can replace the four filters and the two multipliers by a single filter tuned to the symbol rate; this leads to the equivalent circuit shown in Fig. 7.15a. The desired shape of the passband of this filter is difficult to calculate, but, if it is approximated by a single tuned section, equating the delays at f_s shows that its bandwidth must be the same as that of the original arm filters.

The second sampling is essentially a demodulation, so the BPF can, in turn, be replaced by its low-pass equivalent as shown in Fig. 7.15b. Finally, the sampler can be drawn as a phase detector and the sampled VCO represented as a forward Euler integrator; the conventional PLL illustrated in Fig. 7.15c results.

In order for the LPF, which is switched at f_s, to have the same bandwidth as the BPF which is switched at Lf_s, the distance of the LPF pole from the unit circle must be L times that of the BPF poles. Also, the LPF must have a zero at $z = -1$, and its gain at d.c. must equal that of the BPF. Therefore, if we write $r = 1 - \delta$, the transfer function of the equivalent LPF is

$$F_{LP}(z) = \frac{L}{2 \sin \phi_+} \frac{z + 1}{(z - 1 + L\delta)}$$

This filter acts as a loop filter,† and there is no need for the loop filter per se to be anything more than a constant gain G. The closed-loop error transfer function is therefore

$$\frac{\theta_i - \theta_o}{\theta_i} = \frac{(z - 1)(z - 1 + L\delta)}{z^2 - (2 - L\delta) z + 1 - L\delta + [GAL (z + 1)/(2 \sin \phi_+)]} \qquad (7.20)$$

For the desirable condition of critical damping, we find that

$$G \simeq \frac{L\delta^2 \sin \phi_+}{4A}$$

and the loop behaves just like a conventional sampled-data PLL [Li1] as far as trade-off between acquisition time and output jitter is concerned.

This PLL has a true sinusoidal phase detector, so it will suffer from hang-up just like any similar PLL. It is recommended that when a positive zero crossing occurs, the state of the final divide-by-4 circuit or microprogram counter should be set appropriately, as discussed in Section 7.1.3. Since the signal has already been well filtered, the input to the phase detector (sampler) has very little jitter, and an up/down counter is not needed; nevertheless, the usual

---

† In a conventional PLL it is common for the loop filter to have a pole a $z = 1$, so that the loop can track frequency offset with no phase error. This cannot be done here; the two BPFs must have finite Q values in order to constrain their outputs.

precaution of allowing this setting only during acquisition (that is, turning it off during data reception) is recommended.

Relationship Between the Timing Recovery and the Phase Splitter. As Godard pointed out, in order to improve the transient response of the loop, it is desirable to keep the delay through the phase splitter outside the timing recovery loop. This means that the timing should be recovered before the phase splitter, and therefore that the timing signal from the VCO must be delayed before being used to sample the complex data signal. If it is arranged that the delay through the splitter is an integer multiple of T/L, the simple circuit shown in Fig. 7.16 can be used; this may also include some *mild* bandpass filtering of the receive signal, as shown by the dotted box.

However, serious problems arise if the receive signal is very contaminated by ACI (e.g., in a V.22 bis modem) and a lot of filtering is needed before the phase splitter. If the clock recovery were done before this filtering as shown in Fig. 7.16, the narrow filters at $(f_c \pm f_s/2)$ would have to reject the ACI, and their Q value would have to be very high. Also, the amplitude of the in-band part of the input to the narrow filters would vary over a wide range, and the stability of the loop would be questionable.

On the other hand, the delay of such a large filter should not be included in the loop, and interpolation is probably a much better approach.

7.5.2.3 *Interpolation.*

The problem of interpolation can be defined as follows. Given a set of values of a function, $f(t_m) \triangleq f_m$ for $m = 1$ to M at equally spaced values of its variable t, find an approximation to $f(t)$, where t is inside the range of the t_m.

The simplest solution is based on fitting a polynomial of degree $(M-1)$ through the observed $f(t_m)$ and calculating the value of the polynomial at t. The calculations are best done by creating a table of successive differences:

t_m	f_m			
t_1	f_1	$\delta_{11} = (f_2 - f_1)$		
t_2	f_2	$\delta_{12} = (f_3 - f_2)$	$\delta_{21} = \delta_{12} - \delta_{11}$	$\delta_{31} = \delta_{22} - \delta_{21}$
t_3	f_3	$\delta_{13} = (f_4 - f_3)$	$\delta_{22} = \delta_{13} - \delta_{12}$	
t_4	f_4		etc.	

and then

$$f(t) = f_1 + \frac{(t - t_1)}{(t_2 - t_1)} \delta_{11} + \frac{(t - t_1)(t - t_2)}{2!(t_2 - t_1)} \delta_{21}$$
$$+ \frac{(t - t_1)(t - t_2)(t - t_3)}{3!(t_2 - t_1)} \delta_{31} + \cdots \quad (7.21)$$

where $t_2 - t_1 = t_3 - t_2 = t_4 - t_3$.

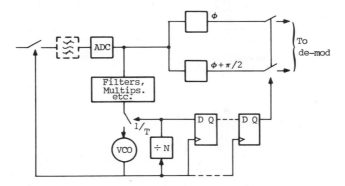

Figure 7.16 Allowing for delay through phase splitters.

The required value of M for a given maximum or RMS interpolation error obviously decreases as the degree L of oversampling increases. For the common cases of $L = 3$ and 4, it appears that $M = 4$ and 3, respectively, are completely adequate (RMS interpolation error $< 0.5\%$) even for pulses with small excess bandwidths ($\alpha < 0.25$).†

The parameter $(t - t_1)/(t_2 - t_1)$ can be considered as an "interpolation phase" — equivalent to the output phase of a VCO but expressed in periods rather than radians. If this "phase" is defined as x, (7.21) becomes

$$f(t) = f_1 + \delta_{11} x + \frac{\delta_{21} x(x + 1)}{2} + \frac{\delta_{31} x(x + 1)(x + 2)}{6} + \cdots \qquad (7.22)$$

In order to minimize the delay through the interpolator, f_1 should be defined as the most recent sample, and the interpolation should be made between it and the previous sample; that is, $0 < x < 1$.

Just as in a conventional PLL, x is generated by integrating the filtered phase error. This might be calculated in several ways, but the BECM method of Section 7.5.2.2 is especially suitable; the phase error is then the sample of the signal at f_s that must be driven to zero. For $M = 4$, the complete loop would be as shown in Fig. 7.17.

As the receive clock precesses relative to the transmit clock, x will occasionally slip out of the range $0 < x < 1$. If it goes below 0, it must be incremented by one and the data in the register must not be shifted; similarly, if it goes above one, it must be decremented by one and the data in the register must be shifted twice.

---

† Pulses with larger excess bandwidths are smoother and therefore easier to interpolate. For $\alpha = 1$ the RMS error is about half of what it is for α small.

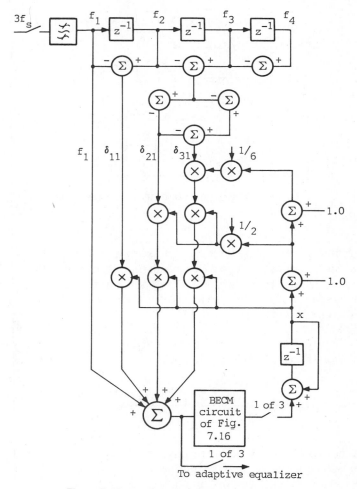

Figure 7.17 Interpolator included in BECM loop.

7.6 TIMING RECOVERY FOR FRACTIONALLY SPACED ADAPTIVE EQUALIZERS

In Section 7.5.2 we discussed the aliasing effects that occur when a passband signal with very asymmetrical amplitude and delay distortion is sampled at the symbol rate. In Chapter 8 we shall see that these effects can be avoided by adequately sampling the signal and using an equalizer with a fractional tap spacing of T' where $1/T' > (1 + \alpha)f_s$. It is most convenient if $T' = MT/L$, where M and L are small integers.† Then the initial sampling and A/D conver-

† The only ratios that have been considered are $\frac{1}{2}$, $\frac{2}{3}$, and $\frac{3}{4}$.

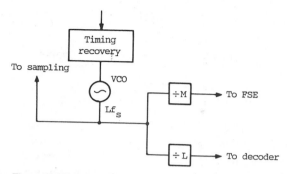

Figure 7.18 Splitting a frequency divider for an *L/M* FSE.

sion must be done at some integer multiple of Lf_s, and the final stages of the frequency divider are split to divide by both L and M as shown in Fig. 7.18.

A Fractionally Spaced Equalizer (FSE) can augment the delay of a signal to match any phase of the symbol-rate sampling, so its performance is insensitive to the timing phase, and, strictly speaking, the receiver need only be frequency-locked or symbol-locked to the incoming data so that the data can be decoded and clocked out at the correct rate. Nevertheless, we will discuss both phase-locked and frequency-locked methods.

7.6.1 Phase-Locked Methods

The forward-acting BECM method can still be used; in this case the narrow filtering around the band edges is used, not to achieve a particular phase for the timing signal, but to reduce the data-related jitter that the equalizer could not follow. The only modification to the previous methods that may be needed is the splitting of the frequency divider. If sampling must be performed at the receiver input, the sampled output of the f_s signal can also be used as the control for an interpolator, as described in Section 7.5.2.3 for T-Spaced Equalizers (TSEs).

Since phase is unimportant, it is tempting to try to avoid the rather cumbersome BECM method, and to reduce the amount of processing by using the FSE to generate a jitter-free control signal. For a one-dimensional signal, the first trailing tap, c_{NM+1}, would be a promising candidate, but for a QAM signal, it is difficult to find a parameter that is independent of the demodulating phase. Furthermore, the delay through the equalizer would be inside the timing loop, and would impair its transient response.

7.6.2 A Frequency-Locked Method

The following is a slight modification of the method suggested by Ungerboeck [Un3], in which the information for frequency lock is derived from the equalizer taps, c_n ($n = 1$ to N). The sampling time is adjusted according to

$$\tau_{k+1} = \tau_k + \Delta_T D \tag{7.23}$$

where

$$D = \sum_{n=1}^{NM-1} |c_i|^2 - \sum_{n=NM+1}^{N} |c_i|^2 \tag{7.24}$$

and NM is the index of the main tap. Since impulse responses typically have more trailing than leading echoes, NM is made slightly greater than $(N-1)/2$. The modification of omitting the NMth tap from the calculation of D would seem to allow this tap to stay near an initialized value of 1.0 and thereby to reduce slightly the convergence time of the equalizer.

The amount of processing needed for this feedback method is not much less than for the forward-acting methods, but there may be some other benefits. Qureshi [Qu2] showed that sometimes the MSE can be significantly reduced by adaptively adjusting the position of the main tap. This is difficult to do with a symbol-rate equalizer because the MMSE goes through many alternating local minima and maxima as the sampling phase is changed, but it should be much easier with an FSE because its MMSE is a monotonic function of the sampling phase. One possible strategy would be to modify (7.24) to

$$D = \sum_{n=1}^{N_1} |c_n|^2 - \sum_{n=N-N_1}^{N} |c_n|^2 \tag{7.25}$$

where now $N_1 \simeq N/4$. This should have the effect of balancing the tails of the equalizer by allowing the main tap to move within the range $N/4$ to $3N/4$ and might allow the length of the equalizer to be reduced by about 20% without increasing the MMSE.

A Caveat. It should be noted that the timing may have to move several symbol periods; to see the implication of this, let us consider further the analogy suggested at the beginning of Section 7.5. There are now several markers, spaced a symbol period apart, set up on the first moving walkway, and the observer must walk toward a particular one. In contrast to the jittery situation discussed in Section 7.1.1, the information that the observer receives is now consistent for a long time, and that individual will walk at maximum speed; consequently, the observer may move, relative to the first walkway, at more than twice the difference speed of the walkways.

This is a much more stringent requirement than any we have encountered so far. Consider, as an example, a V.32 modem with $f_s = 2400$ s/s; if the maximum difference between the clocks is 200 ppm and the insert or delete capability of the PLL is 300 ppm,† the loop will move the timing by $T/2$ — requiring a complete retraining of the equalizer taps — in 416 ms.

---

† Note that it makes no difference whether this is achieved with a first- or second-order loop.

Transient Response. The delay through the equalizer appears in the timing recovery loop in an extremely complicated way, and it would be very difficult to optimize the transient response of the timing. Fortunately, as Ungerboeck pointed out, the equalizer will converge with any timing phase, and there is no need for the timing to converge (i.e., for D to reach zero) before data transmission starts. If the equalizer can keep up with the relatively fast and prolonged migration of the timing, the transient response of the timing is not important.

7.6.3 Relative Merits of Phase-Lock and Frequency-Lock Methods

As we saw in the previous section, in the feedback, frequency-lock, method the adaptive FSE has to follow the changing of the impulse response as the receive clock precesses relative to f_s. There is an exact parallel to carrier recovery here; a complex equalizer (TSE or FSE) can compensate for any demodulating carrier phase, but it is easier to deal with frequency offset by using a separate circuit or algorithm that, because it deals with only one variable, carrier phase, can move faster without causing jitter. The relative frequency offset seen between clocks is typically much less than that between carriers, but the advantages of tracking the changing timing phase of a signal separately from the equalizer may still be important.

7.7 TIMING RECOVERY FOR PARTIAL-RESPONSE SIGNALS

The principal ways in which timing recovery for partial-response systems differs from that for systems that use some excess bandwidth are:

1. Because the signal is band-limited to $f < f_s/2$, no clock component can be generated by a squaring of the signal.
2. As we shall see in Chapter 8, the performance of an infinite equalizer is insensitive to timing phase, and that of a finite one is only slightly sensitive.
3. All the recovery methods that were tested generated a steady-state timing phase that was acceptable (< 1 dB reduction of the SNR) for an unequalized signal; because of condition 2 (above), this was also acceptable for a signal that was to be equalized.
4. The amount of timing information contained in a random data signal is much less than for full-response systems, and, correspondingly, the jitter resulting from both random data and noise is much greater.

Consequently, in any comparison of the different methods we do not need to consider steady-state phase, or to differentiate between equalized and unequalized systems; we need only compare the jitters. All the forward-acting methods described in this section were simulated for Classes I and IV (the only two of

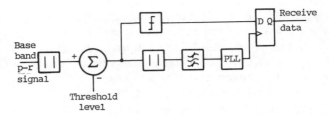

Figure 7.19 Rectifier/level shifter/rectifier/BPF combined with detector.

interest) with no distortion, and also with enough delay distortion to half close the eye. The feedback method described in Section 7.7.2 is so closely coupled with the equalizer that no separate quantitative evaluation is possible; we will have to limit ourselves to a qualitative description only.

7.7.1 Forward-Acting Methods

Rectifier/BPF. This is the simplest method of all; the circuit comprises just a rectifier and a BPF (or, of course, a BPF/PLL); it has been simulated by D'Andrea and Mengali [D&M]. As noted above, soft rectification will not work; a hard rectifier (an absolute-value circuit) is best but may be difficult to implement in hardware appropriate for high-speed data.

Rectifier/Level Shifter/Rectifier/BPF. The magnitude of the clock component can be increased and the jitter reduced by rectifying the signal twice. Obviously, if the second rectifier immediately followed the first, it would have no effect; a level shifter must be interposed. Ideally, in order to make the second rectifier most effective, the shifted signal should have no d.c. component, but the performance is not very sensitive to the amount of the shift. In an unequalized receiver the shift can be made equal to the slicing level used for detection and decoding; the economical configuration shown in Fig. 7.19 results. When an adaptive equalizer is used, the detection thresholds are not needed at the input to the equalizer,† so the rectifier and level shifter must be provided just for the timing recovery.

Since there are now two rectifiers, each one needs only to double the frequency range of its input. It appears, from the limited amount of simulation done so far, that either hard or soft rectifiers can be used.

Rectifier/Unipolar Limiter/Level Shifter/Rectifier. Ridout [Ri1] scrutinized a rectified Class IV eye pattern (with undistorted sampling levels of ± 1 and 0) and observed that the inner band (0 to 0.5) contains very little timing information but the middle and outer bands (0.5 to 1.0 and > 1.0 respectively) contain

† Either they are calculated at the output of the equalizer, or the main tap of the equalizer is adjusted to match some fixed decision thresholds.

significant clock components that are of opposite phase. The sequence of operations he proposed is, in more familiar terms, full-wave rectify, shift by -0.5,
half-wave rectify, shift by -0.5, and full-wave rectify.

Again, it would seem that the hardness of the rectifiers is not important.

Rectifier/BPF/Rectifier/BPF. The d.c. can also be removed from the rectified
signal by passing it through a BPF centered at $f_s/2$. Since a rectified partial
response eye somewhat resembles a shifted full-response eye, it would seem that
this method is similar to using a prefilter for BECM, but the results of the
simulations were surprising:

1. The method failed completely for Class IV signals, even though level
   shifting (which is equivalent to very sharp high-pass filtering) worked very
   well. I can offer no explanation of this.
2. For Class I signals, one would expect that the jitter would decrease mono-
   tonically as the Q of the prefilter is increased. Surprisingly, there appears
   to be a fairly sharp minimum of the jitter at a Q of about 5.

Threshold-Crossing Pulse Generator/BPF. In Section 7.3.1, when considering
full-response signals, we said that because the relationship between the data
signal and a (nonzero) threshold level is a function of the state of the AGC,
threshold-crossing techniques should be used only for two-level signals. However, stabilization of the AGC and generation of threshold levels are essential
parts of the detection process for partial-response signals, and the timing recovery can take advantage of them.

The basic circuit shown in Fig. 7.20*a* is an extension to three levels of that of
Fig. 7.5. It can be simplified to that in 7.20*b*, which is the circuit in Fig. 7.19 with
the second rectifier replaced by a zero-crossing detector.

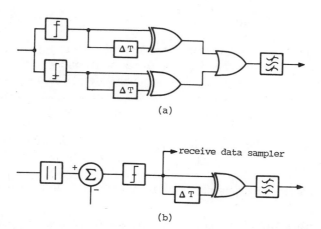

(a)

(b)

Figure 7.20 Threshold-crossing detectors/BPF: *(a)* two detectors; *(b)* simplified form.

TABLE 7.4   Jitters for Partial-Response Timing Recovery by Several Methods

	Class I			Class IV		
	Pattern Jitter without Distortion	Noise Jitter without Distortion	Total Jitter with Distortion	Pattern Jitter without Distortion	Noise Jitter without Distortion	Total Jitter with Distortion
R/F	0.0041	0.0087	0.025	0.034	0.021	a
R/LS/R/F	0.0018	0.0052	0.023	0.0068	0.0089	0.059
R/UL/LS/R/F	a	a	a	0.0056	0.0089	0.020
R/F/R/F	0.0038	0.011	0.016	a	a	a
TC/F	0.0042	0.0078	0.026	0.0081	0.0084	0.046

[a] Unacceptably large—typically 0.2.

Results of the Simulation. The mean-squared jitters resulting from noise and undistorted random data are approximately additive, so, for the sake of versatility, the two effects are shown separately in Table 7.4. The Q value of the f_s filter was set at 100 to facilitate comparison with the results in [D&M] for the basic rectifier/BPF circuit. It was assumed that the spectral shaping of the signal was split equally between transmitter and receiver; the noise was thus band-limited to $f < f_s/2$ and approximately shaped to the square root of the appropriate partial-response spectrum; an SNR of 20 dB was used throughout.

The combined effects of noise and random data can be calculated for any Q and noise variance σ^2 by scaling the mean-square jitter resulting from noise by $1/Q$, and that from random data by $1/Q^2$.† In any particular system the Q should be chosen for the best compromise between jitter and speed of response.

As an example, the jitter when using the double-rectify/filter method with a Q of 20 and an SNR of 26 dB on an undistorted Class I signal would be

$$\text{RMS jitter} = [(0.011)^2(\tfrac{1}{4})(\tfrac{100}{20}) + (0.0038)^2(\tfrac{100}{20})^2]^{1/2} \approx 0.023$$

When there is severe delay distortion, and the pattern jitter becomes large, noise and pattern jitter are no longer additive (presumably because so many of the operations are nonlinear). Table 7.4 shows the total jitter for just one case of $Q = 100$ and SNR $= 20$ dB; it is more useful for ranking the methods than for predicting what the jitter would be in other circumstances.

Comparison of the Methods. The following conclusions can be drawn from Table 7.4:

1. The single rectifier circuit varies from just acceptable for Class I (about

---

† As for full-response systems, the pattern noise is very dependent on frequency, and a BPF is very effective at eliminating it.

twice the jitter of the best of the other methods) to totally unacceptable for Class IV.

2. For Class I signals, the R/LS/R/F and R/F/R/F configurations are comparable (the first gives the least jitter with no distortion, the second when there is distortion); R/LS/R/F is preferable overall because of its simpler circuitry.

3. For Class IV signals, Ridout's circuit gives by far the least jitter under all conditions; if the amount of circuitry required is prohibitive, the TC/F circuit is a much simpler alternative.

Dependence on Phase of the Demodulating Carrier. For an SSB Class IV signal Ridout reported only a 5 dB decrease in the clock component when the carrier phase error was maximum (90° for SSB); this is comparable to the results shown in Fig. 7.8 for full-response signals, so this augurs well for QPRS signals.

However, the other simpler double-rectification methods do not appear to work. Although simulations similar to those that resulted in Fig. 7.8 have not been made for any of the circuits described in this section, preliminary analyses suggest that R/LS/R/BPF with hard rectification actually generates the wrong timing information if the carrier phase error is maximum (45° for QPRS), and with squaring it generates zero information. As pointed out in [Ri1], when there is a large phase error, timing information is contained only in the envelope of the signal; this suggests that methods such as R/UL/R/F, which have the effect of ignoring the inner levels, may be the only ones that will allow carrier and timing acquisition to proceed simultaneously.

7.7.2 Feedback Methods

Without an Adaptive Equalizer. Simple criteria based on learned terms of the impulse response are intuitively more reasonable than they are for full-response signals. For Partial-Response (PR) Class I, the two main terms of the impulse response can be made equal by feeding back either $(h_1 - h_0)$ for a baseband system, or $(hp_1 - hp_0)$ for a QPRS system; for PR Class IV (baseband or SSB), the h_0 term† is an obvious choice because the slope of the impulse response is maximum at $t = 0$.

With an Adaptive Equalizer. The factors involved in the choice between feedforward and feedback methods are different for full and partial-response signals. Since for partial response signals there is no excess bandwidth and no foldover, the choice of sampling phase for an infinite equalizer is not critical. However, as shown in [Qu3], the residual ISI becomes more dependent on the sampling phase as the equalizer is shortened. Qureshi showed that the derivative of the MSE with respect to the sampling phase could be found by putting the output of the adaptive equalizer z_k into another Tapped Delay Line (TDL)

---

† Recall that nominally $h_{-1} = 1$, $h_0 = 0$, and $h_1 = -1$.

and averaging the output. He recommended using a stochastic gradient method, and updating the timing phase each symbol time according to

$$\tau_{k+1} = \tau_k - \Delta_T \operatorname{Re} \left[\epsilon_k^* \sum_{m=1}^{M} \frac{(-1)^m}{m} (z_{k-m} - z_{k+m}) | \right]$$

where ΔT is the timing step size and ϵ_k^* is the complex conjugate of the output error. Qureshi suggested that a very short TDL ($M = 1$) might be adequate, but I have not seen any results of investigations into this question.†

The timing phase could be changed by any of the methods discussed previously (digital PLL, interpolation, etc.). The delay through the equalizer is inside the timing loop, and care would have to be taken in defining the gain and transient reponse.

† Partial-response methods seem to have been much less popular in the decade since Qureshi's paper.

Chapter 8

Linear Adaptive Equalizers

High-speed data transmission through channels with severe distortion can be achieved in several different ways:

1. By designing the transmit and receive filters so that the combination of filters and channel results in an acceptably low mean squared error (from the combination of ISI and noise) at the slicer(s) in the receiver.
2. By designing an equalizer in the receiver that counteracts the channel distortion.
3. By accepting the distorted impulse response of the channel and using some fairly sophisticated maximum-likelihood detection technique to recover the data.

If, as is usually the case, the distortion of the channel is unknown at the beginning of transmission, the chosen method must be made adaptable; the adaptation should, preferably, be automatic. For the first method, adaptation of both the transmit and receive filters would require a lot of complicated feedback from the receiver to the transmitter, and it is rarely attempted; it will not be considered here.

The suboptimum second method is the most commonly used; it is the subject of this chapter (on linear equalizers) and the next (on nonlinear equalizers). Adaptive equalizers have received an enormous amount of attention since they were first invented by Lucky in 1966 [Lu1]; a survey article by Qureshi [Qu6] cited 141 references, and, as with icebergs, these are probably just the most visible 10 percent.

The third method—often called *Maximum-Likelihood Sequence Estima-*

tion (MLSE)—requires much more processing than the second and is used only if the distortion is very severe and/or it is necessary to achieve the absolute optimum performance in the presence of noise; it is discussed in Chapter 9.

Structure of the Equalizer. A linear equalizer could have any regular structure with a number of variable parameters, but in order to simplify the adaptation, the preferred form is a tapped delay line. The delay line may be either continuous (consisting of lumped or distributed delay elements) or sampled (consisting of switched analog storage devices or digital registers); it is usually depicted as shown in Fig. 8.1*a*. Sometimes—particularly at very high data rates—the individual elements may be poor approximations to an ideal delay (constant amplitude and linear phase), but this is generally ignored in devising the adaptation algorithms; at worst, it would result in some coupling between the taps and a slight slowing of the adaptation. With the assmption that the elements are perfect, the transfer function of the equalizer can be written as a function of the delay operator $\exp(-j\omega\tau)$ or z^{-1}.

The delay line nearly always has feedforward taps; if it has only these, the transfer function of the equalizer is a polynomial in z^{-1} (viz., its poles are at $z = 0$), and it is called *nonrecursive, Finite Impulse Response* (FIR), a *Transversal Equalizer* (TE), or simply a *Tapped Delay Line* (TDL). If it also has feedback taps as shown in Fig. 8.1*b*, its transfer function is a rational function of

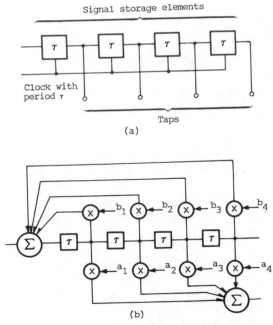

Figure 8.1 Tapped delay lines: *(a)* basic TDL; *(b)* a TDL with both feedforward and feedback taps.

z^{-1} (has both zeros and poles), and it is called *recursive* or *Infinite Impulse Response* (IIR).

Some channel distortions are caused by only a few echoes after the main pulse (trailing echoes), and the impulse responses can be accurately characterized as short polynomials in z^{-1} that have all their zeros inside the unit circle in the z plane. An example would be

$$H(z) = h_1 + h_2 z^{-1} + h_3 z^{-2} = \frac{h_1 z^2 + h_2 z + h_3}{z^2}$$

where $h_1 > h_2 > h_3$. These would be "naturals" for equalization by simple recursive functions with only one feedforward term.

However, if for any reason, an echo occurs before the main pulse (called a leading "echo" even though that seems like an oxymoron) then one zero of $H(z)$ would be outside the unit circle; consequently, one pole of the equalizer transfer function would also be outside, and the equalizer would be unstable; that is, its response to noise would grow without bounds.

Since most channel distortions involve both leading and trailing echoes, and *very few* can be guaranteed to have no leading echoes, recursive *linear* equalizers are rarely used,† and for the rest of this chapter we will be concerned with only TDLs.

Real or Complex Equalizers. An equalizer for a one-dimensional baseband system (and its passband equivalents, SSB and VSB) has real input signals and tap coefficients. For a two-dimensional QAM system, both are complex, and all operations must use complex arithmetic—usually requiring four times as many multiplications. When the number of multiplications needed for a particular algorithm is stated, it should be understood that these are real or complex multiplications as appropriate.

In most of the discussions of equalizers the real or complex character of all the quantities need not be specified; to simplify the notation, the symbol for complex conjugate will not be used, but readers should be aware that in vector and matrix operations such as $\mathbf{y}^T \mathbf{y}$ and $\mathbf{H}^T \mathbf{H}$ the complex conjugate of the transpose should be used. We will need to distinguish between real and complex operations only when we discuss (1) the interaction between a complex equalizer and an adaptive rotator in Section 8.2.2 and (2) "two-phase" initialization signals for QAM systems in Section 8.4.3.

Sampling Rate of the Equalizer. Sometimes it is advantageous to sample the signal that is input to the TDL at a rate higher than the symbol rate f_s (i.e., τ in Fig. 8.1 is less than T). These Fractionally Spaced Equalizers (FSEs) are considered in Section 8.7, but in order to establish the basic principles of equalization

---

† However, Decision-Feedback Equalizers (DFEs), which are recursive and *nonlinear,* are frequently used; they are discussed in Chapter 9.

TABLE 8.1 Summary of Vector and Matrix Notation

Vector or Matrix	Dimensions		Description
	Rows	Columns	
$\mathbf{H}$	$N + M - 1$	N	Impulse response matrix: defined by (8.2)
$\mathbf{c}$	N	1	Column vector of tap coefficients
$\mathbf{d}$	$N + M - 1$	1	Column vector of ideal impulse response; with only one or two nonzero terms for full- and partial-response signals, respectively
$\boldsymbol{\varepsilon}_k$	$N + M - 1$	1	Output IR error vector at time kT, $= \mathbf{Hc}_k - \mathbf{d}$
$\mathbf{e}_k$	N	1	Coefficient error vector at time kT, $= \mathbf{c}_k - \mathbf{c}_{\text{opt}}$
$\mathbf{y}_k$	1	N	Input signal contents of TDL at time kT
$\mathbf{Y}$		N	Input signal matrix; kth row is $\mathbf{y}_k$
$\mathbf{v}_k, \mathbf{w}_k$	1	N	Transformations of $\mathbf{y}_k$ used in orthogonalizing

and adaptation, we will confine ourselves for the moment to symbol-rate TDLs (for which $\tau = T$).

Summary of the Vector and Matrix Notation of This Chapter. Most of the ideas of this chapter are best expressed in terms of vectors and matrices, and easy assimilation of the ideas may be helped by the summary of the notation that is given in Table 8.1.

The Basic Equalizer Problem. Consider a channel that has an impulse response $h(t)$. Let the samples of this at $t = mT + \tau$ be designated as h_m, and let the response be truncated to M significant samples. Also, for the case of full-response signals, which will be our main concern for a while, let the main sample, which is not necessarily the middle one, be designated h_{MM}.

A TDL equalizer can be considered as equalizing either the channel or the receive signal.† In order to equalize the channel, the tap gains, c_n for $n = 1$ to N, should ideally satisfy the matrix equation

$$\mathbf{H} \cdot \mathbf{c} = \mathbf{d} \tag{8.1}$$

where the elements of $\mathbf{H}$ are

$$h_{i,j} = \begin{cases} h_{i-j+1} & \text{for } 1 < i - j + 1 < M \\ 0 & \text{otherwise} \end{cases} \tag{8.2}$$

{Note that we are using h for both the terms of the impulse response and the elements of $\mathbf{H}$; the use of single subscripts for the former and double for the latter should make the distinction clear.}

As an example, let us consider the short impulse response with just four

---

† By contrast, DFEs are dependent on the form of the receive signal and cannot equalize the channel per se.

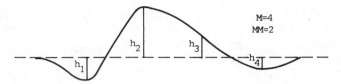

Figure 8.2 A short impulse response with four samples.

significant samples that is shown in Fig. 8.2 ($M = 4$ and $MM = 2$); this is input to a TDL that has five taps with the main one in the middle ($N = 5$ and $NM = 3$). The set of equations (8.1) can be written out explicitly as

$$
\begin{array}{ccccc}
c_1 & c_2 & c_3 & c_4 & c_5
\end{array}
$$

h_1	0	0	0	0	$= 0$	(8.3a)
h_2	h_1	0	0	0	$= 0$	(8.3b)
h_3	h_2	h_1	0	0	$= 0$	(8.3c)
h_4	h_3	h_2	h_1	0	$= 1.0$	(8.3d)
0	h_4	h_3	h_2	h_1	$= 0$	(8.3e)
0	0	h_4	h_3	h_2	$= 0$	(8.3f)
0	0	0	h_4	h_3	$= 0$	(8.3g)
0	0	0	0	h_4	$= 0$	(8.3h)

This is an overdetermined set of equations — there are only N (five in the example) unknowns to control the $N + M - 1$ (eight in the example) terms resulting from the convolution of the input impulse response with that of the equalizer — and it can only be solved approximately; the methods of *zero-forcing* and *Least Mean Squares* (LMS) are described in Sections 8.1 and 8.2, respectively.

If, on the other hand, the equalizer has to equalize a received signal that carries continuous data, the tap gains should satisfy the infinite set of equations

$$\mathbf{Y} \cdot \mathbf{c} = \mathbf{a} \tag{8.4}$$

where $\mathbf{a}$ is the infinite vector of transmit data, $(a_1, a_2, \ldots)$, and the elements of $\mathbf{Y}$ are $y_{i,j} = y_{i-j+1}$, where, again, we are using single and double subscripts to denote the components of a vector and a matrix, respectively. The term $y_{i,j}$ can be expanded as

$$y_{i,j} = \sum_{m=1}^{M} h_m a_{i-j-m+2} + \mu_{i-j+1} \tag{8.5}$$

where the μ_{i-j+1} are samples of the noise.

If the continuous data is truly random, then solution of the two sets of equations (8.1) and (8.4)—using the same criterion (zero-forcing or LMS) for both—yields the same tap gains. A third method that uses periodic short sequences of data for the initial training of the equalizer results in a slightly different setting of the taps; this is discussed in Section 8.5.

In general, the impulse response of the channel is not known in advance. It could be learned and the equations solved "offline," but this is usually more tedious than an iterative inversion; methods of iteration toward the two types of solution are also described in Sections 8.1 and 8.2.

Initialization of the Tap Coefficients. Before we consider how the tap coefficients get there, we must briefly discuss where they start from.

All the iterative methods we shall consider are data-directed—that is, they rely on knowledge, or at least an estimate, of the transmitted data. A *decision-directed* algorithm, in which the equalized signal is sliced to form the estimate, is the simplest and is used for channels that have only a moderate amount of distortion.

If, however, the initial eye is closed, the error rate of the data derived by slicing may be quite high and might seriously impair convergence. In this case the early training of an equalizer will go faster if the data is known a priori in the receiver. Since the sequence can be easily agreed on beforehand by transmitter and receiver, the only problem is synchronizing the sequence generator in the receiver to the received signal; depending on the training pattern used and the initial setting of the taps, this synchronization may have to be precise, approximate, or may not be needed at all. Methods of synchronization are discussed in Section 8.5.1.

It is usually best to use any knowledge of an average channel to design a fixed compromise equalizer that precedes the adaptive equalizer (this is often incorporated into the receive filter); then the "best" starting values for the tap coefficients—in the sense that, on the average, they will have the least distance to go—are . . . $0, 0, 1, 0, 0$. . . . By setting the main tap thus, we have defined the delay through the equalizer, and the synchronization will have to be precise. {Note that this should also be the initial setting when decision-directed training is used because it minimizes the initial error rate.}

On the other hand, if all the taps are set initially to zero, the convergence will take longer, but the synchronization need not be precise because the main tap can "grow" anywhere it is needed. The only limitation is that if the main tap is forced too far from its optimum position (usually near the middle of the equalizer), truncation of the coefficient vector at one end or the other will cause an increase in the Minimum MSE (MMSE). An error in synchronization of plus or minus two symbol periods would usually be just acceptable.

In the cyclic equalization method described in Section 8.5.2 wrap-around occurs instead of truncation, and synchronization is not needed at all.

Noise Enhancement. In the process of equalizing attenuation distortion in the received signal, linear equalizers will enhance† the noise. Because the sampled input noise is assumed to be white, the total noise enhancement can be defined by

$$\text{Noise power enhancement} = \frac{\text{input SNR}}{\text{output SNR}} = \sum_{n=1}^{N} c_n^2 \qquad (8.6)$$

8.1 ZERO-FORCING

In this method [Lu1] the solution of the equations (8.1) for a full-response signal is approximated by solving exactly a subset defined by

$$\sum_{n=1}^{N} c_n h_{MM+i-n} = \delta_{i,NM} \qquad \text{for } i = 1 \text{ to } N \qquad (8.7)$$

where δ_{ij} is the Kronecker delta (= 1 if $i = j$, and zero otherwise). Lucky and colleagues [LS&W] showed that if the original binary eye was open (i.e., $\Sigma |h_m| < 2h_{MM}$), then this method minimizes the equalized peak distortion.

8.1.1 Iterative Adaptation

This is done by a modification of the Newton method. That is, at each stage of the iteration, the new tap vector is calculated from the old by

$$\mathbf{c}' = \mathbf{c} - \frac{\Delta}{d\boldsymbol{\varepsilon}_{ZF}/d\mathbf{c}} \qquad (8.8)$$

where Δ is the step size and $\boldsymbol{\varepsilon}_{ZF}$ is the shortened impulse-response error vector defined by its components

$$\varepsilon_{ZF,i} = \sum_{n=1}^{N} c_n h_{MM+i-n} - \delta_{i,NM} \qquad \text{for } i = 1 \text{ to } N$$

It can be seen from (8.3) that all the derivatives in (8.8) are equal to h_{MM}, the main sample. If the eye is open and the total energy in the pulse has been adjusted to unity by the AGC, then $h_{MM} \simeq 1$, so that the updating simplifies to

$$\mathbf{c}' = \mathbf{c} - \Delta\boldsymbol{\varepsilon}_{ZF} \qquad (8.9)$$

---

† We will use the word "enhance" rather than amplify in order to avoid any suggestion that the increase in noise is constant across the band.

If a single impulse (data symbol) is transmitted, it can be seen from (8.3) that the terms of ε_{ZF} can be observed directly at the output while the main sample is within the span of the equalizer. Therefore, each tap can be incremented by an amount proportional to the error in the output at the time that the main sample is at that tap position.

A single pulse takes $(N + M - 1)$ symbol periods to pass through the equalizer, so for this simple method of iterating, a pulse could be transmitted, and the taps updated, only at that interval. The main disadvantage of this is that the signal power is reduced from the maximum possible by a factor of $(N + M - 1)$ (a typical reduction might be 15 dB) and the resultant SNR for tap adjustment is very low; consequently, the tap values would be very jittery.

Iteration with Continuous Random Data. If random data is transmitted, equation (8.8) can still be used for updating, but the components of ε_{ZF} cannot be observed directly; they can only be estimated from the signal error vector, $(\mathbf{Yc} - \mathbf{a})$, where we are not yet being precise about the phase of the data vector, $\mathbf{a}$. The output of the equalizer at time k is

$$z_k = \sum_{n=1}^{N} c_n \, y_{k-n+1} \tag{8.10}$$

which, from (8.5)

$$= \sum_{n=1}^{N} c_n \sum_{m=1}^{M} h_m \, a_{k-m-n+2} + \mu_{k-n+1} \tag{8.11}$$

and since the delay through channel and equalizer is $(MM + NM - 2)T$, this is intended to convey the data symbol a_{kd}, where the "delayed" subscript $kd = k - MM - NM + 2$. If decision-directed training is used, then a_{kd} should, strictly speaking, be replaced by its estimated value, $\hat{a}_{kd}$, but we will drop the estimate sign from now on.† The component of the signal error vector is then $\epsilon_k = (z_k - a_{kd})$.

If the data is binary and random, then

$$E[a_k \, a_j] = \delta_{kj} \tag{8.12}$$

Therefore, the components of ε_{ZF} can be estimated by correlating the signal error with past, present, and future received data:

$$\varepsilon_{ZF,n} = E[a_{k-n+MM+2} \epsilon_k] \qquad \text{for } n = 1 \text{ to } N \tag{8.13}$$

If decision-directed training is used, the "future" received data is not yet available, so the error signal must be delayed by $(NM - 1)T$ before correlation

---

† The effects of decision errors on the rate of convergence of an equalizer are controversial; for example, estimates of the maximum error rate allowable for guaranteed convergence vary from 2 to 20%.

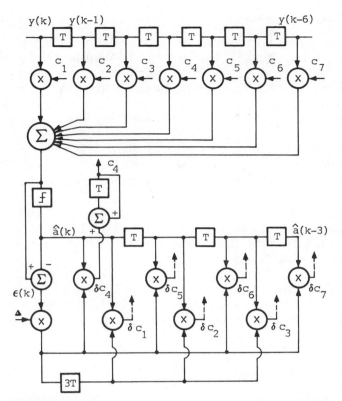

Figure 8.3 Zero-forcing equalizer with seven taps — in decision-directed mode.

with the present and past received data; a typical structure for this is shown in Fig. 8.3.

Deterministic Updating. In order to find ε_{ZF} accurately, with very little jitter, we would have to average (8.13) over many symbol periods before **c** could be updated by (8.8); this is obviously very slow and it is seldom done.

Stochastic Updating. Several writers have discussed the question of what is the best number of symbols to average over in order to achieve the fastest convergence with a given acceptable amount of jitter in the taps; the conclusion has been that the best number is one. That is, averaging should be abandoned — thereby saving one storage location per tap — and the updating performed each symbol period according to

$$c_{n,k+1} = c_{n,k} - \Delta a_{k-n+MM+2}\epsilon_k \tag{8.14}$$

A structure that implements stochastic updating toward a zero-forcing solution is shown in Fig. 8.3.

8.1.2   Uses of Zero-Forcing

Zero-forcing equalization is often used for microwave digital radio signals that are subjected to multipath fades. For these, the following conditions exist: (1) data rates are very high, (2) signal processing is difficult and expensive, and (3) the noise level is low. Consequently, the aim is to get the eye open with the minimum amount of processing, and the zero-forcing solution is best.

However, for other media, this solution has two disadvantages:

1. If the binary eye is initially closed, the solution is no longer guaranteed to be optimum; in the process of forcing the central terms of the output impulse response to be zero, the equalizer may generate significant terms outside its range.

2. It cannot take account of the noise; the frequency response of a long equalizer will be approximately the inverse of that of the input impulse response, and it may therefore peak considerably at some frequencies and thereby seriously enhance the noise.

These disadvantages become very important in modems for the telephone network, where both distortion and noise levels are high; the somewhat more complicated LMS solution is nearly always used in these modems.

8.2   LEAST-MEAN-SQUARES SOLUTION

In this method [Lu2] the sum of the squares of the right-hand sides of (8.1) is minimized. In order to explain the theory most easily, we will consider the noiseless case first, and then show the slight modification that is needed to minimize the sum of the squares of distortion and noise.

The squared modulus of the error is

$$
\begin{aligned}
|\varepsilon|^2 = \varepsilon^T \varepsilon &= (c^T H^T - d^T) \cdot (Hc - d) \\
&= c^T H^T Hc - d^T Hc - c^T H^T d + d^T d
\end{aligned}
\tag{8.15}
$$

where the superscript "T" indicates the transpose of a vector. This squared modulus can be expanded by the matrix equivalent of the scalar method of completing the square; that is,

$$
\begin{aligned}
|\varepsilon|^2 = (c^T - d^T H (H^T H)^{-1}) \cdot H^T H \cdot (c - (H^T H)^{-1} H^T d) \\
- d^T H (H^T H)^{-1} H^T d + d^T d
\end{aligned}
\tag{8.16}
$$

and this is minimized by equating the first, quadratic, term to zero. That is,

$$
c_{opt} = R^{-1} H^T d
\tag{8.17}
$$

where $\mathbf{R}$ ($=\mathbf{H^T H}$) is called the *auto-correlation* or *covariance matrix*.† The elements of $\mathbf{R}$ are $r_{i,j} = r_{i-j}$, where the singly subscripted rs are the terms of the auto-correlation function of the samples of $h(t)$; that is,

$$r_k = r_{-k} = \sum_{m=1}^{M} h_m \, h_{m+k} \tag{8.18}$$

[It is worthwhile emphasizing again that for simplicity all these equations are written for the case of $h(t)$ and $c(t)$ being real. For a QAM system, for which $h(t)$ and $c(t)$ are complex, all the transposes must be replaced by their complex conjugates, and the products in (8.18) must be written as $h_n^* \, h_{n+k}$.]

Minimization of Mean-Square Distortion and Noise. If white Gaussian noise of variance σ^2 is added to the input distorted impulse, the extra contributions to the output MSE can be taken into account by augmenting equations (8.1) by

$$c_n \mu_n = 0 \qquad \text{for } n = 1 \text{ to } N$$

which, for the example, can be written out as

c_1	c_2	c_3	c_4	c_5			
μ_1	0	0	0	0	=	0	(8.3i)
0	μ_2	0	0	0	=	0	(8.3j)
0	0	μ_3	0	0	=	0	(8.3k)
0	0	0	μ_4	0	=	0	(8.3l)
0	0	0	0	μ_5	=	0	(8.3m)

The LMS solution to this expanded set of equations is still given by (8.17), but now the terms on the main diagonal of the matrix $\mathbf{R}$, $r_{i,i}$ must be increased by σ^2. A program to solve simultaneous equations (with this specialized form of impulse-response input, and added noise) is given in Appendix I. Intersymbol interference is not Gaussian, but the residual ISI of an equalizer output has many terms, and if its mean-squared value is small in comparison to the noise (the situation aimed for in equalizer design), it can be reasonably added to the enhanced noise to form a total MSE. That is,

$$\text{MSE} = \text{MSISI} + \sigma^2 \cdot \Sigma c^2$$

or
$$E[\epsilon^2] = \langle a^2 \rangle |\varepsilon|^2 + \sigma^2 \cdot \Sigma c^2 \tag{8.19}$$

---

† This is also often designated as $\mathbf{A}$ or $\mathbf{\Phi}$, but $\mathbf{R}$ is used in this book to emphasize the correlation feature and to preserve as and As for data.

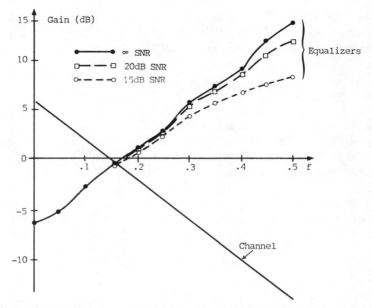

Figure 8.4 Amplitude responses of an attenuation distorted channel and three equalizers.

A simple example of the effects of adding White Gaussian Noise (WGN) to an input distorted pulse may be useful. A baseband signal with $\alpha = 0.25$ is passed through a channel as defined by (7.1) with $A = 20$ and $\beta = 0$ (i.e., linear attenuation distortion and no delay distortion); the resultant eye is well closed. The amplitude responses of the channel and three 15-tap equalizers designed for input SNRs of ∞, 20, and 15 dB are shown in Fig. 8.4.† It can be seen that as the noise increases the equalizer does not try so hard in the frequency region of large attenuation; the mean-squared distortion of the output pulse therefore increases, but the noise enhancement decreases.

In many practical situations an equalizer trains during fairly long periods of low noise and then has its hardest work to do during fairly short periods of high noise. Therefore, it is interesting to compare the noise power enhancement NPE (measured in the whole band and defined by (8.6)) and the overall performance (measured by the total MSE defined by (8.19)) of the best equalizer for each noise level with that of the equalizer trained without noise:

Input SNR	NPE	Minimum RMSE	With "No Noise" Equalizer
∞	4.95	0.0053	0.0053
20 dB	3.98	0.211	0.223
15 dB	2.79	0.341	0.396

† The input power is normalized to unity, and the equalizer is adjusted so that the output power is also unity. The SNR of 15 dB was chosen because it is about the lowest that can be tolerated when linear equalization is used.

Thus, at the high noise level, there is a difference of 1.3 dB between the minimum RMSE and that achieved with an equalizer adjusted without noise. This is an extreme example (high attenuation distortion and very high noise), and the effect will usually be much less than this; it is probably not important for linear symbol-rate equalizers. However, as we shall see in Sections 8.7 and 9.1.2, it may be more serious for fractionally spaced linear equalizers and some types of decision-feedback equalizers.

8.2.1 Iterative Adaptation

The basic principle of iterative adaptation toward an LMS solution of (8.1) is to calculate (or at least, estimate) the gradient $\mathbf{g}$ of the squared modulus of the error and adjust the taps by the method of *steepest descent*. That is, the new value of the tap vector $\mathbf{c}$ is found from the old value by

$$\mathbf{c}' = \mathbf{c} - \Delta\mathbf{g} \tag{8.20}$$

where Δ is the step size.

The gradient can be found by differentiating (8.15):

$$\mathbf{g} = \frac{d|\varepsilon|^2}{d\mathbf{c}} = 2(\mathbf{R}\mathbf{c} - \mathbf{H}^{\mathrm{T}}\mathbf{d}) \tag{8.21a}$$

$$= 2\mathbf{H}^{\mathrm{T}}(\mathbf{H}\mathbf{c} - \mathbf{d}) \tag{8.21b}$$

and it can be seen from (8.21) and/or from geometric considerations that the minimum error is achieved when the gradient becomes zero.

Adaptation with Isolated Pulses. If a single impulse (in practice a single data symbol) is transmitted through the channel and the equalizer, the signals appearing at the nth tap at successive times form the elements of the nth column of $\mathbf{H}$. Also, the error signals (the actual outputs minus the desired outputs) form the elements of the column vector $(\mathbf{H}\mathbf{c} - \mathbf{d})$. Therefore, according to (8.20) and (8.21), after the passing of a single pulse, each tap could be incremented by an amount proportional to the sum of the products of the output errors and the signals at that tap at each time. This is one version of the *deterministic gradient* method—so called because the gradient is calculated exactly (if there is no noise) before updating.

These single impulses (data symbols) could be transmitted every $(N + M - 1)$ symbol periods without interfering with each other, so the taps could be updated at that rate. However, as with the zero-forcing method, isolated pulses are seldom used for adaptation because of their low signal power and the resultant large effects of noise.

Adaptation with Continuous Data. If continuous "random" data is used, then $\mathbf{H}$ and $\mathbf{d}$ are not observable, and (8.21) must be replaced by

$$\mathbf{g} = 2E[\mathbf{Y}^{\mathrm{T}}(\mathbf{Y}\mathbf{c} - \mathbf{a})] \tag{8.22}$$

{Note that by expanding the elements of $\mathbf{Y}$ according to (8.5) and invoking the randomness of the data, it can be shown that

$$E[\mathbf{Y}^\mathsf{T}\mathbf{Y}] = \mathbf{R} \quad \text{and} \quad E[\mathbf{Y}^\mathsf{T}\mathbf{a}] = \mathbf{H}^\mathsf{T}\,\mathbf{d}$$

so that substitution into (8.22) confirms (8.21)}

Thus the gradient is a measure of the correlation between the signal error and the signals delivered to each tap multiplier. This is intuitively reasonable because, as pointed out previously, the adaptation stops when the minimum error is achieved and the gradient becomes zero — that is, when there is no longer any correlation between the errors and the signals at the taps, and changing the weightings applied to these signals (the tap coefficients) cannot reduce the error any further.

The deterministic gradient method could be implemented by averaging (8.22) over many symbol periods. Compared to the implementation with isolated pulses, this has the advantage of greater signal power and the disadvantage of imperfect estimation of the gradient because the data cannot be completely "random" over a finite period, and the components interfere with each other.

Stochastic Gradient. Just as with the zero-forcing method, deterministic updating is slow and seldom used. It is faster to update the taps each symbol time according to

$$\mathbf{c}_{k+1} = \mathbf{c}_k - \Delta\epsilon_k\,\mathbf{y}_k^\mathsf{T} \tag{8.23}$$

which can be expanded in terms of the individual tap coefficients as

$$c_{n,k+1} = c_{n,k} - \Delta y_{k-n+1}\,\epsilon_k \tag{8.24}$$

A structure that implements stochastic gradient iteration toward an LMS solution is shown in Fig. 8.5.

This method has two serious disadvantages:

1. If the amplitude spectrum of the sampled signal is not flat — because either the channel has attenuation distortion or partial-response shaping is used — the path of steepest descent is not a direct one, and convergence is slowed. This problem is generic to all gradient methods; the problem and various solutions are discussed in Sections 8.3 and 8.4.

2. The "random" data used for training the equalizer is usually generated by an K-bit maximum length sequence generator; this has a period $2^K - 1$ that is much greater than the expected convergence time. Although the long-term auto-correlation of these sequences is satisfactory, the short-term ones are usually not, and at each stage of the iteration the estimation of each component of the gradient is perturbed by random functions of the other terms. This problem is discussed in Section 8.5.

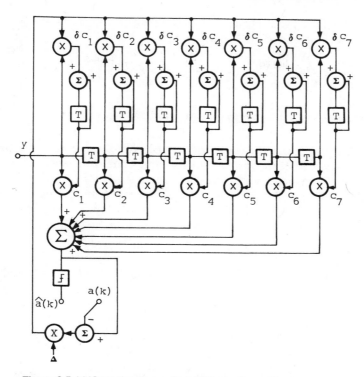

Figure 8.5 LMS stochastic gradient (SG) equalizer with seven taps.

Nevertheless, despite these disadvantages, the simplicity of the random data/stochastic gradient method makes it the preferred method for continuous reception from one transmitter† via channels that do not change very quickly. For other cases (e.g., fast-fading radio channels), where fast adaptation of the equalizer is required, the methods discussed in Sections 8.4 and 8.5.2 may be needed.

8.2.2 Interaction of a Complex Equalizer with an Adaptive Rotator

A combination of linear complex equalizer and adaptive rotator described in [Fa1] is shown in Fig. 6.12. The equalizer could adapt to any rotation of the received signal, but it is better to let the single-minded rotator do the job. The only feature of the arrangement that has not been discussed already is the derotator. Because the output of the equalizer is rotated by θ_e (i.e., multiplied by $e^{j\theta_e}$), the error signal must be multiplied by $e^{-j\theta_e}$ before it can be correlated with

---

† In this case the receiver needs to be trained only once at the beginning of transmission, and, even with slow training, the proportional reduction of through-put is small.

the contents of the TDL. The reason for drawing the loop filter in Fig. 6.14 with a discrete delay is now clear.

The sequence of operations is as follows: (1) calculate equalizer output, slice, and calculate error; (2) derotate error, and update taps; and (3) calculate phase error, filter, and update θ_e.

8.3 THE PROBLEM OF EIGENVALUE SPREAD

This problem arises in the iteration toward both the zero-forcing and LMS solutions; however, it is most serious for large equalizers, and since these are nearly always designed for an LMS solution, that is what we will consider here.

As before, we will first consider the deterministic gradient method — impractical though it is — because the theory is simpler and can serve as an introduction to the practical stochastic gradient method.

8.3.1 Deterministic Gradient Method

At the kth stage of the iteration, the error in the tap vector can be defined as

$$\mathbf{e}_k = \mathbf{c}_k - \mathbf{c}_{\text{opt}} \tag{8.25}$$

so that from (8.20) and (8.21)

$$\mathbf{e}_{k+1} = \mathbf{c}_k + \Delta(\mathbf{Rc}_k - \mathbf{H}^{\text{T}}\mathbf{d}) - \mathbf{c}_{\text{opt}} \tag{8.26}$$

where, for convenience, the factor of two has been absorbed into the step size. {Note that we do not have to be concerned at this stage with how many symbol periods constitute an iteration period; we need only assume that there are enough to accurately define the gradient. Later, in the stochastic update method, when the number shrinks to one, the subscript k will denote both iteration and symbol periods.}

Then substituting for $\mathbf{H}^{\text{T}}\mathbf{d}$ from (8.17), we obtain

$$\begin{aligned}
\mathbf{e}_{k+1} &= \mathbf{c}_k - \Delta\mathbf{Rc}_k + \Delta\mathbf{Rc}_{\text{opt}} - \mathbf{c}_{\text{opt}} \\
&= (\mathbf{I} - \Delta\mathbf{R})\mathbf{e}_k
\end{aligned} \tag{8.27}$$

where $\mathbf{I}$ is the identity matrix. The ideal situation occurs when the amplitude spectrum of the sampled signal is flat; then, if the pulse has been normalized so that $r_0 = 1$, $\mathbf{R} = \mathbf{I}$, and the choice of $\Delta = 1$ allows convergence in one step.

If the spectrum is not flat, and $\mathbf{R} \neq \mathbf{I}$, then the rate of convergence depends on the spread of eigenvalues of $\mathbf{R}$. This can be analyzed by transforming the coefficient error vector $\mathbf{e}_k$ to $\mathbf{d}_k$†:

$$\mathbf{d}_k = \mathbf{P} \cdot \mathbf{e}_k \tag{8.28}$$

† Note that this $\mathbf{d}$ is not the previously defined ideal IR vector.

where the columns of $\mathbf{P}$ are the eigenvectors of $\mathbf{R}$. Then the decay of the squared modulus of this transformed coefficient error vector can be described by

$$|\mathbf{d}_{k+1}|^2 = \sum_{n=1}^{N} (1 - \Delta\lambda_n)^2 |d_{n,k}|^2$$

and the cumulative effect by

$$|\mathbf{d}_k|^2 = \sum_{n=1}^{N} (1 - \Delta\lambda_n)^2 |d_{n,0}|^2 \tag{8.29}$$

where the λ_n are the eigenvalues of $\mathbf{R}$ ordered so that

$$\lambda_1 (\lambda_{\min}) < \lambda_2 < \ldots < \lambda_N (\lambda_{\max})$$

Then the convergence of the signal MSE, ϵ^2,† can be described by

$$\epsilon_k^2 = \epsilon_{\min}^2 + \sum_{n=1}^{N} \lambda_n (1 - \Delta\lambda_n)^{2k} |d_{n,0}|^2 \tag{8.30}$$

That much is fairly standard matrix theory; further analysis becomes very specialized and complicated, and we will only report and interpret the main results. The reader is referred to the original papers [Ge], [G&W2], [Ma4], [P&M], and [Un1], or to the excellent summary of them in [Qu6], for details.

Maximum Step Size. It can be seen from (8.30) that convergence is ensured if

$$\Delta < \frac{2}{\lambda_{\max}} \tag{8.31}$$

Rate of Convergence. If the step size is chosen as

$$\Delta = \frac{2}{\lambda_{\max} + \lambda_{\min}} \tag{8.32}$$

then the two extreme components of ϵ_0^2, $d_{1,0}$, and $d_{N,0}$, will decay at the same rate, $(\rho - 1)/(\rho + 1)$, where $\rho = \lambda_{\max}/\lambda_{\min}$. All other components will decay faster, but if $d_{1,0}$ and $d_{N,0} \neq 0$, then, strictly speaking, the convergence will eventually be controlled by the slowest decay rate; if there is a lot of amplitude distortion, this may be very slow indeed. However, this argument leads to an unduly pessimistic estimate of the time needed to converge to an acceptably small error; the decay of ϵ^2, given by (8.30), depends on N different time constants, and the extreme eigenvalues usually have only a small effect on the convergence.

---

† For simplicity of notation, we will omit the expectation function from hereon.

We will consider the convergence in more detail for the more practical stochastic gradient method.

8.3.2   Stochastic Gradient (SG) Method

Analysis of the convergence in this case is complicated by the fact that the estimates of the gradient that are used to update the tap coefficients are very noisy, and their short term averages depend on the particular data sequence transmitted. Consequently, the random departures from the path of steepest descent are very difficult to predict. Furthermore, as pointed out in [Ma4], the assumption of independence of successive signal vectors that was used to simplify all the early analyses is certainly not valid (in fact, at each symbol time only one of the N components of the signal is new — all the others are merely shifted one position in the storage register). Nevertheless, some useful results have been obtained.

Maximum Step Size. It has been generally agreed that, in order to ensure convergence for *all* sequences, the step size must be bounded by

$$\Delta_{max} < \frac{2}{N\lambda_{max}} \tag{8.33}$$

This has proved to be a fairly tight bound because, for each of several different impulse responses tested by simulation, some "random" sequence has been found that causes divergence with a step-size only slightly greater than $2/\lambda_{max}$. {Note that this step size is only one Nth of that allowed when using the deterministic gradient.}

Maximizing the Rate of Convergence. If the channel has only a small amount of amplitude distortion, the initial rate of convergence is maximized by using a step size of

$$\Delta_0 = \frac{1}{N\lambda_{max}} \tag{8.34a}$$

If the channel has a large amount of amplitude distortion, the simple form of (8.34a) is too conservative, and it appears that it can be generalized to

$$\Delta_0 = \frac{2}{N\left(\lambda_{max} + \lambda_{min}\right)} \tag{8.34b}$$

{Note that these step-sizes are again $1/N$ of those recommended for the deterministic gradient case.}

If $\rho = \lambda_{max}/\lambda_{min} \gg 1$, however, then Δ_0 given by (8.34b) is dangerously close to Δ_{max} and potential divergence; therefore, a compromise between ensured convergence, (8.34a), and fastest convergence, (8.34b), is recommended. One simple compromise, which reverts to (8.34a) for $\lambda_{max} = 1$ and has worked well in the small number of cases tested so far, is

$$\Delta_0 = \frac{1}{N \sqrt{\lambda_{max}}} \qquad (8.34c)$$

Estimating the Rate of Convergence. Sometimes, in order to decide whether any type of accelerating procedure is necessary, it is desirable to predict the rate of convergence. This is affected by (1) the spread of eigenvalues, which makes the path of steepest descent a long one and (2) the wandering around that path because of stochastic updating with random data.

Some writers have suggested that the spread of eigenvalues only slows down the convergence (from that achieved with all the eigenvalues equal to unity) by a factor of 2, but for channels with a lot of amplitude distortion, this is much too optimistic. For a better estimate during the most important early stages of convergence, a wide range of eigenvalues can be replaced by a single $\lambda_{eq} = \sqrt{\lambda_{min}}$. {Note that this formula has the correct form if all the λ_n are unity and that it also recognizes that λ_{min} plays a role—albeit reduced—in the convergence.}

The second effect can be modeled by simply multiplying the convergence time by a *random-data-slowing-down factor* μ. For slow convergence (with $\rho \gg 1$), the randomness of the path tends to average out, and for different data sequences μ may vary only from 1.5 to 2.5; as suggested in [Qu6], $\mu = 2$ is a reasonable average. However, fast convergence (achieved when $\rho \approx 1$) is more dependent on the particular data sequence, and μ may vary from 3 to 5.

These suggestions can be summarized by a simplification of (8.30) in which ϵ_{min}^2 (typically very small, and therefore unimportant during the early stages of convergence) is ignored:

$$\epsilon_k^2 \approx \epsilon_0^2 (1 - \Delta \lambda_{eq})^{2k/\mu} \qquad (8.35)$$

where $\qquad \lambda_{eq} = \sqrt{\lambda_{min}}$

$\qquad \mu \approx 2$ if there is a large spread of eigenvalues

$\qquad \approx 4$ if there is a small spread of eigenvalues

If the initial stepsize is chosen according to (8.34c), this becomes

$$\epsilon_k^2 \approx \epsilon_0^2 \left(1 - \frac{1}{N\sqrt{\rho}}\right)^{2k/\mu} \qquad (8.36)$$

8.3.3   Estimation of the Maximum and Minimum Eigenvalues

Before tackling the case of general amplitude distortion, it is useful to consider two special cases for which the eigenvalues can be calculated precisely.

Partial-Response Shapings. Chang (in [Ch2]) showed that for a PR Class IV signal

$$\lambda_{max} = 1 - \cos\left[\frac{(N+1)\pi}{N+3}\right] \tag{8.37a}$$

$$\lambda_{min} = 1 - \cos\left(\frac{2\pi}{N+3}\right) \tag{8.37b}$$

His method can also be used for a PR Class I signal, with the generalized result:

$$\lambda_n = 1 - \cos\left(\frac{n\pi}{N+1}\right) \quad \text{for } n = 1 \text{ to } N \tag{8.38}$$

so that

$$\lambda_{max} = 1 - \cos\left(\frac{N\pi}{N+1}\right) \tag{8.39a}$$

$$\lambda_{min} = 1 - \cos\left(\frac{\pi}{N+1}\right) \tag{8.39b}$$

{Note that the eigenvalue spread is much greater for Class I than for Class IV; for example, for $N = 15$, $\lambda_{min} = 0.0192$ compared to 0.0603.}

Equation (8.37) can be written more informatively as

$$\lambda_n = 2 \sin^2\left[\frac{n\pi}{2(N+1)}\right] \quad \text{for } n = 1 \text{ to } N$$
$$= \left|S\left[\frac{nf_s}{2(N+1)}\right]\right|^2 \tag{8.40}$$

Thus the eigenvalues are equal to the power spectral density at N frequencies spaced evenly throughout the band 0 to $f_s/2$.

General Amplitude Distortion. It was shown in [Ge] and [Ko1] that

$$|H_{min}(f)|^2 < \lambda_n < |H_{max}(f)|^2 \quad \text{for } n = 1 \text{ to } N \text{ and } 0 < f < \frac{f_s}{2} \tag{8.41}$$

so that, if N is large, $\lambda_{min} \simeq |H_{min}|^2$ and $\lambda_{max} \simeq |H_{min}|^2$.

If the maximum and/or minimum attenuations occur at the edges of the band and, as in the cases of the partial-response shapings, the slopes of the attenuation there are high, a more accurate approximation is needed. The expressions for λ_{min} and λ_{max} depend on the shape of $|H(f)|$; for example, for the common case of $|H(f)|$ monotonically decreasing, we obtain

$$\lambda_{min} \simeq \left| H\left[\frac{f_s}{2(N+1)} \right] \right|^2 \text{ and } \lambda_{max} \simeq \left| H\left[\frac{Nf_s}{(N+1)} \right] \right|^2 \quad (8.42)$$

8.3.4 Three Simple Examples

Three examples will be used to illustrate the effects of eigenvalue spread; they are summarized in Table 8.2. Even though most practical signals and channels are bandpass and their impulse responses are complex, baseband signals are used here for the examples because it is difficult to define a "typical" bandpass channel—it is easier to examine some simple stereotype baseband channels and then extrapolate to any particular bandpass case.

In each case the channel distortion was sufficient to just close the eye, the signal level was adjusted so that r_0 and $\lambda_{av} = 1$, and the order of the equalizer N was chosen to achieve a final RMSE of approximately 0.01.

The convergence of LMS equalizers for these three signals—using the same step size throughout each run, and averaging $|\epsilon_k|^2$ over several different runs of pseudo-random data—is shown in Figs. 8.6a–8.6c together with the simplified prediction given by (8.35). Some comments about these results may be useful:

1. The signal with only delay distortion converges much more slowly than simple theory would predict; the random-data-slowing-down factor μ is about 3.5.

2. The signal with severe amplitude distortion converges very slowly, requiring about 1000 symbols to reach an acceptable RMSE; some method of acceleration would probably be needed. The nearly linear behavior of log(MSE) for the first 25 dB or so confirms that the model with a single equivalent eigenvalue is a valid one.

3. It can be seen that for the partial-response signal, the convergence time becomes intolerably long; acceleration would certainly be needed.

TABLE 8.2 Three Examples Demonstrating Convergence of the Stochastic Gradient

				λ_{min}		λ_{max}			
α	A^a	β^a	N	Estim.^{c}	Actual^d	Estim.	Actual	Δ_{max}	$\Delta_{opt}{}^b$
0.25	0	1.75	9	1.0	1.0	1.0	1.0	0.22	0.11
0.25	15	0	15	0.14	0.12	2.88	2.99	0.044	0.022
0^e	0	4.0	15	0.019	0.019	1.98	1.98	0.067	0.033

[a] See Section 7.2 for definitions of these channel parameters.
[b] This "optimum" step size was calculated from (8.34a).
[c] These eigenvalues were estimated from (8.42).
[d] These were found by calculating the zeros of Det($R - \lambda I$), using the SIMEQ program listed in Appendix I.
[e] This is a PR Class I signal.

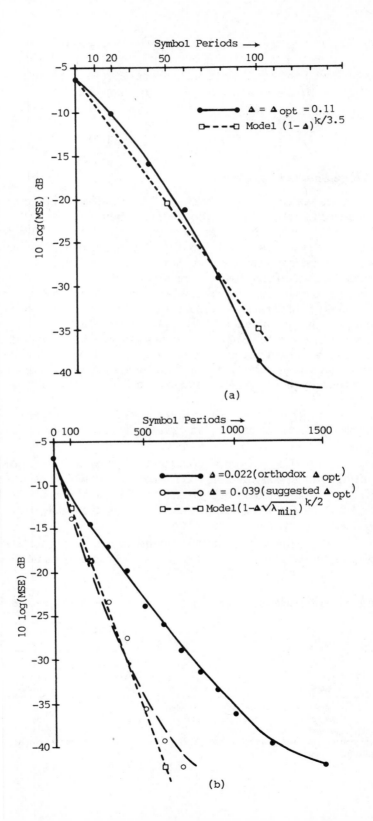

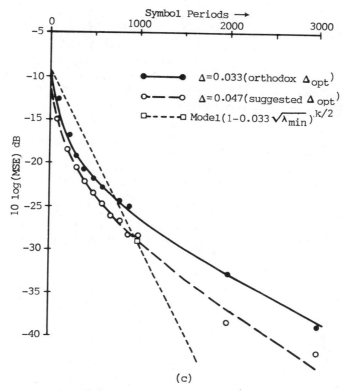

Figure 8.6 SG convergence of MSE: *(a)* delay-distorted full-response signal; *(b)* attenuation-distorted full-response signal; *(c)* delay-distorted PR Class I signal.

8.4 METHODS OF COUNTERING THE EFFECTS OF EIGENVALUE SPREAD

There are two basic methods of doing this, which, for reasons that will become clear presently, have been called [Bi6] *Double-Sided Orthogonalization* (DSO) and *Single-Sided Orthogonalization* (SSO). In order to study the different characteristics of DSO and SSO, we will first consider the case of a partial-response signal for which most of the spectral shaping — or at least, the part of it that we hope to compensate for — is predictable because it is performed in the transmit and receive filters. Consequently, the compensation can be precalculated and need not be adaptive.

If a channel has a lot of amplitude distortion, and either (1) very fast initial training is needed or (2) the channel characteristics change more rapidly than a conventional TE can follow, then some form of adaptive orthogonalization is needed. Both approaches, DSO and SSO, have received much attention, and the reader is referred to the literature for details. We will discuss briefly only the main DSO method; the many different SSO methods must wait until more work has been done and a consensus has been reached on their use. {Note that there is another important difference between the partial- and full-response cases besides the fixed/adaptive orthogonalization difference: for a full-response signal, the equalizer must counteract the shaping, but for a partial-response signal, the shaping must be retained in the output, and the operations are only performed to speed the adaptation.}

8.4.1 Fixed Orthogonalization of Partial-Response Signals

Four different methods of compensating for the effects of partial-response shaping on the adaptation have been described:

1. Chang [Ch2] transformed the N outputs of a TDL to produce N other signals that were used both as inputs to the tap multipliers and for correlating with the error in order to calculate the updates of the taps; this is DSO. Chang's implementation requires $\sim N^2$ multiplications per symbol time.
2. Mueller [Mu1] suggested an alternative in which the transformed outputs were used only to calculate the updates and the tap multipliers were applied to the untransformed outputs of the TDL as in a conventional TE; this is SSO. His implementation also requires $\sim N^2$ multiplications per symbol time.
3. Makhoul [Ma3] suggested a lattice structure to implement Chang's transformation using only $2N$ multiplications.
4. Bingham [Bi6] described a modified ladder structure that implements SSO using only $2N$ multiplications.

The relationship between the four methods can be summarized thus:

	$\sim N^2$ Multiplications	$2N$ Multiplications
DSO	Chang	Makhoul
SSO	Mueller	Bingham

Double-Sided Orthogonalization. In order to orthogonalize the auto-correlation matrix $\mathbf{R}$ "from both sides," the input vector $\mathbf{y}_k$ is transformed to $\mathbf{v}_k = \mathbf{y}_k \cdot \mathbf{P}$,† which is then used to calculate the output and tap updates according to modifications of (8.10) and (8.23), respectively:

$$z_k = \mathbf{v}_k \cdot \mathbf{c}_k \tag{8.43}$$

and

$$\mathbf{c}_{k+1} = \mathbf{c}_k - \Delta\epsilon_k \mathbf{v}_k^{\mathsf{T}} \tag{8.44}$$

† Note that this is not the same $\mathbf{P}$ as was used in Section 8.3.1.

As a result, equation (8.27), which defined the new coefficient error vector, is modified to

$$\mathbf{e}_{k+1} = (\mathbf{I} - \Delta \mathbf{P}^{\mathrm{T}}\mathbf{RP})\,\mathbf{e}_k \qquad (8.45)$$

so that if $\mathbf{P}^{\mathrm{T}}\mathbf{RP} = \mathbf{I}$ and $\Delta = 1$, convergence will occur in one step. DSO thus has the effect of almost removing the partial-response spectral shaping with the transforming network and then restoring it *and* equalizing the channel with the taps.

Makhoul's implementation of this transform will become important when we consider adaptive DSO in Section 8.4.2, but for the fixed case SSO is preferable.

Single-Sided Orthogonalization. In order to orthogonalize $\mathbf{R}$ from the left only, $\mathbf{y}$ is retained for the calculation of the output, and the transformed vector, $\mathbf{v} = \mathbf{y} \cdot \mathbf{Q}$, is used only for the updating. Equation (8.23) is therefore modified to

$$\mathbf{c}_{k+1} = \mathbf{c}_k - \Delta E[(\mathbf{y}_k \cdot \mathbf{Q})^{\mathrm{T}} \cdot \boldsymbol{\epsilon}_k] \qquad (8.46)$$

so that
$$\mathbf{e}_{k+1} = (\mathbf{I} - \Delta \mathbf{Q}^{\mathrm{T}} \cdot \mathbf{R}) \cdot \mathbf{e}_k \qquad (8.47)$$

and ideally
$$\mathbf{Q}^{\mathrm{T}} \cdot \mathbf{R} = \mathbf{I}$$

With SSO the taps function as in a conventional TE and equalize the channel only. It is generally agreed that this is a better-conditioned process than DSO, and that the sensitivity of the transfer function to the tap coefficients is less than for DSO. Another advantage of SSO over DSO is that it preserves the delayed samples of the input in its TDL; this may be useful when doing the prefiltering for nonlinear equalizers (see Section 9.1).

We will summarize the implementation of SSO only for the most important case of a PR Class I signal and refer the reader to [Bi6] for details of the method for both Class I and IV.

The orthogonalizing matrix $\mathbf{Q}$ can be factorized into

$$\mathbf{Q} = \mathbf{U} \cdot \mathbf{D} \cdot \mathbf{U}^{\mathrm{T}} \qquad (8.48)$$

where $\mathbf{U}$ is upper triangular and $\mathbf{D}$ is a normalizing diagonal matrix. The elements of $\mathbf{U}$ and $\mathbf{D}$ can be shown to be

$$u_{i,j} = \frac{(-1)^{i-j}i}{j} \qquad \text{for } i < j \qquad (8.49)$$

and
$$d_{i,i} = \frac{2i}{i+1} \qquad (8.50)$$

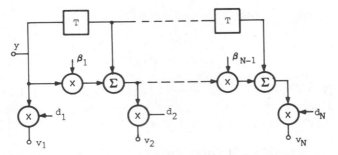

Figure 8.7 Forward ladder to implement DSO of a PR Class I signal.

As a result of this particularly simple form of its elements—which comes about because the original auto-correlation function has only two nonzero terms—**U** can be implemented by the forward ladder shown in Fig. 8.7, and the full transforming matrix, **Q**, by the combination of a forward ladder, a scaling of its outputs, and a complementary backward ladder.

This would require three multiplications per tap, but, because $\beta_n = -d_n/2$, the number can be reduced to two by combining the pairs of multiplications as shown in Fig. 8.8 and absorbing the constant -2.0 multiplier on all the outputs into the step size. The output of the forward ladder w_n, which need be stored only temporarily (i.e., not from one iteration period to the next), can be calculated from the equations:

$$w_1 = y_1 \tag{8.51a}$$

and $\qquad w_n = y_n + \beta_{n-1} w_{n-1} \qquad$ for $n = 2$ to N $\qquad$ (8.51b)

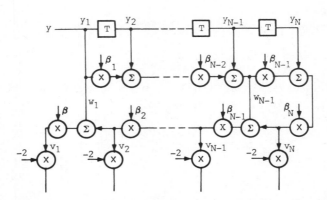

Figure 8.8 Double ladder implementing full Q matrix for SSO of a PR Class I signal.

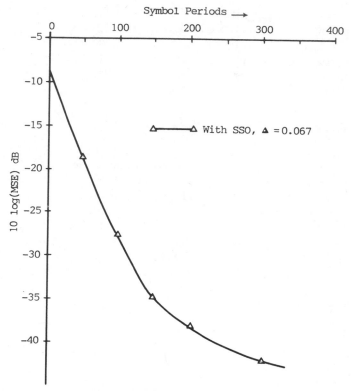

Figure 8.9 SG convergence of MSE for delay-distorted PR signal using SSO.

and then the elements of the transformed vector $\mathbf{v}$ can be calculated from

$$v_{N+1} = 0 \tag{8.52a}$$

and
$$v_n = \beta_n (v_{n+1} + w_n) \quad \text{for } n = N \text{ to } 1 \tag{8.52b}$$

and
$$\beta_n = -\frac{n}{n+1} \tag{8.53}$$

An Example. The partial-response signal defined in Table 8.2 was transformed according to (8.3); the convergence of ϵ^2 — averaged over several different sequences of data — is shown in Fig. 8.9. It can be seen that orthogonalization has accelerated the convergence by a factor of about 14, and that it is now comparable to that of the full-response system with only delay distortion shown in Fig. 8.6a.

A channel with only delay distortion was chosen for this example in order that there should be no further spreading of the eigenvalues beyond that caused by the partial-response shaping and to thereby demonstrate most dramatically the effects of the orthogonalization. With a channel that has amplitude distortion, there will still be some residual spread after the transformation, and the acceleration factor will not be so great; the rate of convergence for the "net" system (partial-response shaping, channel distortion and partial-response orthogonalization) can then be reliably predicted by the same rules as given for a full-response system.

8.4.2   Adaptive Lattice (DSO) Algorithms

The cascade of lattices shown in Fig. 8.10 was originally described in [I&S], and its use in an adaptive equalizer was first described in [S&A]. There have been several variations — some simpler, some more complex — of the basic method, [S&P], [M&G], [Pr2], and [Qu6], but we will study only the original. We will first describe the method and then discuss some of its features and shortcomings.

The input signal $y(k)$ is transformed to a set of N intermediate output signals $b_n(k)$, which are used as the input to the tap multipliers *and* to the calculation of the updates of the multiplying coefficients. The two outputs of each stage of the lattice are calculated from the inputs by

$$f_1(k) = b_1(k) = y(kT) \tag{8.54a}$$

$$f_n(k) = f_{n-1}(k) + \kappa_{n-1}(k)\, b_{n-1}(k-1) \tag{8.54b}$$

and $\qquad b_n(k) = b_{n-1}(k-1) + \kappa_{n-1}(k)\, f_{n-1}(k) \qquad$ for $n = 2$ to N $\qquad$ (8.54c)

The b outputs of the lattice are then used like those of a TDL as inputs to the tap multipliers as shown in Fig. 8.10; the output of the equalizer is given by

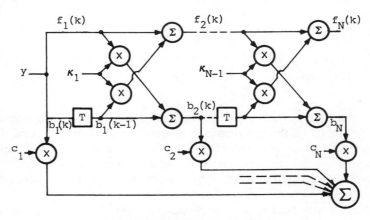

Figure 8.10 Tapped lattice equalizer.

$$z(k) = \sum_{n=1}^{N} c_n(k)\, b_n(k) \tag{8.55}$$

{Note that several different combinations of notation, subscript conventions and polarities of the κ values have been used in the literature. The use of f and b for the lattice signals does not fit the convention in the rest of this book, but they indicate the forward and backward residuals of linear prediction theory and are too well entrenched to be discarded. The present subscript convention is that at time kT the $\kappa(k)$ and $c(k)$ are used to calculate the output, and then the new $\kappa(k+1)$ and $c(k+1)$ are calculated ready to be used next time.}

In order to understand the goal of the adaptation process, it is useful to go back for a moment to the fixed lattice described in [Ma3] for decorrelating a PR Class I signal. The lattice coefficients are

$$\kappa_n = \frac{(-1)^n}{n+1} \qquad \text{for } n = 1 \text{ to } N-1$$

If the input signal is the basic partial-response duplet . . . . 0, 0, 0, 1, 1, 0, 0, 0 . . . [temporarily scaled to a pulse energy of 2.0 in order to avoid having to write many $(1/\sqrt{2})$s], then the outputs of the stages at successive times are as shown in Table 8.3; it can be seen that

$$E[b_n\, b_j] = \delta_{nj} \qquad \text{for } n = 1 \text{ to } N$$

In order to achieve this uncorrelated state adaptively for a general, unknown distortion, the average power level at the input of the nth stage of the lattice $p_n(k)$ is first learned by putting successive observations of the powers into a set of first-order recursive low-pass filters with poles at $z = 1 - \gamma$ and d.c. gains of unity. That is,

$$p_n(0) = p_0 \tag{8.56a}$$

and
$$p_n(k) = (1 - \gamma)\, p_n(k-1) + \frac{\gamma[f_n^2(k) + b_n^2(k-1)]}{2} \tag{8.56b}$$

Then the κ values are updated according to

$$\kappa_{n-1}(k+1) = \kappa_{n-1}(k) - \Delta_{\kappa,n}(k)\, \frac{[f_n(k) b_{n-1}(k-1) + f_{n-1}(k) b_n(k)]}{2\dagger} \tag{8.57}$$

where the step size for the nth lattice coefficient at time k, $\Delta_{\kappa,n}(k) = \gamma/p_n(k)$.

† Few authors use the factor of 2.0; it is included here to preserve the idea that the p_n are powers and to maintain compatibility with q_n, defined later.

TABLE 8.3 Outputs of the Orthogonalizing Lattice for a PR Class I Signal

	b_1	b_2	b_3	b_4	$b_5 \ldots$
$t = 0$	1	$-1/2$	$1/3$	$-1/4$	$1/5 \ldots$
T	1	$1/2$	$-1/3$	$1/4$	$-1/5 \ldots$
$2T$	0	1	$1/3$	$-1/4$	$1/5 \ldots$
$3T$	0	0	1	$1/4$	$-1/5 \ldots$
$4T$	0	0	0	1	$1/5 \ldots$
.	.	.	.	.	.
.	.	.	.	.	.
.	.	.	.	.	.

Two methods of updating the tap coefficients were described in [S&A]. We will introduce both methods first and discuss them later. In the first (a generalization of the conventional method used for the LMS SG adjustment of a TE) the coefficients are updated so as to drive to zero the correlation between the output error $\epsilon(k) = z(k) - a(k)$ and the b_n; that is,

$$c_n(k + 1) = c_n(k) - \Delta_{c,n}(k) b_n(k) \epsilon(k) \tag{8.58}$$

The step-size for the nth tap-coefficient at time k, $\Delta_{c,n}(k)$, $= \gamma / q_n(k)$, where the $q_n(k)$, which are the learned powers of the b_n alone, are the outputs of a second set of low-pass filters with the same bandwidths as the first. That is,

$$q_{n-1}(0) = q_0 \tag{8.59a}$$

and
$$q_{n-1}(k) = (1 - \gamma) q_{n-1}(k - 1) + \gamma b_n(k) \tag{8.59b}$$

In the second method the (eventual) orthogonality of the b_n is exploited, and the increment of each coefficient is made proportional to the output error, not of the full equalizer, but of only the equalizer up to that tap. If the transmit data is known — that is, reference data is being used for training — the successive "partial errors" and the tap updates can be calculated by

$$\epsilon_0(k) = -a(k) \tag{8.60a}$$

$$\epsilon_n(k) = \epsilon_{n-1}(k) + c_n(k) b_n(k) \tag{8.60b}$$

and
$$c_n(k + 1) = c_n(k) - \Delta_{c,n}(k) b_n(k) \epsilon_n(k) \tag{8.61}$$

If the transmit data is not known (during decision-directed training or when receiving customer data), the output of the last stage and the estimate of the transmit data $a(k)$ must first be calculated in the conventional way, according to (8.55); then the intermediate errors and tap updates can be calculated from (8.60) and (8.61). It can be seen that the complete process requires $13N$ multiplications and $2N$ divisions per symbol time.† Other variations reported have reduced this to about $10N$ and $2N$. {Note that when the lattice has orthogonalized the input signal, each stage of it does a progressively better equalization of the *amplitude* distortion; the taps are then mainly concerned with equalizing the original delay distortion of the channel and that introduced by the lattice.}

Initial Step Size. The initial step sizes for the lattice and the taps are $\gamma/p(0)$ and $\gamma/q_n(0)$. For the fastest ensured convergence, the latter should be set equal to $1/N\lambda_{max}$, but not much has been written about the former. In [S&A] fast convergence was shown with $\Delta_{\kappa,n}(0)$ as large as 1.0, but this is rather worrying and is not recommended without a lot of reassuring simulation!

8.4.2.1 *The Problem of Self-Noise (Jitter).*
The penalty paid in an adaptive lattice equalizer for orthogonalization is jitter of both the lattice and tap coefficients and consequent increase of the final MSE. These effects have not been adequately studied, and I can attempt only a qualitative analysis.

Jitter of the Lattice Coefficients. It can be seen from (8.57) that when the lattice coefficient κ_n has converged, then

$$E[f_n(k)\, b_{n-1}(k-1) + f_{n-1}(k)\, b_n(k)] = 0$$

However, unlike all the previous adaptation algorithms, the "error" that is multiplied by the step size in order to calculate the increments of the κ_n does not tend to zero as the iteration progresses — only its average does. The jitter can be reduced only by making the step size $\gamma/p_n(k)$ small.

If, as we have assumed throughout, the AGC operates to make the input power unity, then the steady-state value of the first power, $p_1(\infty)$, is unity. For most impulse responses encountered in practice, the power does not decrease very much through the lattice,‡ so that $p_n(\infty) \simeq 1.0$ for $n = 1$ to N. Therefore, the final step-size for all lattice stages is approximately equal to γ, which must be chosen accordingly.

Jitter of the Tap Coefficients. The tap coefficients must follow the lattice coefficients, and even though, in the first method of tap updating, the error should finally become small, the tap jitter may still be very large.

The second method of updating the taps is an attempt to minimize this effect — each tap coefficient depends on only the sections of the lattice that precede it. However, this modification of the algorithm introduces another problem; even in the converged state, the error signals that are used to update

---

† The basis for comparison is the SG method applied to a conventional TE; this requires $2N$ multiplications per symbol per symbol time.

‡ In the extreme case of partial-response shaping, it can be seen from Table 8.3 that the power decreases by a factor of 2 from the input to the end of the lattice.

the tap coefficients only finally become small at the end of the lattice. Consequently, the amounts by which the taps (particularly those before the main tap) are incremented have much larger random factors than with the conventional SG update method.

This effect can be simply and dramatically illustrated by considering an undistorted signal input to a lattice in which, so as to focus attention on tap jitter only, the lattice coefficients are frozen at their correct values (zero). For taps before the main tap (with a nominal tap coefficient of unity), the error signal will always have a magnitude of unity, and the increments of the coefficients will jitter accordingly. The final MSE—for a signal for which a simple TE would hold the MSE at zero—would be quite large!

Final Step Size. The final values of the powers of the b_n, $q_n(\infty)$ are also close to unity,† and so the final step sizes for all lattice stages and taps are approximately equal to γ. This must be chosen small enough to make the final MSE (resulting from jitter and noise) acceptable and should also allow the equalizer to track any anticipated changes in the channel.

Unfortunately, all the simulations reported in the literature appear to have been done with binary data and a fairly low SNR; under the circumstances the large amount of final MSE was acceptable, but it is not easy to discover how much of it is due to lattice and tap coefficient jitters and how much to noise.

If the data signal is a multilevel, low-noise signal that can tolerate only a very small MSE, γ will have to be made very small, and the tracking capability may well become inadequate. Some improvements such as are suggested in the next section may be necessary.

8.4.2.2 *Proposed Simplifications and Improvements.* Readers should be aware that the following ideas have been only partially tested (i.e., simulated), and have not been exposed to any peer criticism.

Fixed Step Size. Satorius and Alexander [S&A] showed satisfactory convergence with $p_n(0)$ set equal to unity. As discussed above, this is tantamount to using an almost constant step size throughout the lattice and for the entire iteration. If this is acceptable (perhaps for binary data in high-noise environments), then the algorithm can be simplified by discarding both (8.56) and (8.59); this eliminates the divisions and reduces the number of multiplications to $6N$.

Freezing the Lattice. For the reception of multilevel signals from low-noise channels that do not change substantially in the duration of a connection, it

† In fact, several authors have suggested discarding either equation (8.56) or (8.59) and using q_n or p_n as the weight for the update of both κ_n and c_n.

may be better to freeze the *lattice* coefficients after they have "converged" and then to revert to the conventional, first, method of updating the taps. This has the effect of reducing the MSE much further than is possible with an adapting (jittering) lattice, and yet, because the gross orthogonalizing effect of the lattice is retained, it still preserves more tracking capability than a pure TE. {Note that this approach would not work with a channel that has an amplitude dip, caused by multipath interference, that moves across the band; for this the power spectrum changes radically, and a frozen lattice would be worse than useless.}

Shortening the Lattice. The purpose of the lattice is, of course, to remove the correlation between successive inputs to the tap multipliers, that is, to equalize the amplitude spectra of these signals. The lattice coefficients typically decrease through the lattice, and each stage merely refines the equalization performed by the previous ones. This suggests that most of the benefits of decorrelation could be obtained with just a very few lattice stages followed by a TDL (equivalent to setting the coefficients of the remaining "lattice" stages to zero). Simulations with the attenuation-distorted signal defined in Table 8.1 have shown that the eigenvalue spread is reduced from 25 : 1 to about 3 : 1 — and the convergence accelerated accordingly — with just three stages. {These amplitude-equalizing stages should probably be frozen after a short "pretraining" period.}

The combination of all of these measures would reduce the number of multiplications for a 15-tap equalizer to about one fifth of those required for a fully adaptive lattice.

8.4.3 Kalman, Fast-Kalman, FAEST, and FTF (SSO) Algorithms

These methods have been described in [Go1], [F&L2], [CM&K], and [C&K], respectively; they all calculate the updates of the multiplying coefficients applied to the outputs of a TDL. The number of multiplications per symbol time for the basic Kalman algorithm is of the order of N^2; for the three later algorithms, it is about $7N$. No easily digestible description or comparison of the methods has appeared, and I am not qualified to attempt one; the reader is referred to the original papers for details.

Adaptive orthogonalizing algorithms are needed in order to allow fast adaptation (either initial or during rapid changes of the channel) in the presence of severe amplitude distortion. However, as we have already seen — and will consider in more detail in Chapter 9 — linear equalization of this distortion would result in considerable noise enhancement; therefore, nonlinear equalization (particularly DFE) is preferable. As we shall see, it is best if the prefilter for a DFE applies the tap multiplications directly to the stored samples of the input signal. Any adaptive orthogonalization should therefore be done using an SSO method, but there have been no reports of the application of any of the above methods to DFEs.

8.4.4 Adaptive Slope Equalization

This is a crude form of adaptive DSO. It is sometimes used in conjunction with baseband adaptive equalizers for high-speed data on microwave radio. The reader is referred to the specialized literature [G&C] and [T&H] for more details.

8.5 DATA USED FOR TRAINING

At the beginning of this chapter we briefly considered training an equalizer with isolated pulses, but mostly we have been concerned with training with "random" data that is of the same form (e.g., multilevel) as is subsequently used for transmission of data. Since we have ignored the possible effects of differences between the transmit and receive data (errors), we have implicitly assumed that the transmit data is known a priori at the receiver — that is, that the receiver can generate the same data sequence and can synchronize it to the received signal.

We will briefly discuss a method of synchronizing and then consider other training signals and methods that can be used (1) to speed convergence and (2) in cases where synchronization of the receiver is difficult or impossible. It is worthwhile stressing at this stage that the techniques discussed here are useful only during training; once information transmission has started, further adaptation (to a changing channel) can be accelerated only by the orthogonalization methods discussed in the previous section.

8.5.1 Synchronizing a Reference Data Generator

In order to start the reference data generator in the receiver at the right time, a marker should be included in the simple signal that is used to train the AGC and the carrier and clock-recovery "circuits." For QAM modems, the most common form of this signal is, as we have already discussed, alternating $+90°$ and $-90°$ phase changes; these can be interrupted by a single extra $+90°$, which should be easily recognized in the receiver. As an example, let us consider the sequence recommended in V.32.

A 4-point subset *ABCD* of the 16-point constellation shown in Fig. 8.11 is used, and the transmitted sequence is a large number of *AB* pairs† followed by eight *CD* pairs and then random data.

In the receiver, in order to simplify the carrier recovery, the set will probably be temporarily rotated by 26.6° (plus, of course, some unknown integer multiple of 90°) to a conventional 4-point set; the sequence can then be more reliably decoded as dibits. . .00 11 00 00 00 11 00. . . , and the extra 00 can be used to locate the beginning of random data from some defined state 16 symbol periods later.

† The number need not be precise because the receiver will have to train its AGC and carrier and clock recovery and will miss some of the data anyway.

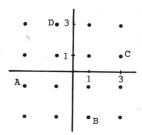

Figure 8.11 Subset of 16QAM constellation used for training.

The "defined state" applies primarily to the serial binary data. If this could be transformed to a defined state of the line signal, and thence of the received signal, the equalizer could be preset with the main tap coefficient equal to unity. However, if, as is usually the case, the polarity of the transfer function of a baseband (real) channel is unknown, or the phase of the carrier in a QAM system is rotated by an unknown integer multiple of 90°, the equalizer must be started from the all-zero state and the main tap allowed to grow in either direction.

8.5.2 Training with Periodic Sequences

Pseudo-random data that is used for training usually has a repetition period that is much longer than the expected convergence time of the equalizer. Ideally, $E[a_i a_j] = \delta_{i,j}$, so that the gradient can be estimated accurately according to (8.22). However, in the short term this may be far from true, and the tap-coefficient vector may wander a long way from the path of steepest descent. As we have seen, the use of a noisy estimate of the gradient may slow the convergence by a factor of between 2 and 4 relative to that that would be achieved if the gradient were known exactly for each update.

One method [C&H] of reducing the variance of the estimate and thereby speeding convergence of an equalizer with N taps is to transmit a repetitive sequence that has a period NT and is such that

$$\sum_{n=1}^{N} a_n a_{n+j} = \delta_j \qquad \text{for } j = 0 \text{ to } N-1 \qquad (8.62)$$

{There are slight differences needed in the explanation for N even or odd; we will consider the more common case of N odd.}

With such a sequence the gradient could be calculated exactly from (8.22) after each set of N symbols (averaged update). However, in practice it is more convenient to update all the taps each symbol time (stochastic update); Qureshi [Qu6] showed that this is as fast as the averaged update method if the channel has no amplitude distortion and faster if it has distortion.

Within the Nyquist band (0 to $f_s/2$) the sampled version of this sequence can be defined by N Fourier series coefficients; the signal can therefore be used to adjust the real and imaginary parts of the equalizer spectral response at $(N + 1)/2$ discrete frequencies (including zero) evenly spaced across the band. The adjustment can be exact; that is, the combination of channel and equalizer will have constant amplitude and delay at these frequencies, and *for this particular repetitive sequence* the ISI is made zero.

How closely the equalizer found with these sequences approximates that that minimizes the MSE with truly random data can be best understood qualitatively in the frequency domain, but it can be most easily quantified in the time domain. The closeness depends on how accurately the sampled spectrum of the channel can be modeled by the $(N + 1)/2$ samples; if there are no rapid changes in either the amplitude or delay, the samples will adequately characterize the spectrum. Rapid changes in the frequency spectrum would result in long impulse responses, so Qureshi suggested the simple criterion that if the time span of the equalizer is long enough to contain at least 95% of the energy of the channel impulse response, the "periodic" equalizer will converge to an acceptable approximation to the MMSE solution. (In practice, these periodic sequences are used for only a short time, and then, for refinement of the equalizer, the data is switched to pseudo-random with a much longer period.)

The eigenvalue spread of the received signal, and thence the convergence rate of an equalizer, depend on the amplitude spectra of both the channel *and* the transmitted sequence. Therefore, it is usually desirable that all $(N + 1)/2$ components of the Fourier series expansion of the periodic training signal have the same magnitude. Such sequences, which also satisfy (8.62), are called *Constant-Amplitude Zero Auto-Correlation* (CAZAC) sequences. Unfortunately, for $N > 4$, they can be constructed only by using data symbols a_i that are not in the simple set $\pm 1, \pm 3 \cdots \pm(L - 1)$. Although such symbols might be acceptable with some very flexible methods of implementation, in general they would be very inconvenient.

8.5.2.1 *Short Pseudo-Random Sequences.*

Another solution, suggested by Forney [Fo3] is to use a pseudo-random binary sequence (viz., $a_i = \pm 1$) of length $N = 2^M - 1$.† For these

$$\sum_{n=1}^{N} a_n a_{n+j} = \begin{cases} 1 & \text{for } j = 0 & \text{(8.63a)} \\ -\dfrac{1}{N} & \text{for } j = 1 \text{ to } N - 1 & \text{(8.63b)} \end{cases}$$

These were first used with initial tap values of . . . 0, 0, 1, 0, 0 . . . , and very fast convergence was observed. However, these starting values would require

---

† The M values 4 and 5 are the most common; these are included in Table 12.1, which is a list (up to $M = 25$) of all the most easily generated sequences used for scramblers.

either decision-directed training or precise synchronization of the transmitter and receiver sequences.

Cyclic Equalization. Forney [Fo3] suggested starting with all zero taps while using any phase for the receiver sequence generator (i.e., without trying to synchronize) and allowing the main tap to "grow" in any position. Because the data is periodic with a period equal to the length of the equalizer, the taps are themselves "periodic." Only one period of the equalizer can be observed with any particular phase of the receive generator, but changing that phase (e.g., by j symbol periods) preserves all the coefficients and merely rotates, or wraps them around; that is, the new taps are derived from the old by

$$c'_n = \begin{cases} c_{n-j} & \text{for } n = j+1 \text{ to } N \\ c_{N+n-j} & \text{for } n = 1 \text{ to } j \end{cases}$$

When the equalizer has converged enough to firmly establish the position of the main tap, the taps and the receive sequence generator can be rotated to bring the main tap into the desired position (usually the middle or slightly before the middle) and thereby minimize the error when random data begins.

Unfortunately, when all zero starting values are used instead of setting the main tap coefficient to unity, not only is the initial error larger, but it often decreases much more slowly. This is because although $(N-1)/2$ of the discrete spectral components of the data signal have equal amplitude, the component at d.c. is greatly reduced. Mueller and Spaulding [M&S1] showed that if the channel has no distortion, the eigenvalues of **R** are

$$\lambda_1 = \frac{1}{N}$$

and
$$\lambda_n = 1 + \frac{1}{N} \qquad \text{for } n = 2 \text{ to } N$$

Consequently, if the initial tap error vector (e.g., the all zero vector) has a significant d.c. component, then that component, and hence the total error, will decay very slowly. For most channels, the d.c. value of the equalizer transfer function is approximately unity,† so that the starting vector . . .0, 0, 1, 0, 0 . . . has only a small d.c. error. It was shown in [M&S1] that "phase impartiality" and low d.c. error of the initial tap vector can be achieved together by choosing

$$c_n(0) = \frac{1}{N} \qquad \text{for } n = 1 \text{ to } N$$

Maximum and Optimum Step Sizes. Because the maximum length of any adverse data patterns is closely controlled by using these short periodic sequences,

---

† A notable exception is a radio channel subjected to a midband fade (see Section 1.5.2)

it might seem that the step size could be larger than that used with random data. Unfortunately, this does not appear to be the case; the step sizes for guaranteed and fastest convergence are still given by (8.31) and (8.32).

An Example of a Delay-Distorted Signal. Figure 8.12 shows the decay of the MSE for the delay-distorted channel defined in Table 8.1, when using all three possible sets of starting taps. The MSE in each case is the average of that obtained by using two different 15-bit pseudo-random sequences (derived from shift registers with taps at 1, 4 and 3, 4). It can be seen that all three behave the same; the log(MSE) decays rapidly for about two periods and then decreases linearly and slowly (indicating that the decay is controlled by a single, small eigenvalue) toward a steady-state value. {The steady-state MSE is unrealistically small because noise was not included, and the equalizer was much longer than it need be.}

For the all-zero initial tap vector, the d.c. component of the initial error is

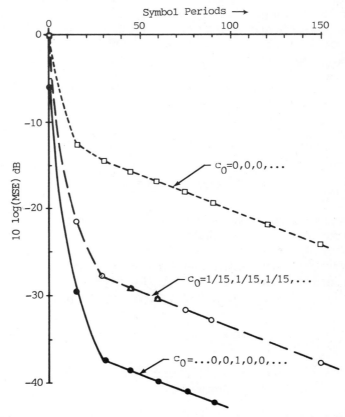

Figure 8.12 Decay of MSE for delay-distorted full-response signal using cyclic equalization.

large, and it very soon dominates the convergence. Comparison with Fig. 8.6*a* shows that when used this way, short sequences are slower than random data! For the other two initial vectors, simple theory suggests that the d.c. error should be zero, and that convergence should be controlled only by the multiple eigenvalues around unity. However, during the first two periods of N symbols the short-term randomness of the data drives the tap-vector away from the condition of zero d.c. error, and eventually the convergence is controlled by the small eigenvalue at d.c. Because the initial MSE is larger for the $1/N$, $1/N$ . . . starting vector than for . . . 0, 0, 1, 0, 0 . . . , the accumulated d.c. error is larger, and the convergence slower.

This shows that if there is only delay distortion, then these sequences should be used for only one or two periods; thereafter convergence would be slower than for random data. When the short sequences are used this way, initial convergence is speeded up by a factor of 2 – 3.

Use with Amplitude-Distorted Signals. If the channel has amplitude distortion, and thereby disperses the eigenvalues of **R**, the use of short training sequences still speeds the initial convergence by approximately the same ratio. If, as in most channels, there is more attenuation at high frequencies than at d.c., the effect of the low amplitude of the d.c. component of the data signal is much diminished; there is no longer the distinct break between the initial and final convergences that can be seen in Fig. 8.6*a*, and the faster convergence obtained with the short sequences continues for much longer. It becomes more difficult to decide how many of the $(2^M - 1)$ periodic sequences of data should be used; simulation of the convergence over all combinations of line distortions is probably the only way to choose the training signal.

Equalizers with Lengths Not Equal to $2^M - 1$. Sometimes it is not convenient to use an equalizer of length $2^M - 1$ (e.g., the jump from slightly more than 15 to 31 taps might have serious hardware consequences). If the length of the equalizer is chosen only to achieve an acceptable MMSE, nearly all the benefits of equalization with short periodic sequences can be obtained by using the next longer pseudo-random sequence. The steady-state solution (which is never reached because the data is switched to random long before then) becomes an LMS solution instead of an exact one, but that is not important. The tap coefficients no longer wrap around completely, and some form of crude synchronization will be needed to ensure that the principal taps do not roll off one end.

8.5.3 Decision-Directed Training

Even if the signals for synchronizing a local reference are provided, the circuitry or code to do it are not trivial, and it is always worthwhile to consider ignoring synchronization and using decision-directed training.

The limiting factor is, of course, the initial error rate; if it is too high, the adaptation will diverge or, at least, hang-up (i.e., the MSE will neither increase nor decrease significantly). Unfortunately, no consensus has been reached on what is "too high"; [M&S1] showed a channel with an initial error rate of about 22% that could be made to converge with some step sizes; other experts have suggested that more than about 5% would be too high for guaranteed convergence. The explanation for this disagreement seems to be that the channel in [M&S1] had very little eigenvalue spread ($\lambda_{min} \simeq 0.7$); consequently, the large movements of the taps when the data decisions were correct had the effect of quickly opening the eye — thereafter convergence was assured. For channels with a large eigenvalue spread the few wrong decisions seem to have a proportionately larger effect, and the equalizer is never able to move into the error-free region. For any particular system, careful simulation using the channel with the largest possible attenuation distortion, and many different data sequences is probably the only way to decide whether decision-directed training will work.

The initial error rate should be reduced by compromise equalization, and for QAM, the use of a two-phase training signal can also reduce the initial distortion and error rate.

Two-Phase Training Signals for QAM. During normal transmission of a two-dimensional signal, both channels are subjected to the in-phase and quadrature distortion. That is, the MSE on each channel is

$$\text{MSE} = \sum_{m=1}^{M} |h_m - d_m|^2 = \sum_{m=1}^{M} (h_{pm} - d_m)^2 + (h_{qm})^2$$

where the complex impulse response

$$h_m = h_{pm} + j \, h_{qm}$$

and the desired impulse response $\mathbf{d}$ (. . 0, 0, 1, 0, 0 . . . for a full-response system) has an appropriate delay included.

If, however, the input data is encoded as only one bit per symbol (simple binary differential coding), one channel could be modulated and the other held at zero. In the receiver the desired signal on the two channels would be the data signal and zero, and the ISI would be the *separate* real and imaginary parts of the impulse response, respectively. Or, considered in the frequency domain, one channel would have to deal with the signal produced by the sum of the upper and lower sidebands, and the other, with that produced by the difference.

In order that the AGC can also train while receiving this two-phase signal, it is desirable that the power in the pair of points be approximately equal to (certainly within a few tenths of a decibel) the average power of the multipoint set. However, as can be seen from Fig. 4.6, none of the commonly used constellations contains a pair of integer points on the real axis that satisfy this requirement. Therefore, it is easier in the transmitter to use two points on the integer grid as shown in Fig. 4.6; for 16QAM, for example, the points *A* and *C* shown in

Fig. 8.11 could be used. Then in the receiver the carrier recovery must be biased to rotate the constellation (by $-18.4°$ for 16QAM) so that the nominal points are again real.

To test the possible usefulness of two-phase training the impulse responses were calculated for a V.27† signal ($f_c = 1.8$ kHz, $f_s = 1.6$ ks/s and $\alpha = 0.5$) transmitted through the four different "worst-case" channels defined in [Q&F] and then passed through a simple compromise equalizer as discussed in Chapter 5. In each case the real eye was just open or only just closed (but the complex eye was well closed); the error rate for random two-phase data would be low enough for reliable decision-directed SG training.

However, it should be noted that a few end-to-end connections on the PSTN have more distortion than these particular "worst cases" and, furthermore, that the more recent, higher-speed modems, which use $f_s = 2.4$ ks/s, are much more susceptible to distortion than a V.27. For these, receiver-reference training is probably essential.

8.6   EQUALIZER LENGTH, STEP SIZE, AND COEFFICIENT PRECISION

So far we have discussed equalizers rather generally; now we must be more specific and consider how each of these quantities affects the final peak distortion or MSE.

8.6.1   Equalizer Length

If one knows the exact source of a channel's distortion and can model the impulse response precisely, one can sometimes calculate the order of the equalizer transfer function (length of the TDL) needed to reduce either the MSE or the peak distortion to an acceptable level. One case where this might be possible is that of radio signal subjected to a multipath fade.

In general, however, the only practical method of deciding the length of the equalizer is calculation of several "worst-case" impulse responses and then offline solution for each of them of the appropriate set of simultaneous equations‡. Usually the equalizer needs to be slightly longer than a truncated version of the impulse response for which the truncation error is comparable to the acceptable final error. In some special cases this may lead to either a very pessimistic or a very optimistic guess, but at least it provides a simple starting point for exploratory shortening or lengthening (via simulation) of the equalizer.

Usually the precision of the general-purpose computer used for this will be much higher than that of the processor to be used in the modem. Therefore, the

---

† The recommendation includes a segment of random two-phase in the sequence of training signals.
‡ Sample programs that can be used for this are presented in Appendix I.

peak distortions or MSEs thus calculated must be considered as theoretical minima that will be increased by all the cumulative effects of finite precision and finite step-size in the processor.

A Caveat. The simple models we have used so far for "typical" channels — linear attenuation and parabolic delay, or more general *smooth* curves for attenuation and delay that pass through the limits of a particular specification (e.g., the basic 3002 conditioning shown in Table 1.4) — are not suitable for estimating the required length of an equalizer; they are too smooth, and the impulse responses calculated from them are too short; any conclusions about the required length of the equalizer would be too optimistic.

It is necessary to use fairly precise representations of some worst-case channels, paying particular attention to the small high-frequency ripples (e.g., see Fig. 1.12), and also to include the transfer functions of the transmit and receive filters (theoretically perfect, but not so in practice). These filters are often designed with a few percent of RMSE, using the justification that "the adaptive equalizer will take care of that"; however, if the filters are of a high order — as would be needed, for example, in an FDM full-duplex modem such as a V.22 bis — then the errors in the end-to-end impulse response will extend over a long time.

8.6.2 Step Size

In the discussions of convergence so far we have been concerned mainly with reducing the MSE† to some minimum as fast as possible, and we chose the step size accordingly. However, since the residual minimum error is not zero and is, by definition, not correlated with the signals at the taps, the small random updates cause all the tap coefficients to jitter; this, in turn, causes an increase of the MSE that is nearly proportional to N and Δ.

The Orthodox View. Gitlin and Weinstein [G&W2] showed that the final MSE‡ when using a step size of Δ is given by

$$\epsilon_\infty^2 = \frac{\epsilon_{\min}^2}{1 - \Delta N \lambda_{\max}/2} \tag{8.64}$$

so that if Δ is maintained at $\Delta_0 = 1/N\lambda_{\max}$, the value suggested in (8.34a) to maximize the initial rate of convergence, then $\epsilon_\infty^2 = 2\epsilon_{\min}^2$; such a 3 dB increase in MSE would not generally be acceptable. {Note that for channels with a lot of amplitude distortion, the larger initial step size of $1/N\sqrt{\lambda_{\max}}$, suggested in (8.34c), would theoretically result in an even greater excess MSE.}

---

† We will confine our discussion to the LMS solution from now on since it is the more commonly used.

‡ Recall that, strictly speaking, the MSE should be written as $E[\epsilon^2]$.

For the 0.3 dB increase suggested in Table 5.1, the step size would have to be reduced at some stage to $0.13/N\lambda_{max}$. However, as also shown in [G&W2], the step size that maximizes the rate of convergence after k iterations depends on the ratio of the MSE at that stage ϵ_k^2 to the minimum MSE ϵ_{min}^2:

$$\Delta_{k,opt} = \frac{1}{N\lambda_{max}} \frac{(\epsilon_k^2/\epsilon_{min}^2) - 1}{(\epsilon_k^2/\epsilon_{min}^2)}$$

From this it can be seen that $0.13/N_{max}$ is the optimum step size only when the MSE has decreased to within 0.6 dB of the minimum. Therefore, several authors have suggested that three different step sizes be used in succession: the first, Δ_0, to maximize the initial rate of convergence; an intermediate, $\Delta_0/2$, for convergence from 3 dB above MMSE to about 1 dB above; and a final one of about $\Delta_0/8$ to polish the equalizer.

In order to decide when to change the step sizes, we must look again at equation (8.35). If, for example, an eventual reduction of the initial MSE, ϵ_0^2, by 33 dB is required, and the convergence model of (8.36) is used, the first change of step size should be made after a 30 dB reduction; the number of symbols required k_1 is given by

$$(1 - \Delta\lambda_{eq})^{k/\mu} = \left(1 - \frac{1}{N\sqrt{\rho}}\right)^{k/\mu} = 0.0316$$

whence $k_1 \simeq 3.5N\mu\sqrt{\rho}$.

For the delay-distorted example in Table 8.2, the eigenvalue ratio $\rho = 1.0$, the number of taps $N = 9$, and the random-data-slowing-down factor $\mu \simeq 3.5$. Consequently, the first change of step size should be made after approximately 110 symbols. For the second, amplitude-distorted example, $\rho = 25$, $N = 15$, and $\mu \simeq 2$. Therefore, the first gear shift should be made after approximately 500 symbols.

During the second stage the convergence speed is half of what it was previously, but the MSE need decay further by only about 2 dB; $N\mu\sqrt{\rho}$ more symbols should be adequate before switching to the final step size.

For any expected range of amplitude distortion that must be dealt with by any particular equalizer (typically much less than encompassed by these two extreme examples), the duration of the stages should be chosen for the worst distortion, because reducing the step size too late on a fast-converging channel with little distortion is less harmful to the average convergence time than is reducing it too early on a slow-converging channel.

A Simpler but Unorthodox Approach. Most of the analysis that led to (8.64) was based on the assumption that the eigenvalue spread ρ ($= \lambda_{max}/\lambda_{min}$) is small. Theoretical analysis for the more interesting cases where $\rho \gg 1$ would be extremely difficult; simulation is the only practical alternative. LMS equalizers for the attenuation-distorted signal (second example in Table 8.2) were adapted using the same step-size throughout. The results were (1) with $\Delta = 0.039$, the suggested optimum initial step size, RMSE = 0.0050; and (2) with $\Delta = 0.010$, small enough to give the minimum error, RMSE = 0.0046. Thus, even with a step size larger than the "theoretical" optimum, the final RMSE was only insignificantly greater than the minimum.

On the other hand, a similar experiment with the delay-distorted signal (first example in Table 8.2) and a short seven-tap equalizer resulted in: (1) with $\Delta = 0.143$, the optimum initial step-size, RMSE = 0.0270; and (2) with $\Delta = 0.005$, very small, RMSE = 0.0193. Thus the RMSE with $\Delta = 1/N$ was exactly $\sqrt{2}$ times the minimum RMSE.

It is clear that (8.64) is valid for $\rho \simeq 1$ but is very pessimistic for $\rho \gg 1$. The reason for this became clear when the jitter of the tap coefficients was observed in the first example; the tap jitter did, indeed, increase by a factor of about 4 in going from $\Delta = 0.01$ to 0.039, but this had very little effect on the RMSE. This is a beneficial consequence of the "ill-conditioning" caused by eigenvalue spread — during adaptation the taps typically change a lot without reducing the error very much, but then when they jitter because of a large step size and short-term correlations in the data, the jitter does not increase the error very much.

This suggests — again issuing the caveat that this idea has not been adequately tested or exposed to peer criticism — a simple strategy for choosing the step size. Calculate λ_{max} for the channel with the worst possible attenuation distortion, and use $\Delta = 1/N\sqrt{\lambda_{max}}$ *throughout* the adaptation. For those channels with less spread of eigenvalues, the final MSE may be about 2 dB higher than the minimum possible *for that channel,* but that is not important because that minimum will be much less than the minimum for the worst channel on which the design of the equalizer was based.

8.6.3 Coefficient Precision

This subject has been thoroughly explored in [GM&T] and [G&W2] and excellently reviewed in [Qu6]. This discussion is based on their results.

In an analog or infinite-precision implementation, the MSE ϵ^2 could eventually be reduced to the minimum ϵ_{min}^2 by reducing Δ asymptotically to zero. However, if each coefficient is represented in store by a finite number of bits, adaptation will stop when $\Delta\epsilon_k y_k$ is less than half of the Least-Significant Bit (LSB).[†]

The error signals ϵ_k are randomly distributed, so that eventually some will cause a change in the coefficients and some will not. Nevertheless, on the average, adaptation will stop when

$$\Delta\epsilon_k < 2^{-B} \tag{8.65a}$$

[†] This is assuming that, as is most common, the arithmetic is done with greater precision and the coefficients are rounded before being stored.

where each coefficient is represented over the range of -1 to 1 by B bits (including the sign) and it is assumed that the input signals y_k have unit power. For a QAM signal, this power is divided between two channels, and (8.65a) must be modified to

$$\Delta\epsilon_k/\sqrt{2} < 2^{-B} \tag{8.65b}$$

This equation may seem strange because it says that for a given precision of the coefficients the final error is inversely proportional to the step size — the opposite of the infinite precision case! This occurs because occasional large errors will cause the MSE to increase until (8.65) is satisfied *on the average*. This is an inadequate explanation of the phenomenon, but the practical result is that (8.64) should be used to calculate the final step size — assuming that the precision will be adequate — and then the tap precision that is needed to reach the desired ϵ_∞ should be calculated from (8.65). That is,

$$\Delta_\infty = \frac{2(1 - \epsilon_{\min}^2/\epsilon_\infty^2)}{N\lambda_{\max}} \tag{8.66}$$

and

$$B = -\log_2(\Delta_\infty\epsilon_\infty) \tag{8.67a}$$

or, for a QAM signal,

$$B = -\log_2(\Delta_\infty\epsilon_\infty) + \tfrac{1}{2} \tag{8.67b}$$

Using fewer bits would stop the adaptation before it reached ϵ_∞, and using many more would be wasteful because the extra bits would merely jitter, and the MSE would be determined solely by the step size.

In practice, the randomness of the ϵ_k means that the adaptation does not stop abruptly, and when it continues in response only to large errors, the correlation that exists in the data at these times will cause some further tap jitter.† In anticipation of this, any fractional value of B found from (8.67) should be rounded *up* in order to provide a little extra precision.

Application of equations (8.66) and (8.67) to equalizers for high-speed telephone modems that were designed in the early 1980s usually resulted in the conclusion that a 12-bit representation of the coefficients was sufficient, but this probably involved some hindsight that was dependent on the DSP hardware then available. Similar calculations in the late 1980s will probably conclude that 16 bits are available and (therefore?) necessary.

---

† The extra 0.6 dB degradation of the MSE reported in [G&W2] was probably due to this effect.

8.7 FRACTIONALLY SPACED EQUALIZERS

The main disadvantage of symbol-rate sampling and equalization is that a sampled spectrum is created by foldover from the excess band $f_s/2 < f < f_s$ into the main band $f < f_s/2$† *before* the equalizer, so that there is no opportunity to operate separately on the components in the two bands. By contrast, a fractionally spaced equalizer (FSE) uses samples at a rate f_{samp} ($> (1 + \alpha)f_s$) that is sufficient to characterize the whole band; an FSE is therefore able to operate separately on the two bands.

It is most convenient if f_{samp}/f_s is the ratio of two small integers; in this discussion we will consider only the simplest ratio of 2/1. Ratios of 3/2 and 4/3 have also been used; they reduce the amount of computation and increase the amount of storage — the basic principles are the same. An FSE has the following *possible* advantages over a symbol-rate (T-spaced) equalizer (TSE):

1. It can match the attenuation of the signal in both bands.
2. It can equalize the delay over the whole band.
3. It can synthesize a flat delay of any fraction of T.
4. With the first three as "preliminaries", it can then equalize the folded-over spectrum, and thereby implement the classical optimum linear receiver.

As we shall see, the benefits of the first two capabilities — eliminating the loss of sampled energy due to aliasing — are insignificant for low-pass signals. However, for passband signals that have been subjected to severe phase distortion, the ability to equalize the delay before foldover occurs can become paramount. For both bands the benefit of the third capability — allowing the phase of the eventual symbol-rate sampling (to recover the data) to be fixed at anything convenient, without incurring any aliasing penalties — depends on the method of timing recovery that is used.

Steady-State Solution of an FSE.   Consider an impulse response sampled at $2f_s$ that has M terms, input to an FSE with N taps. There would be $(M + N - 1)$ output terms, but only every alternate term would be of interest. With no noise there would therefore be only either $(M + N)/2$ or $(M + N - 2)/2$ equations to define N taps. Since it is usual to make $N > M$, there would be fewer equations than taps, and there would be an infinite set of solutions. In the frequency domain it can be seen that if $\alpha < 1.0$, the band from $(1 + \alpha)f_s/2$ to f_s has no signal energy in it, and the equalizer transfer function can assume any values in that band.

---

† We will begin this discussion with one-dimensional baseband signals and consider complex signals later.

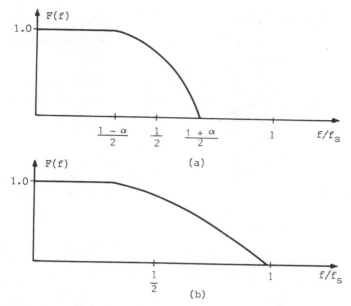

Figure 8.13 Responses required of band-limiting filters: *(a)* for a TSE; *(b)* for a T/2 FSE.

However, the addition of only a small amount of noise to the signal is sufficient to make the solution unique.† The FSE must then filter out the out-of-band noise in addition to performing its usual task of achieving a compromise between equalization and enhancement of in-band noise.

Some type of continuous filter will still be needed to band-limit the signal to $f < f_s$. The response of the low passes used for a TSE and an FSE are shown in Figs. 8.13a and 8.13b. The first one could also be used for an FSE, but the second has two advantages:

1. It has a more gentle roll-off and is easier to implement.
2. It is easier for an FSE to provide high attenuation beyond $(1 + \alpha)f_s/2$ if it is allowed to provide most of the roll-off in the transition band from $(1 - \alpha)f_s/2$ to $(1 + \alpha)f_s/2$.

The relative merits of TSEs and FSEs are very different for baseband and passband signals, so we will consider them separately.

† When FSEs were first described [Gu], [Un3], [Q&F], and [G&W3] there was considerable worry about their stability, and several more complicated adaptation algorithms [GM&W] and [U&S] were proposed to ensure this, but it now seems to be generally agreed that these are unnecessary.

8.7.1 FSEs for Baseband Signals

It may be helpful to consider two simple examples of signals with severe distortion.

Attenuation Distortion. A signal with 25% excess bandwidth is passed through a channel with an attenuation distortion that increases linearly with frequency and is 20 dB at $f_s/2$ (a 0.25/20/0 signal in the terminology given in Section 7.2). Since there is no delay distortion, sampling at the right time results in no aliasing loss. TSEs and $T/2$ FSEs were designed using the SIMEQ program listed in Appendix I. The shortest ones that will do a reasonable job—ensuring that the residual ISI reduces the output SNR by only a few tenths of a decibel—have 9 and 19 taps, respectively; that is, the time spans of the two equalizers are the same. The resultant RMSEs are almost equal:

	T	$T/2$
RMSE with 36 dB input SNR	0.034	0.035
RMSE with 17 dB input SNR	0.275	0.277
RMSE of 36 dB equalizer at 17 dB SNR	0.301	0.305

The amplitude responses of the two 17 dB SNR equalizers are shown in Fig. 8.14.

Delay Distortion. The same signal is passed through a channel with parabolic delay so that $d\phi/d\omega = 2T$ at $f_s/2$ (a 0.25/0/2.0 signal). Aliasing due to symbol-rate sampling produces a minimum of the sampled spectrum of -0.06 dB and

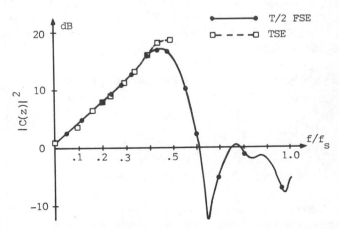

Figure 8.14 Amplitude responses of two FSEs for linear amplitude distortion.

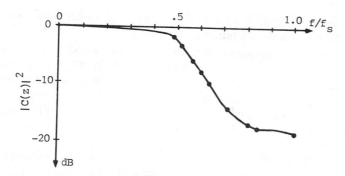

Figure 8.15 Amplitude response of an FSE for delay distortion.

a total loss of sampled energy of only 0.003 dB. For the residual ISI to reduce an input SNR of 20 dB by 0.2 dB† the TSE and FSE each need 11 taps.

The amplitude response of the FSE is shown in Fig. 8.15. It achieves a fair approximation to the square root of a raised-cosine in the transition band.

Relative Merits of TSEs and FSEs. It appears that the relative number of taps required for TSEs and FSEs depends on the type of distortion. For pure delay distortion, the $T/2$ FSE needs no more than the TSE, but for pure attenuation, it needs twice as many. For channels that have both distortions, the ratio is undoubtedly between 1 and 2.

The conclusions for low-pass signals are that (1) the price of the ability to synthesize any delay is always greater complexity and (2) the benefits of equalizing the delay before sampling are negligible.

8.7.2 FSEs for Passband Signals

As shown in Section 7.5, aliasing distortion can be much more severe in an asymmetrical passband than in the baseband. It will be useful to look further at the problem of equalizing the V.27 and V.32 signals distorted by a 99 percentile channel as defined by (7.19).

1. Table 7.3 shows that when a V.27 signal (with 50% excess bandwidth) is sampled at the symbol rate, aliasing reduces the sampled power by only 0.15 dB. If this signal is then equalized by a TSE of infinite length, noise enhancement reduces the effective SNR by 2.2 dB. An FSE, on the other hand, because it equalizes the delay before sampling and thereby prevents

---

† Since there is no noise enhancement in equalizing only delay distortion, this means that the output SNR is 19.8 dB.

aliasing, reduces the SNR by only 1.3 dB. The 0.9 dB difference might be significant in some situations.

2. Symbol-rate sampling reduces the power of a V.32 signal (with 12.5% excess bandwidth) by only an insignificant 0.01 dB. Equalization by an infinite TSE enhances the noise by 4.4 dB, and the use of an an infinite FSE only reduces this enhancement to 4.3 dB; it is clear that very little of the enhancement is caused by equalization of a spectral dip resulting from aliasing.

It is clear, therefore, that for high-speed modems that make efficient use of an overall band by using small excess bandwidths, the theoretical benefits of FSEs (assuming infinite length equalizers) are negligible. On the other hand, as shown in [Q&F], the situation changes when one considers practical, finite equalizers. The dip(s) in the sampled spectrum caused by aliasing is (are) often much narrower than any in the unsampled spectrum,† so that the TSE must have a much greater time span than the FSE. The aim usually is to reduce the ISI to a level where it only adds a few tenths of a decibel to the enhanced noise, and for channels with very asymmetrical delay characteristics, this requires more taps in a TSE than in an FSE.

8.7.3 Adaptation

An FSE can be designed adaptively in the same way as a TSE. For a $T/2$ FSE, the samples are shifted through the TDL at a $2f_s$ rate, but the output need be calculated only· at an f_s rate—that is, at alternate times. The error then is correlated with all the stored values and the products used to update the taps as in a conventional SG LMS method. All published reports show that an FSE converges at approximately the same rate as a TSE.

8.7.4 Relation to Timing Recovery

The ability of an FSE to synthesize any delay, and the consequent insensitivity of the MSE to timing phase, may be a big advantage in all-digital receivers in which the sampling (at a rate Lf_s) is done at the input to the receiver. If—and it is a big "if"—the FSE can keep up with the precessing clocks, there is no need to change the phase of that sampling or do any explicit interpolation.

---

† If there were narrow dips in the unsampled spectrum—although it is hard to think of what might cause them—then the time spans of the two equalizers would have to be equal, and a $T/2$ FSE would need twice as many taps.

Chapter 9

Nonlinear Equalizers

The only nonlinear operation that is commonly used in the equalization and detection of *linearly modulated* signals is a slicing (thresholding) to produce a decision — either preliminary or final — about the transmitted data. If the slicing is used only in the detection, the equalization itself is linear; if the result is used to change subsequent signals before slicing, the equalization is nonlinear. We will consider three types of nonlinear equalizer: Decision-Feedback Equalizers (DFEs), ISI Cancelers, and Maximum-Likelihood Sequence Estimators (MLSEs).

In most cases the motivation for the nonlinear processing is to avoid the noise enhancement that occurs when an *attenuation-distorted* signal† is linearly equalized. The aim is always to approach the performance (error rate) that would be achieved with an undistorted received signal with the same SNR.

In some cases such as the multipath distorted channel, because the ISI is caused mainly by a zero in the channel transfer function, it can be equalized with many fewer feedback taps than linear (feedforward) taps. One such channel is used as an example throughout the discussion of DFEs in Section 9.1.

Prefiltering. In most cases the nonlinear part of the equalizer must be preceded by some linear processing to put the impulse response into a form that can be handled by the particular non-linear processor. As might be expected, a matched filter is often a theoretically essential part of this prefilter. However, as

---

† At the risk of boringly belaboring a point, it must be emphasized that a signal that has only delay distortion can be linearly equalized perfectly — that is, without any noise enhancement.

pointed out by Clark [Cl2] the primary function of the matched filter† — that of equalizing the delay so that no further attenuation distortion is introduced by aliasing when the signal is sampled at the symbol rate — becomes much less important when using band-edge timing and a nonlinear equalizer. This is because (1) the sampled energy is maximized by the choice of timing and (2) a DFE or MLSE does not try to equalize the distortion per se and therefore enhances the noise much less than a linear equalizer does.

Therefore, we will use the concept of a matched filter only occasionally — first when we briefly discuss fractionally spaced prefilters, and then to establish the theoretical basis for practical adaptive prefilters; most of the discussion will assume that the signal has been passed through a fixed filter matched to the *transmit signal* and a compromise *delay* equalizer.‡ Thereafter all the processing will be at the symbol rate.

Terminology. Many authors have used the terms "decision-feedback equalizer" and "maximum-likelihood sequence estimator" to mean the combination of a basic DFE or MLSE with the prefilter of their choosing. Since the performance of each depends very strongly on the type of prefilter used, we will distinguish between the various combinations by using a prefix (e.g., SSTDL) to describe the prefilter.

9.1 DECISION-FEEDBACK EQUALIZERS

The idea of using previous decisions to reduce the distortion of data signals is very old; Price [Pr1] listed 43 references dating back to 1919.

A basic adaptive DFE is shown in Fig. 9.1. Decisions (binary or multilevel as appropriate) are made about the received data; these are stored in a shift register(s) and are then multiplied by the learned values of the trailing echoes of the impulse response (h_m for $m = MM + 1$ to M) and subtracted from the received signals at subsequent times. That is,

$$y'(k) = y(k) - \sum_{m=MM+1}^{M} \hat{a}(kd - m + MM)\, h_m(k - 1) \qquad (9.1)$$

where $\hat{a}(kd)$ is the sliced value of $y(k)$. The error $\epsilon(k)$ is defined as usual as $[y'(k) - \hat{a}(kd)]$, and the estimate of each trailing echo is updated by correlating the error with the stored received data; that is,

$$h_m(k) = h_m(k - 1) + \Delta\epsilon(k)a(kd - m + MM)$$
$$\text{for } m = (MM + 1) \text{ to } M \qquad (9.2)$$

† The other function — matching the attenuation of the channel in the excess band, and thereby reducing the noise slightly — becomes insignificant for high-speed modems that use only a very little excess bandwidth.

‡ Note that we must not use a compromise attenuation equalizer because that would enhance the noise — precisely what we are trying to avoid.

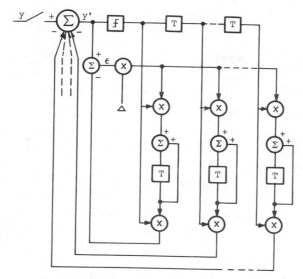

Figure 9.1 A basic adaptive DFE.

It is important to note that the adaptation is not trying to invert the channel response—only learn it. It is therefore much less affected by a spread of eigenvalues, and the adaptation of a basic DFE is typically much faster than that of a linear equalizer.

Minimum and Maximum Phase, Trailing and Leading Echoes. A minimum-phase transfer function is defined as one whose phase shift at each frequency is a minimum for the given overall amplitude response. This means that the transfer function expressed as a rational function of $p\,(=\sigma + j\omega)$ has no zeros in the Right Half-Plane (RHP). Strictly speaking, there is no such thing as a maximum-phase function, but the term is sometimes used to describe one that has no zeros in the Left Half-Plane (LHP); we will use the term occasionally, but put it within quotation marks to emphasize its imprecision.

A non-minimum-phase transfer function, resulting from channel and filters, cannot be made minimum-phase by cascading it with a prefilter because cancellation of the RHP zeros would require RHP poles that would make the prefilter unstable. This has led to some confusion about the role played by matched filters and delay equalizers. A signal that has been perfectly delay-equalized is often shown as having an impulse response that is symmetrical about $t = 0$. Such a signal has zero phase, and surely zero is less even than minimum? However, such a signal would be noncausal, and in practice the signal can be made (nearly) symmetrical only about some delayed time; this "pseudo-minimum" phase is then (nearly) linear and, because of the quadruplets of poles and zeros that do not affect the amplitude response, it is greater even than the "maximum" phase.

The combination of channel and transmit and receive filters (often delay-equalized) is seldom either minimum or "maximum" phase. Nevertheless, it is useful to begin our discussion of prefilters by discussing these two classes. When the received and filtered signal is sampled to generate an impulse response represented as $H(z)$, the requirement for minimum (maximum) phase is that none of the zeros of $H(z)$ lie outside (inside) the unit circle. One of four special types of $H(z)$ may occur:

1. Minimum-phase, and the first impulse response term is the largest (i.e., $|h_1| > |h_m|$ for all $m \neq 1$): in this case no prefilter is needed, and a pure DFE is adequate.
2. Minimum-phase but with severe amplitude distortion, so that some later term (h_2 except in the most extreme cases) is the largest: in this case it might seem that the early, smaller term(s) should be canceled somehow and the largest one used for detection. However, as we shall see, this is not advisable, and there is little that the prefilter can or should do.
3. Maximum-phase, and the last term is the largest: in this case the best solution is for the prefilter to "time-reverse" the impulse response without changing its amplitude response. That is, it should add delay so that the resultant differential delay (total delay minus some arbitrary constant amount) is the negative of the input delay.
4. Maximum-phase with severe amplitude distortion, so that the last term is not the largest. Again, the prefilter should operate only to time-reverse the IR and create a "pseudo" minimum-phase IR of type 2.

A more general type of $H(z)$ usually has both leading and trailing echoes, with the number of each equal to the number of zeros outside and inside the unit circle respectively.

Error Propagation. If noise causes one of the decisions $\hat{a}(kd)$ to be wrong, then while it is stored in the shift register it will be fed back wrongly to all subsequent received signals and will thereby reduce their noise margin and increase their probability of error. If the data is transmitted in blocks, and any block that is found to contain an error is retransmitted, then any secondary errors that may occur because of incorrect feedback from the primary error are inconsequential. Even if the data is character-formatted, the multiple errors that occur because of long scramblers and differential phase modulation (see Sections 12.3 and 5.6) will typically be much more serious than those resulting from incorrect feedback.

The conclusion of most workers has been that, except in the most extreme cases, and for certain data formats, error propagation in DFEs is not a serious problem.

Moving the "Main" Sample. Extreme cases of error propagation arise when the attenuation is very severe; then the numbers of leading echoes and of zeroes of

$H(z)$ outside the unit circle may differ by one.† The prefilter cannot remedy this, and its output will have one large leading echo, which must be considered as the main sample and used for recovering the data. This will have two effects: (1) the probability of a primary error will be increased because the power in the trailing echoes is discarded by a DFE, so that there is less power left in the "main" sample to overrride the noise, and (2) feedback of an "echo" larger than the "main" sample will certainly cause error propagation.

9.1.1 A Single-Sided TDL (SSTDL) as a Prefilter

The original [Au] and [NM&M] and simplest way of dealing with leading echoes is to use a TDL that has taps only before its main tap. If decision-directed training is used, this can be ensured by presetting the taps to . . . 0, 0, 0, 1; if a receive reference is used, the delay is calculated as described previously, but no taps are provided after the main one.

The very simple impulse response and equalizer example considered in Section 8.1 ($M = 4$, $N = 5$) can be modified to illustrate an SSTDL/DFE by dividing the five taps into three feedforward and two feedback. Then the equations that must be solved for an LMS solution with noise are

$$\begin{array}{ccc} \underline{c_1} & \underline{c_2} & \underline{c_3} \end{array}$$

$$\begin{array}{cccll} h_1 & 0 & 0 & = 0 & \text{(9.3a)} \\ h_2 & h_1 & 0 & = 0 & \text{(9.3b)} \\ h_3 & h_2 & h_1 & = 0 & \text{(9.3c)} \\ h_4 & h_3 & h_2 & = 1 & \text{(9.3d)} \\ \mu_1 & 0 & 0 & = 0 & \text{(9.3e)} \\ 0 & \mu_2 & 0 & = 0 & \text{(9.3f)} \\ 0 & 0 & \mu_3 & = 0 & \text{(9.3g)} \end{array}$$

and then
$$h_1' = h_4 c_2 + h_3 c_1 \qquad \text{(9.3h)}$$
and
$$h_2' = h_4 c_3 \qquad \text{(9.3i)}$$

An SSTDL linearly equalizes only the leading echoes and leaves the trailing ones, only slightly changed, to be handled by the DFE; that is, it approximately equalizes the distortion caused by those zeros of the impulse response that are outside the unit circle. If a channel has only attenuation distortion, its impulse response is symmetrical about some delayed time reference, the zeros of $H(z)$ occur in pairs that are such that $z_{2i+1} = 1/z_{2i}$ for $i = 1$ to $(M-1)/2$, and the attenuation distortion is caused equally by the leading and trailing echoes. For

---

† This is a generalization of types 2 and 4 of $H(z)$ referred to above.

such a channel, an SSTDL/DFE will cause approximately half as much noise enhancement as a linear TE.

On the other hand, if a channel has only delay distortion, linear equalization of only the zeros outside the unit circle would cause attenuation distortion and consequent noise enhancement. By contrast, such a channel could be linearly equalized without any noise enhancement. Thus SSTDL/DFEs sometimes perform worse than conventional TEs, and they are not used for voice-band telephone channels or any other channels that may combine delay distortion with high noise.

However, for very high speed data signals that have been subjected to multi-path fades and have high SNRs, the main aim in the equalizer is to open the eye (often multilevel) with the minimum amount of processing; for these, an SSTDL/DFE combination is recommended.

9.1.2 A Double-Sided TDL as Prefilter

If a conventional TE with both leading and trailing taps is used as the prefilter, then since the prefilter by itself is capable of fully equalizing the signal, it might seem that the feedback taps are superfluous. If there are NF forward and NB backward taps, and there is no noise, then the $(NF + NB)$ variables must satisfy $(M + NF - 1)$ equations. It can be seen that if $NB > M - 1$, there is just one exact, but trivial and useless, solution in which the prefilter is transparent (i.e., one $c_n = 1.0$ and all the rest are zero) and the DFE uses the first sample of the impulse response — no matter how small it is — for detection, and treats all the rest as trailing echoes.† If, on the other hand, $NB < M$, the solution will be unique, but some of the tap values may be very large.

When noise is added to the input signal there is always a unique LMS solution to the equations, and the sets of taps play fairly separate and distinct roles: the leading taps deal with the leading echoes, the trailing taps follow the leading taps in such a way as to minimize the output noise, and the feedback taps deal with both the original trailing echoes and all those created by the prefilter.

If the simple linear equalizer used as an example in Section 8.1 (five forward taps to equalize an impulse response with four terms) is augmented by two feedback taps, the forward taps must satisfy the reduced set of equations (8.3a–8.3d, 8.3g–8.3m), and the feedback taps must cancel the first two modified trailing echoes

$$h_1' = c_2 h_4 + c_3 h_3 + c_4 h_2 + c_5 h_1$$

and

$$h_2' = c_3 h_4 + c_4 h_3 + c_5 h_2$$

$$(9.4)$$

---

† This extreme case and its useless solution should be borne in mind as we consider DSTDL/DFEs further.

Adaptation. The taps of the DSTDL are updated just as they are for a completely linear equalizer—using any of the methods discussed in Chapter 8. However, the matrix whose eigenvalues control the rate of convergence is no longer the simple auto-correlation matrix. Because the feedback taps typically train much faster than do the forward ones, they have the effect of removing a block of equations (8.3e and f in the example) from the adaptation, so that the matrix is no longer necessarily positive definite. Even if the noise level is high, the eigenvalues may be dispersed considerably more than they would be by the channel alone, and the rate of convergence becomes extremely difficult to predict.

The terms of the modified impulse response are learned just like those of the original one—according to (9.2).

Amount of Computation Required. The DSTDL usually has equal numbers of leading and trailing taps, so for calculating its output and updating all the forward taps, it will require twice as many multiplications as for an SSTDL. The multiplications for the DFE have the data (two or three bits at most) as one of the multiplicands and thus can often be performed much more easily. The total amount of computation will therefore be between one and a half and two times as much as for an SSTDL/DFE.

Two Examples. In order to estimate the value of this mixed equalization method compared to linear equalization, we will consider again the two examples of distortion that we have already linearly equalized in Chapter 8:

Example 1. For the attenuation-distorted signal (0.25/20/0), four DSTDL prefilters were designed for input SNRs of ∞, 40, 20, and 15 dB. As was expected, the equalizer designed for no noise went berserk (see Fig. 9.2a), but as little as -40 dB of noise was sufficient to keep the tap values under control. Therefore, as we did for the linear equalizer, the RMSE of the best equalizer for each noise level was compared to what would be achieved if the equalizer were trained with this low noise level; the results are shown in Table 9.1.

If these figures are compared with those in Section 8.2 for the linear equalizer, it can be seen that the MSE has been reduced by between 1.8 and 2.0 dB. Also, at the high noise level, there is a difference of only 0.3 dB (compared to 0.6 dB for the linear equalizers) between the MMSE and that achieved with the

TABLE 9.1   RMSEs for an Attenuation-Distorted Impulse Response with Various Prefilters

SNR (dB)	NPE	Minimum RMSE	RMSE with -40 dB Equalizer
40	1.60	0.013	(0.013)
20	1.52	0.125	0.127
15	1.38	0.217	0.225

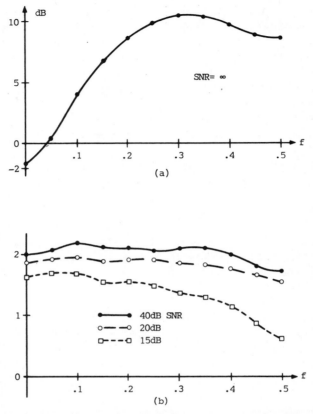

Figure 9.2 Amplitude responses of four DSTDL prefilters for a 0.25/20/0 signal: *(a)* infinite SNR; *(b)* finite SNRs.

equalizer adjusted for low noise. Thus this equalizer can be trained satisfactorily at almost any noise level *except zero*.

The amplitude responses of these three equalizers are shown in Fig. 9.2*b*; the significance of the fact that they are all approximately all-pass will be discussed in Section 9.1.4.

Example 2. For the QAM signal subjected to a 20 dB midband fade [T&S] the impulse response (real because the passband response is symmetrical about the carrier frequency) is 0.0256, −0.0912, 0.8167, −0.5630, 0.0811, −0.0224 for a "Minimum-Phase" Fade (MPF)† and the time inverse of this for a Non-Minimum-Phase Fade (NMPF).

---

† Unfortunately, the terminology appropriate to passband radio signals is not consistent with that used for sampled baseband signals. The "minimum-phase" refers to the relationship between the main and the interfering paths; however, as can be seen from Fig. 9.3, which shows the five zeros of $H(z)$, only the dominant zeros of the impulse response polynomial are inside the unit circle, and the sampled response is not truly minimum phase.

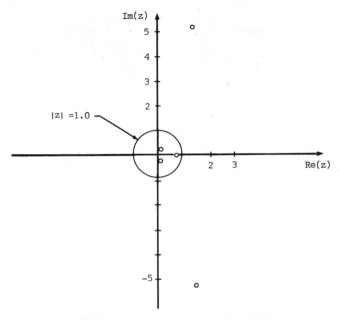

Figure 9.3 Zeros of $H(z)$ for a minimum-phase fade.

The model used to calculate this impulse response was developed for micro-wave radio [Ru1] and [Ru2] and as discussed in [S&M] and the references cited therein, present technology does not allow any more than the simplest equalization at the very high data rates used in that medium. Nevertheless, in the expectation that digital processing will become faster, narrower bands and lower data rates will be used, and/or the model will be valid in other media, we will consider a more complicated equalizer (15 forward and 7 feedback taps), which might be used to reduce the MSE to a level appropriate for an SNR of, say, 20 dB.

For the MPF, since most of the ISI is in the trailing echoes and can be dealt with by the DFE, almost any prefilter will do a good job. We would expect very little difference between the performance of an SSTDL, a DSTDL, and the two other prefilters considered in the next sections; simulation confirmed the expectations, so we will concentrate our attention on the NMPF.

As in the previous example, the equalizer adjusted for no noise was useless when any significant level of noise was added, so the equalizer designed for a 40 dB SNR was again used as a reference; the results are shown in Table 9.2.

Comparison with the results given in Chapter 8 shows that with the prefilter *appropriate for the noise level,* the MSE of a DSTDL/DFE is 2.7 dB less than that of a linear equalizer. However, as much as 0.8 dB of that advantage may be lost by using the equalizer trained during quiet periods.

The amplitude responses of the three prefilters are shown in Fig. 9.4; it can be seen that the 40 dB one is already beginning to behave rather strangely! It is not

TABLE 9.2 RMSEs for an NMPF Impulse Response with Various Prefilters

SNR (dB)	NPE	Minimum RMSE	RMSE with Equalizer Trained at 40 dB SNR
40	2.01	0.016	(0.016)
30	1.71	0.043	0.046
20	1.61	0.130	0.142

clear how much noise is needed to control a DSTDL/DFE for any given distortion; consequently, this equalizer, which may converge to some significantly worse-than-optimum solution if the noise level during training is not greater than some unknown amount, should be considered with skepticism and used with caution.

9.1.3 Fractionally Spaced Prefilters

Monsen [Mo1] described a prefilter that used a TDL with a tap spacing at the Nyquist rate of the channel. It was therefore theoretically able to combine all the functions of a matched filter and a filter that cancels leading echoes while minimizing the noise. However, this approach has two disadvantages:

1. The overendowment of parameters—from both the oversampling and the use of trailing *and* feedback taps—may result in some unusable sets of tap values if the equalizer is trained with too little noise.

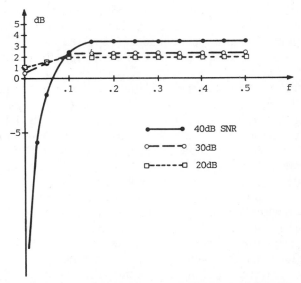

Figure 9.4 Amplitude responses of three 15-tap DSTDL prefilters for a NMPF.

2. If, as is most common, the tap spacing is $T/2$, the number of multiplications required will be between two and four times as many as for the symbol-rate prefilters that are described in the following sections.

Since, as we have already discussed, the matching function is not so important in a prefilter for a DFE (or an MLSE), fractionally spaced prefilters are not recommended.

9.1.4   Justification of an All-Pass as the Best Prefilter

The first thing that must be emphasized in this discussion of all-pass prefilters is that the all-pass is *not a delay equalizer*. In fact, if the input signal has no delay distortion, so that the impulse response is symmetrical about some delayed time, then the prefilter, in equalizing the leading echoes, will act as a delay unequalizer!

It was proved in a series of closely related papers [Mo1], [Pr1], [An], [Sa3], and [H&M] that the optimum prefilter for a DFE is a matched filter (either continuous or sampled at the Nyquist rate for the channel) followed by a sampler and an SSTDL that compensates for the zeros of the impulse response that are outside the unit circle. If the equalizer is infinitely long and the noise level is very low, this combination constitutes a *sample-whitened matched filter,*† and the samples of the noise at the input to the slicer are uncorrelated (white).

If the matched filter (necessarily adaptive) is dispensed with, and a receive filter, matched to the transmit signal, plus a compromise delay equalizer are used, then in order to calculate the transfer function of the sampled prefilter we must combine the (incidental) effects of a matched filter in the main part of the band below $(1 - \alpha)f_s/2$ with those of the SSTDL.

To this end Clark [Cl1] considered the sampled impulse response $H(z)$ to be factorized into the product of $H_1(z)$ and $H_2(z)$, which are of order M_1 and M_2, and have all their zeros inside and outside the unit circle, respectively. Then the sampled, or pseudo-, matched filter would have the transfer function

$$F_1(z) = z^{-M} H\left(\frac{1}{z}\right) = z^{-M_1} H_1\left(\frac{1}{z}\right) z^{-M_2} H_2\left(\frac{1}{z}\right) \qquad (9.5)$$

The product of $H(z)$ and $F_1(z)$ can be written so as to show separately the factors with zeros inside and outside the unit circle:

$$H(z)\, F_1(z) = \left[z^{-M_2} H_1(z) H_2\left(\frac{1}{z}\right) \right]\left[z^{-M_1} H_2(z) H_1\left(\frac{1}{z}\right) \right] \qquad (9.6)$$

In order to equalize the leading echoes, the SSTDL, with transfer function $F_2(z)$,

---

† The reader is referred to the original paper [Fo1] and then to the others for theoretical discussions of the SWMF.

would have to approximate the inverse of the second factor (the precursor part); hence the transfer function of the combination of the pseudo-matched filter and SSTDL would be

$$F_1(z)\, F_2(z) = z^{-M} H_1\left(\frac{1}{z}\right) H_2\left(\frac{1}{z}\right) \mathscr{P}\left\{\left[z^{-M_1} H_2(z)\, H_1\left(\frac{1}{z}\right)\right]^{-1}\right\} \quad (9.7)$$

where the symbol $\mathscr{P}\{X(z)\}$ means an LMS polynomial approximation (of as yet unspecified order) to the function $X(z)$.† Then the product of $H_1(1/z)$ and its polynomial quasi-inverse is approximately unity, so that

$$F_1(z)\, F_2(z) \simeq z^{-M} H_2\left(\frac{1}{z}\right) \mathscr{P}\{[H_2(z)]^{-1}\} \quad (9.8)$$

which is the transfer function of a Quasi-All-Pass (QAP).

Another less mathematical, more intuitive argument for an all-pass characteristic is that since the noise spectrum of the output of the fixed receive filter is white, and the noise input to the slicer should also be white, any processing in between must be all-pass.

It might seem that these arguments are disproved by the fact that none of the equalizer examples given in the last section — each optimized for a particular noise level — is *exactly* all-pass. However, the arguments, like most tidy proofs in data transmission, are strictly valid only for infinite equalizers. This is illustrated by Fig. 9.5, which shows the responses of 25-tap equalizers (much longer than necessary to achieve an acceptable MSE) for the case of the NMPF; comparison with Fig. 9.4 shows that for both high and low noise, as the equalizer gets longer, the responses become more nearly all-pass, and the differences between the best equalizers for the various noise levels become less.

Leading Echoes or Zeros Outside the Unit Circle? Clark's argument may be misunderstood if emphasis is placed on compensating for "leading" echoes rather than zeros outside the unit circle. For all moderate distortions, there is no difference; the numbers of both are equal to $(MM - 1)$,‡ and the argument is indisputable. However, for extreme cases of distortion the all-pass result has not been accepted so readily.

Let us consider an example of a severely attenuated signal with "maximum phase" — special case 4 as defined in Section 9.1.2:

$$H(z) = (1 + 2z^{-1})^3$$
$$= 1 + 6z^{-1} + 12z^{-2} + 8z^{-3}$$

---

† For example, $\mathscr{P}\{[1 + az^{-1}]^{-1}\}$ might be $1 - az^{-1} + a^2z^{-2} - a^3z^{-3}$.
‡ Recall that MM is the index of the main sample (the one with the largest magnitude) of an impulse response that has M terms.

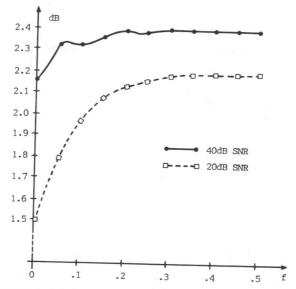

Figure 9.5 Amplitude response of a 25-tap DSTDL prefilter for a NMPF.

which, when normalized to unit energy, becomes

$$0.0639 + 0.3833z^{-1} + 0.7667z^{-2} + 0.5111z^{-3} \qquad (9.9)$$

All three zeroes are outside the unit circle, but there are only two echoes before the main sample. The amplitude response of this impulse response is shown in Fig. 9.6.

Two sets of DSTDL/DFEs were designed for SNRs of 20 and 40 dB. The amplitude responses of the 15-tap and 25-tap prefilters are shown in Figs. 9.7 and 9.8, respectively. As before, the long low-noise prefilter is all-pass, and its output impulse response is the delayed time-inverse of the input. However, the others are far from all-pass, and the short low-noise one is particularly strange. Its (reasonably truncated) output impulse response is

$$1 + 1.98z^{-1} + 1.17z^{-2} + 0.22z^{-3}$$

which has one zero at $z = -1.09$. Thus the prefilter has not fully compensated for one of the zeros that was originally outside the unit circle, and it appears that as the noise is reduced, the prefilter advances its estimate of where the "main" sample is and approaches the useless solution (with its very high noise enhancement) that we considered at the beginning of Section 9.1.2.

For an input SNR of 20 dB, the best short prefilter resulted in an RMSE of 0.177, and this did not change by more than 0.1 dB when the equalizer was

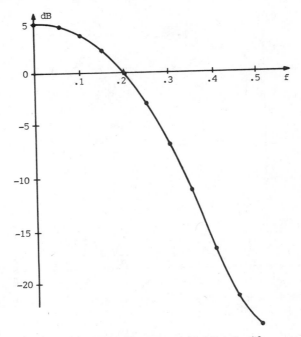

Figure 9.6 Amplitude response of $(1 + 2z^{-1})^3$.

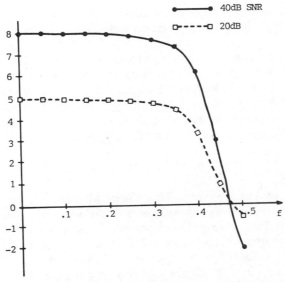

Figure 9.7 Amplitude responses of two 15-tap DSTDL prefilters for $(1 + 2z^{-1})^3$.

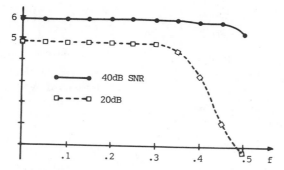

Figure 9.8 Amplitude responses of two 25-tap DSTDL prefilters for $(1 + 2z^{-1})^3$.

made "infinitely" long. If the short, low-noise (40 dB SNR) prefilter and its associated DFE were used at this noise level, the RMSE would be 0.228 (2.2 dB worse!); on the other hand, if the long low-noise (all-pass) prefilter were used, the RMSE would be 0.186 — only 0.4 dB worse than the optimum. It is clear that the all-pass prefilter is a good compromise that achieves nearly the minimum MSE at all noise levels.

9.1.5 An Antimetrically Constrained TDL (ACTDL) as a Prefilter

Clark's and Monsen's methods of calculating the prefilter are offline and not suitable for continuous adaptation. As we have seen, the straightforward approach of using a DSTDL and allowing it to adapt for any particular noise level may become seriously suboptimum when the noise level changes; it also requires much more computation than would seem to be necessary.

A better approach is to adapt only the coefficients needed to equalize the leading echoes and to add some other taps that are constrained so as to make the full structure either all-pass, or at least approximately so. Kurzweil and Bingham [K&B] described an *antimetrically constrained* TDL that has an equal number of leading and trailing taps (i.e., $N = 2NM - 1 = 2NL + 1$) and in which the trailing coefficients are constrained to be the negatives of the leading ones; that is, $c_{N-n+1} = -c_n$. The transfer function can be written in the form

$$F(z) = \sum_{n=1}^{N} c_n z^{-n} \tag{9.10a}$$

$$= z^{-NM} \left\{ c_{NM} + \sum_{n=1}^{NL} c_n (z^{NM-n} - z^{n-NM}) \right\} \tag{9.10b}$$

$$= z^{-NM} \left\{ c_{NM} + 2j \sum c_n \sin[(NM - n)\pi T] \right\} \tag{9.10c}$$

so that

$$|F(z)|^2 = c_{NM}^2 + 4 \left[\sum c_n \sin[(NM - n)\pi T] \right]^2 \tag{9.11}$$

Wheeler [Wh] showed that this form ensures that for small tap values $(|c_n/c_{NM}| < 0.3)$, $|F(z)|$ is approximately constant with frequency (viz., all-pass).

An ACTDL can be most economically implemented by the structure shown in Fig. 9.9. The output (now designated by v, rather than z, to avoid confusion with the variable of the z transform) is given by

$$v = c_{NM} y_{NM} + \sum_{n=1}^{NL} c_n (y_n - y_{N-n+1}) \tag{9.12}$$

Since the leading tap coefficients operate on the difference between the leading and the trailing signals (i.e., the samples stored before and after the main tap), an LMS solution is obtained by correlating the error with these differences and updating the tap coefficients accordingly:

$$c_n(k + 1) = c_n(k) - \Delta_c \epsilon(k) [y_n(k) - y_{N-n+1}(k)] \tag{9.13}$$

No theoretical analysis of the convergence of an ACTDL has been made, and we can only try to extrapolate from the more firmly based findings for linear equalizers. Since the increment of each tap coefficient depends now on the difference between two signals, its RMS value is increased by a factor of $\sqrt{2}$;

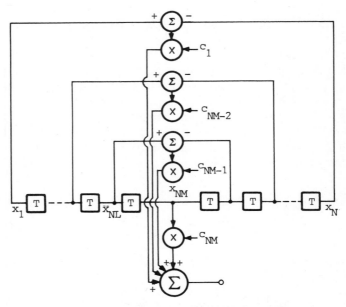

Figure 9.9 Antimetrically constrained prefilter (ACTDL).

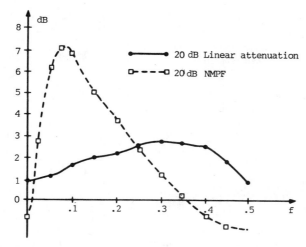

Figure 9.10 Amplitude responses of ACTDL prefilters for attenuation-distorted (0.25,20/0) and NMPF signals.

therefore, the optimum step sizes given in (8.34) should probably be amended to

Orthodox: $\Delta_{opt} = \dfrac{1}{(\sqrt{2}N\lambda_{max})}$ for a moderate spread of eigenvalues (9.14a)

Suggested: $\Delta_{opt} = \dfrac{1}{(N\sqrt{2\lambda_{max}})}$ for a wide spread of eigenvalues. (9.14b)

Convergence with Noise. Because of the constraints on the trailing tap coefficients, the convergence and the final state of an ACTDL are largely independent of the noise level. The only reaction of an ACTDL to an increased noise level is to reduce the overall gain slightly so as to minimize the mean-squared combination of ISI, which is principally contributed by the main sample, and noise. Reducing the gain (and hence the MSE) this way, however, does not, of course, reduce the probability of error; it only readjusts the nominal level of signal plus noise.

Example 1: The Attenuation-Distorted Signal. The tap coefficients were found to be 0.014, −0.0207, 0.0144, −0.06, 0.0647, −0.4204, 1.0, 0.4204, and so on, with $\Sigma c^2 = 1.612$. The amplitude response of this transfer function is shown in Fig. 9.10; it is a moderately good approximation to an all-pass. The RMSE with a 15 dB SNR is 0.226, which is the same as with the "low-noise" DSTDL, and only 0.4 dB worse than with the optimum DSTDL.

Example 2: The NMPF Signal. The tap coefficients were found to be 0.0255, 0.0605, 0.1074, 0.1738, 0.2742, 0.452, 0.797, 1.0, −0.797, −0.452, and so on, with $\Sigma c^2 = 2.922$. The amplitude response of this ACTDL is also shown in Fig. 9.10; it is far from all-pass. The RMSE with 20 dB SNR is 0.145, which is only 0.1 dB worse than with the "low-noise" DSTDL but 0.9 dB worse than with the optimum DSTDL.

ACTDLs for QAM Signals. If both the input signal and the taps of the prefilter are complex, it is not clear what constraint should be put on the imaginary parts of the tap coefficients in order that the transfer function be as nearly all-pass as possible. This was discussed in [K&B], but the conclusions there were questionable. The question has not been pursued further.

9.1.6   A Recursive All-Pass as a Prefilter

The only way to ensure a true all-pass transfer function is to use a recursive prefilter whose transfer function is constrained so that the numerator is the time-reverse of the denominator; that is,

$$F(z) = \frac{c_1 + c_2 z^{-1} + \cdots + c_{NL} z^{-NL+1} + z^{-NL}}{1 + c_{NL} z^{-1} + \cdots + c_1 z^{-NL}} \qquad (9.15)$$

Just as with the ACTDL, the leading coefficients are adjusted by either a zero-forcing or LMS algorithm so as to minimize the leading echoes [which result from zeros of $H(z)$ outside the unit circle]; the zeros of $F(z)$ are therefore outside the unit circle. This alleviates the usual worries about instability that arise when adapting recursive structures, because the poles are constrained to be the inverses of the zeros and are therefore inside the unit circle.

9.1.6.1   *Conventional Recursive Configurations.*   All the conventional recursive digital filter structures described in [Ja], [R&S], and elsewhere in the literature can be made all-pass by constraining the feedback coefficients to be equal to the reversed feedforward coefficients. However, the stored values, which must be correlated with the error in order to update the taps, have each been processed by different transfer functions that involve all the poles of $F(z)$ and various, difficult-to-calculate zeros. It appears, from a preliminary analysis, that the attenuation distortion produced by these intermediate transfer functions always increases the distortion of the input and thereby exacerbates the problem of eigenvalue spread. For this reason these structures are not recommended.

9.1.6.2   *A Modified Recursive Configuration.*   The samples of the input and output can be preserved in separate TDLs and the relationship between the numerator and denominator coefficients exploited in order to halve the number of multiplications; this is shown in Fig. 9.11 in such a way as to emphasize the similarity to the ACTDL of Fig. 9.9. The output at each sample time is calculated from

$$v_{NM} = y_{NM} + \sum_{n=1}^{NL} c_n (y_n - v_{NM-n+1}) \tag{9.16}$$

The leading tap coefficients should be updated in much the same way as was used for the ACTDL; that is,

$$c_n(k+1) = c_n(k) - \Delta_c \epsilon(k) [y_n(k) - v_{NM-n+1}(k)] \quad \text{for } n = 1 \text{ to } NL \tag{9.17}$$

and the step size Δ_c should be chosen according to (9.14).

The gain of $F(z)$ in (9.15) is unity, and it is difficult to make it otherwise while retaining the all-pass characteristic. If the data is multilevel, and, as is usual, the AGC is not able to precisely control the relationship between the level of the equalized signal and the slicing thresholds, this control can be achieved in either of two ways:

1. A separate multiplier can be applied to the output of the AP as indicated by dashed lines in Fig. 9.11. The coefficient c_{NM} can be adapted by correlating the error with the main stored sample y_{NM}, but the adaptation interacts with that of the other taps, and convergence is slower than with the second method.

2. The DFE can learn the magnitude of the main sample of the output of the

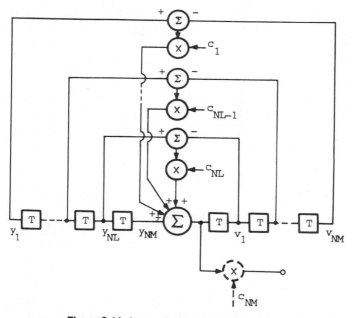

Figure 9.11 A recursive all-pass (AP) prefilter.

prefilter by extending the range of n in (9.5) to 0 to N, and then the learned value of h_0' can be used to control the slicing thresholds.

The choice between these two methods will depend mainly on how the slicing is done in the detector. If some sort of table look-up using predefined regions is used, the first, slower method would probably be the more convenient; otherwise, the second is to be preferred.

Example 1: The Attenuation-Distorted Signal. For comparison with the DSTDL/DFE, an AP/DFE with six leading and seven feedback coefficients was simulated. Using the formula of (9.12b), the "optimum" step size was calculated as 0.027 — compared to 0.039 for the DSTDL. Decay of the MSE, averaged over the same set of runs of random data as was used for the DSTDL, is shown in Fig. 9.12; it can be seen that the convergence rate is the same as for the DSTDL.

The coefficients c_1 to c_6 of the all-pass are 0.0114, −0.0201, 0.0168, −0.0531, 0.0687, and −0.3572, and the terms h_0' to h_7' of its output impulse response are 0.7905, 0.5658, 0.1934, 0.1118, 0.0542, 0.0444, 0.0171, and 0.0015. The MSEs (prorated to the main, first term) at the various noise levels were all within 0.1 dB of the minimum MSEs achieved when a DSTDL was "custom" adapted at each particular noise level. Thus the MSE at 15 dB SNR is 0.3 dB less than that for the DSTDL trained in low noise.

Example 2: The NMPF Signal. The coefficients c_1 to c_7 of the AP were 0.0099, 0.033, 0.0727, 0.1329, 0.2287, 0.3958, and 0.7106, and the terms of its output impulse response were 0.7699, −0.6157, 0.1609, −0.0454, 0.0077, 0.0019, 0.0038, and 0.0020. Again, the MSEs at all noise levels were within

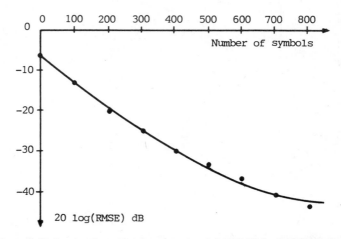

Figure 9.12 Stochastic gradient convergence of AP/DFE for 0.25/20/0 signal.

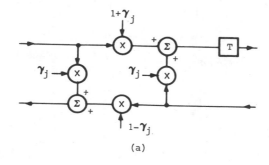

(a)

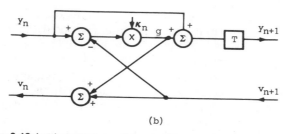

(b)

Figure 9.13 Lattice structures: *(a)* normalized form; *(b)* one-multiplier form.

0.1 dB of the minima achieved with the "custom-adapted" DSTDL and were thus as much as 0.8 dB less than with the low-noise DSTDL.

9.1.6.3 A Short-Circuited Lattice Configuration.

The normalized lattice structure shown in Fig. 9.13a was originally used by Kelly and Lochbaum [K&L] as a model for the vocal tract. Many authors have described variations that require only one multiplication per stage; the one that is most suitable for our purpose was given in [H&M] and is shown in Fig. 9.13b. The forward and backward outputs of each stage are calculated from the inputs by

$$g = \kappa_n (y_n - v_{n+1}) \tag{9.18a}$$

$$y_{n+1} = (y_n + g) z^{-1} \qquad \text{for } n = 1 \text{ to } NL \tag{9.18b}$$

$$v_n = v_{n+1} + g \tag{9.18c}$$

where g is a temporary result that can use the same storage location for all values of n.

The cascade connection of several of these lattices models the incident and reflected waves in a lossless reactive ladder network. Therefore, if a cascade of NL lattices is "short-circuited" at the end by setting $v_{NL+1} = y_{NL+1}$, the transfer

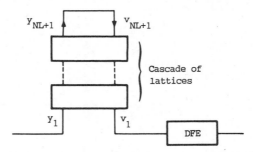

Figure 9.14 Cascade of lattices short-circuited to realize an all-pass.

function $F(z) = v_1/y_1$ is all-pass with a gain of unity, and the cascade can be used as a prefilter as shown in Fig. 9.14. {It should be noted that these lattices are not being used as in Section 8.4; there is no orthogonalizing operation taking place. The lattices are used only because they are an efficient way—*one* multiplication and *one* storage location per stage—of realizing an all-pass.}

The coefficients can be updated by correlating the error with the stored y_n. At the beginning of the adaptation, when all the κ values have been initialized to zero, the y_n are equivalent to the samples of the input that would be stored in a conventional TDL; thus the initial updating is truly steepest descent. However, as the adaptation progresses, each stored y_n is filtered by all the developing poles of $F(z)$ and $(NL - n + 1)$ difficult-to-calculate zeros, and the convergence becomes very difficult to analyze. The zeros appear to have the effect of mitigating the effects of the attenuation distortion introduced by the poles, and convergence is not much slower than when the stored input samples are used for correlation.

One reassuring property of these lattices is that if the magnitudes of all the coefficients are constrained to be less than unity, the transfer function is stable. This "protection" has not been needed in any adaptation simulated so far, but circumstances might arise when it would be useful.

9.1.6.4 Complex All-Passes for QAM.
The requirement for a complex all-pass is that the sum of the squares of the moduli of the real and imaginary parts of the transfer function should be all-pass. This can be ensured by generalizing the transfer function in (9.15) to

$$F(z) = \frac{c_1 + c_2 z^{-1} + \cdots + c_{NL} z^{-NL+1} + z^{-NL}}{1 + c_{NL}^* z^{-1} + \cdots + c_1^* z^{-NL}} \tag{9.19}$$

where the superscript * indicates the complex conjugate. If it is worthwhile to minimize the number of multiplications, (9.19) should be implemented by the complex equivalent of (9.16):

$$v_{NM,r} = y_{NM,r} + \sum_{n=1}^{NL} (y_{n,r} - v_{NM-n+1,r})c_{n,r} - (y_{n,i} + v_{NM-n+1,i})c_{n,i} \qquad (9.20a)$$

$$v_{NM,i} = y_{NM,i} + \sum_{n=1}^{NL} (y_{n,r} + v_{NM-n+1,r})c_{n,i} - (y_{n,i} - v_{NM-n+1,i})c_{n,r} \qquad (9.20b)$$

where the subscripts r and i indicate the real and imaginary parts of the inputs, outputs, and coefficients.

9.1.7 Relative Merits of the Various Prefilters

Before trying to summarize the advantages and disadvantages of each type of prefilter, we should consider further the issue of MSE and the sensitivity of the performance of a DSTDL to the noise level at which it is trained. Let us consider two channels with the same, very severe — probably unrealistic — attenuation distortion (28.6 dB at the band edge):

1. The maximum-phase impulse response considered in Section 9.1.4 was $H(z) = (1 + 2z^{-1})^3$. This is an extreme example where the prefilter has to work very hard. With an input SNR of 20 dB, the MSE of an AP is 2.2 dB *less* than that of a DSTDL trained at 40 dB SNR and equal to that of the DSTDL trained at 20 dB.

2. The minimum-phase impulse response is the time reverse of $H(z)$; in this case the AP and the low-noise DSTDL can only provide a delay, and their MSEs are 0.8 dB *greater* than that of the DSTDL trained at 20 dB SNR.

It appears, therefore, that if the noise level is constant then an AP performs the same as or slightly worse than a DSTDL, but if the high noise occurs only in bursts, then an AP may vary from about 1 dB worse to 2.5 dB better than a DSTDL (with an average improvement of about 1.0 dB).

The MSE performance and many other factors are summarized in Table 9.3,

TABLE 9.3   Ratings (by Several Criteria) of the Various Prefilters

	SSTDL	DSTDL	ACTDL	AP Mod. Recur.	AP Lattice
MSE with steady noise	4	1	3	2	2
MSE with "bursty" noise	3	2	2	1	1
Amount of computation	1	4	2	2	3
Amount of storage	1	2	2	2	1
Assurance of stability	1	1	1	3	2[a]
Speed of convergence	?	2[b]	2	1[c]	2

[a] This assumes that the magnitudes of the lattice coefficients are constrained to be less than unity.
[b] Although the noiseless DSTDL converges at the same rate as the AP (see Fig. 9.12), the last fine adjustment for a particular noise level may be quite slow.
[c] This assumes that the DFE learns h_0'.

which shows the ratings of the various prefilters (1 = best; equal numbers mean either that the methods are the same or are sometimes better and sometimes worse). A designer should select the configuration that best satisfies the design constraints.

9.1.8 Accelerating the Adaptation of a DFE

The heading of this subsection should be made more precise. A pure DFE, which is only learning the trailing echoes, is not much affected by the eigenvalue spread caused by attenuation distortion; the problem lies with the prefilter, which will typically converge as slowly as a linear TE. Ways of applying adaptive DSO to DSTDL/DFEs have been suggested [L&P], but they are extremely complicated and suffer from the tap jitter problem discussed in Section 8.4.2; SSO techniques such as Fast Kalman would seem to be much more suitable.

All orthogonalization methods require a lot of computation (varying between 5 and 11 *extra* multiplications per *adapted* tap), and the fact that the constrained prefilters, ACTDL and AP, adapt only half as many taps as the DSTDL makes them very attractive. Application of adaptive SSO techniques to these — using the modified recursive form of the AP, with its stored samples of the input — has not been attempted but is strongly recommended.

9.1.9 Uses of DFEs

DFEs have two advantages and one disadvantage compared to linear equalizers:

A1. The recursive structure makes them more efficient, in terms of the reduction of MSE for a given number of taps, than linear equalizers for impulse responses that are caused mainly by zeros in the channel response.

A2. With severe amplitude distortion, their noise enhancement and consequent total MSE are much less than those of a linear equalizer.

D. As we shall see in Chapter 10, at the present state of agorithm development, they are incompatible with convolutional coding and Viterbi decoding.

DFEs have not been used very much for systems with multipath interference because of their supposed inability to deal efficiently with the leading echoes caused by a non-minimum-phase fade [S&M, p. 62]. However, it seems that there is a misunderstanding here; a DFE with an SSTDL prefilter will perform slightly better than a linear TE if the fade is NMP and much better if it is MP. Since MPFs are more common than NMPFs, the net gain from using a DFE would be considerable. If noise enhancement is a problem, the extra storage and signal additions required by either an ACTDL or an AP prefilter would be well justified.

DFEs should also be used in medium-speed PSTN modems that do not use convolutional coding (e.g., a V.22 bis). If, as in a V.22 bis, decision-directed training with long-period random data is used, the main tap can be initialized to unity, and an AP can be used as the prefilter with no fears of instability. The improvement in SNR at the slicers may be as much as 2 dB for a severely attenuation-distorted (originate mode) upper channel.

9.1.9.1 *Decision-Feedback Equalization of Partial-Response Signals.* The

partial-response shaping that we will consider here may be produced by the transmit and receive filters or by the channel; in each case it will be assumed that precoding is used in the transmitter to prevent error propagation in the receiver.† The SNR penalty incurred by symbol-by-symbol detection will be less in the first case because of filtering of the noise (assumed white at the receiver input) by the receive filter; nevertheless, for most of the discussion in this section, we can treat the two cases together.

Bergmanns [Be4] derived expressions for the "optimum" prefilter and feedback FIR, but they are impractical; for the simple case of a back-to-back connection (i.e., no channel distortion) where one would expect that the prefilter should be transparent and the feedback coefficients should be zero, he showed that, ideally, the FIR should be infinite in length and have all coefficients equal to ± 1! Even if the FIR were truncated to some reasonable length, the error propagation effects would be unacceptable. We will restrict ourselves to the much simpler DSTDL/DFE and AP/DFE configurations that can be derived by straightforward adaptation, as described in Sections 9.1.2 and 9.1.6.

To assess the value of DFE for partial-response signals, three equalizers— linear, DSTDL/DFE, and AP/DFE (all with a total of 15 taps) were designed for a duobinary signal subjected to the same distortion and noise as considered previously— linear attenuation with 20 dB at $f_s/2$ and 20 dB SNR. The total RMSEs— with those for a full-response, $\alpha = 0.25$, signal included for comparison— were:

	RMSE		
	Linear	DSTDL/DFE	AP/DFE
Duobinary	0.186	0.161	0.165
Full-response	0.211	0.147	0.149

Thus, whereas a DFE achieves either 3.3 or 3.0 dB improvement over the linear equalizer for the full-response signal, it achieves only 1.4 or 1.0 dB improvement for the partial-response signal. Part of the reason for this is clear: the partial-response signal has less power at higher frequencies where the attenuation is greatest, so the linear equalizer does not try so hard in that region and therefore does not enhance the noise so much; on the other hand, it is not so

---

† If the shaping is caused by the channel, it must be anticipated in order to trigger the decision to precode.

clear why the DFE performs slightly worse for the duobinary than for the full-response signal.

Regardless of the reasons, however, it can be seen that most of the motivation for using a DFE — reduced noise enhancement — has been eliminated for the partial-response signal.

Furthermore, as we shall see in Chapter 10, it is fairly simple — and therefore advisable — to use an error detection and correction algorithm (EDCA) for partial response signals to retrieve the 2.1 or 3.0 dB (depending on whether the shaping is provided by the filters or the channel) of SNR that was lost in the shaping. This EDCA assumes uncorrelated noise samples, but if an error is made by the "raw" slicer used in the DFE, the subsequent feedback of the incorrectly weighted trailing echoes will mean that the "noise" seen by the EDCA may be strongly correlated, and the full potential gain will not be achieved.

The conclusion from this must be that decision-feedback equalization *should not be used* for partial-response signals.

9.2 ISI CANCELERS

As we have seen, a DFE can cancel (subtract off) the effects of trailing echos without *explicitly* enhancing the noise, but can do nothing about leading echoes. The much more complicated ISI canceler makes preliminary (tentative) decisions about the received data, delays the signal by a number of symbol periods equal to the number of leading echoes, uses the tentative decisions in a DFE that subtracts from the delayed signal the learned effects of both trailing and leading echoes, and then makes a second, final decision about the data. ISI cancelers have been described in detail in [M&S2] and [Qu6].

Gersho and Lim [G&L2] showed that for optimum performance, the canceler should be preceeded by an adaptive matched filter. Then if — and it is a very important "if" — the tentative decisions are correct, a canceler can achieve an MSE in the presence of noise that is equal to the minimum achieved by an undistorted signal. But that is a rather meaningless result, because if the tentative decisions were correct there would be no need for the rest of the canceler! However, as late as 1985 the "critical question . . . of the effect of tentative decision errors on the final error probability" [Qu6] had not been answered. In the absence of an encouraging answer — and admitting to an intuitive skepticism that there will ever be one — we will not consider ISI cancelers further.

9.3 MAXIMUM-LIKELIHOOD SEQUENCE ESTIMATORS

As we have seen, if there is severe attenuation distortion, a DFE — with a proper prefilter — can do a much better equalization job than a linear equalizer. Nevertheless, since a DFE essentially discards all trailing echoes by merely subtracting their effects from the signal to be sliced, it is apparent that it wastes some of the signal power; we will now consider some other, more complicated, methods that attempt to use all that power.

A theoretically optimum detection technique, which requires knowledge of the impulse response of the channel, is to wait until a complete data signal has been received, correlate the full signal with all possible transmitted messages, and select that message that yields the largest correlation. Such a procedure would be a true Maximum-Likelihood Sequence Estimation (MLSE). Unfortunately, the amount of computation required for this increases exponentially with the length of the message, and the method is impractical.

In the second half of this section we will consider a method, the Viterbi Algorithm (VA), of implementing MLSE that is only very slightly supoptimum, and for which the amount of computation is exponentially dependent on only the length of the *impulse response of the channel.* Nevertheless, for most channels this would still mean a *very* large (usually impractically so) amount of computation. Therefore, nearly all Viterbi detectors are preceded by a prefilter that shortens the impulse response of the channel to some manageable length; the most common lengths are three, four, and five terms, although detectors for as many as nine terms have been described [FMPC].

A generic MLSE is shown in Fig. 9.15. The data feedback to the prefilter (indicated by dashed line in Fig. 9.15) is not used for the symbol-by-symbol generation of the data signal with the Shortened Impulse Response (SIR), but it may be used in the much slower adaptation of the filter and the definition of the SIR itself.

Minimum Distance, and the Probability of Error in MLSE. Before we consider the design of prefilters and the implementation of MLSE by a VA, we must consider the optimum performance of an estimator and what it is that limits that performance.

For simple, symbol-by-symbol decisions, a detector must choose between the two nearest neighbors of the input sample, and the probability of error is the probability that the sample of the noise will exceed half the distance between these nearest neighbors. That is, the detector chooses that transmitted symbol that would have needed the least amount of noise added to it in order to produce the input sample and was therefore most likely to have resulted in the sample. If the signal is undistorted and has nominal levels of $\pm 1, \pm 3 \cdots \pm(2L - 1)$, the distance† between nearest neighbors is 2.

For a distorted signal, instead of considering each sample separately, an MLSE tries to choose the *sequence* of data symbols that was most likely to have resulted in the *sequence* of received samples; since it is assumed that the samples of the noise are uncorrelated, this chosen data sequence is again the one that would have needed the least total noise energy added to it in order to produce

---

† In this chapter and Chapter 10 we will conform to the terminology of coding theory and use the symbols, d and d_{min} to denote distances between two signal *points* or *sequences*. Previously (see, e.g., [LS&W]) d was used for the distance from a point to the nearest *threshold*.

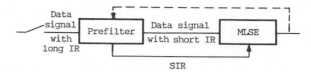

Figure 9.15 Generic MLSE with prefilter to shorten impulse response (IR).

the input sequence of signals. The algorithm will err if some sequence other than the transmitted one would apparently have needed less noise.

Let the transmitted and distorted signal (what would be received if there were no noise) at time kT be

$$x_k = \sum_{m=1}^{M} a_{k-m+1} h_m$$

and the received signal be

$$y_k = x_k + \mu_k$$

where the h_m are the samples of either the impulse response of the channel or the SIR of the channel plus prefilter as appropriate and the μ_k are samples of the noise with variance σ^2. Then the estimates $\hat{a}_k$ of each of the K symbols of the transmitted message are chosen to minimize the total squared Euclidean distance between the y_k and all the possible values of x_k.

This minimization can be thought of in geometric terms by defining the $(K + M)$-dimensional vectors **x** and **y**:

$$\mathbf{x} = \{x_1, x_2, \ldots, x_{K+M}\} \quad \text{and} \quad \mathbf{y} = \mathbf{x} + \mu$$

The vector **y** will be estimated incorrectly as **x**′ if the component of the noise in the direction of the vector $(\mathbf{x} - \mathbf{x}')$ is greater than $|\mathbf{x} - \mathbf{x}'|/2$. It is important to note that even though the variance of the total noise is $(K + M)\sigma^2$, the variance in any one direction in the $(K + M)$-dimensional space—and particularly in the direction of $(\mathbf{x} - \mathbf{x}')$—is only σ^2.

It is possible, of course, for the noise to cross the threshold in the direction of any $(\mathbf{x} - \mathbf{x}')$, but the error rate will usually depend mainly on the number of times that it crosses the nearest threshold(s)—that is, on the ratio of σ to $d_{\min}$, where the latter is defined by

$d_{\min}$ = the minimum over all possible sequences a_k and a'_k of $|x_k - x'_k|$

$$= \min \sum_{k=1}^{K+M} \left| \sum_{m=1}^{M} (a_{k-m+1} - a'_{k-m+1}) h_m \right|$$

Then the asymptotic probability of symbol error (with perfect MLSE) is

$$\mathscr{P}_e = K_1 \, Q \left[\frac{d_{\min}}{2\sigma} \right]$$

where K_1, the *error coefficient,* may be considerably greater than the $2(1 - 1/L)$ that we calculated in Chapter 3 for symbol-by-symbol detection of multilevel signals.

Calculation of the minimum distance in general for any given impulse response is a complicated task that requires searching over all possible sequences of the same length as the impulse response [A&F]. The search is best organized by considering in turn two data sequences that differ in one, two, three, and so on symbols; it often becomes clear that the search can be terminated after only a few sequences have been examined.

If only one symbol of a data sequence is changed into one of its nearest neighbors (i.e., one of the standard set of symbols is increased or reduced by 2), the mth sample of the input signal will be changed by $\pm 2h_m$. Then the squared difference (the squared Euclidean distance) between the two sequences will be

$$d_1^2 = 4 \sum_{m=1}^{M} h_m^2 = 4r_0 \tag{9.21}$$

where r_0 is the first term of the auto-correlation function of $H(z)$.

For sequences that differ in two or more symbols, we will only quote the results and refer the reader to [Fo2] for details. For a difference in two symbols that are separated by j symbols

$$d_2^2 = 4(2r_0 + 2a_j r_j) \tag{9.22}$$

and for a difference in three symbols that are separated by j and k symbols

$$d_3^2 = 4(3r_0 + 2a_j r_j + 2a_k r_k + 2a_j a_k r_{j+k}) \tag{9.23}$$

where $a_j, a_k = \pm 1$, and, as defined previously,

$$r_j = \sum_{m=1}^{M} h_m h_{m+j} \tag{9.24}$$

The minimum distance $d_{\min}$ is then the smallest of all the d_m for $m = 1$ to M and for all possible combinations of as; however, it is seldom necessary to calculate beyond d_3.

For signals with only moderate distortion, d_1 is the smallest, and $d_{\min} = 4r_0$; for these, a VA is able to use all the power in the signal delivered to it. Such a signal will be called *compact.*

For low-pass channels that have high attenuation at the band edge, the smallest d_2 in (9.22) is usually given by $j = 1$ and $a_1 = -1$. Therefore, for these channels, a compact impulse response is defined by

$$r_0 < 2(r_0 - r_1) \qquad \text{or} \qquad r_0 > 2r_1$$

The best known impulse response of this type that is just compact is that of an undistorted PR Class I signal, . . . 0, 1, 1, 0 . . .; this offers a potential gain of 3 dB of a VA over a DFE detector.

Similarly, for channels that have high attenuation at both d.c. and $f_s/2$, d_2 is usually minimized by $j = 2$ and $a_2 = 1$. Then a compact impulse response is defined by

$$r_0 < 2(r_0 + r_1) \quad \text{or} \quad r_0 > -2r_1$$

A PR Class IV impulse response, . . . 0, 1, 0, −1, 0 . . . , is the best known example of this limiting case.

For more severely distorted channels — low, high, or bandpass — the gain of a VA over a DFE may be much greater than 3 dB.

9.3.1 Prefilters for a VA

At the beginning of this chapter we argued that the functions that a matched filter performs (matching the attenuation and equalizing the delay in the transition band) become much less important when the equalizer is nonlinear and when, as in all high-speed modems, the excess bandwidth becomes small. This argument becomes even more convincing for the sorts of channels for which MLSE is used. The attenuation near the bandedge is typically very high, and a matched filter, operating at or above the Nyquist rate of the channel, performs no useful purpose; immediate sampling with subsequent filtering at the symbol rate is completely adequate.

Methods of designing prefilters can be divided into two classes:

1. Those (e.g., [Q&N] and [Be3]) that first calculate some Desired Impulse Response (DIR) and then adaptively design a filter that transforms the channel impulse response to that DIR. In order to reduce the amount of "real-time" computation that must be done during data transmission, the DIR is usually calculated only during the early stages of the training sequence and then held fixed for the remainder of the message, or until another training signal is sent. We will call these *DIR methods*.

2. Those (e.g., [F&M1], [Me2], and later in this section) that do not calculate a desired impulse response, but only generate a *shortened* IR (SIR) in the course of adapting the prefilter according to some LMS criterion. We will call these *SIR methods*.

In general, DIR methods are less attractive because (1) they require more hardware or DSP program code and (2) the DIR may become significantly suboptimum if either the noise level or the channel impulse response changes during transmission. We will therefore concentrate on the SIR methods. However, as we shall see in Section 9.3.3, the second part of the DIR two-stage procedure—transforming to a predefined DIR—is needed in some special-purpose receivers.

It might seem that a modified XXX/DFE, with its highly desirable property of not enhancing the noise, could be used first to eliminate the leading echoes and then to subtract off the effects of all trailing echoes beyond a certain point; variations on this approach have been described in [Qu1] and [L&H]. However, all MLSEs in general, and the Viterbi algorithm in particular, involve a considerable delay before the received data is known with any certainty, and the decisions are not available soon enough to be used by the DFE to cancel the trailing echoes immediately following the SIR. Consequently, nearly all pre-filters that have been used for MLSE have been linear,† and we will confine our attention to these.

Performance Objectives for a Prefilter. The theory of operation of a VA is based on the assumption of uncorrelated (spectrally white) noise. Both of the two-stage, DIR, methods of [Q&N] and [Be3] explicitly recognize this and strive, in different ways, to match the amplitude spectrum of the DIR to that of the input signal; they are thus trying to make the prefilter as nearly all-pass as possible and thereby to maintain the presumed whiteness of the added noise. Having chosen this DIR, they then try to generate it with the minimum MSE.

On the other hand, the fully adaptive, SIR, methods do not predefine a DIR that matches the input impulse response in any way, but use a prefilter (either a DSTDL or an AP/SSTDL) to minimize the MSE subject to a constraint. Since the amount of noise enhancement usually increases with the amount of amplitude equalization, minimizing the MSE and its main component, the enhanced noise, has the indirect effect of making the spectra of the impulse response and SIR similar. There are four possible forms of the constraint; in order of increasing sophistication and complexity;

1. Fix the first term (or, indeed, any one of the terms) of the DIR to be unity [Me2]

2. Fix the total energy in the DIR to be unity [F&M1]. This has the computational disadvantage that it requires a division—a tedious operation in most digital signal processors.

3. Fix the minimum distance of the DIR at some predefined value. If the signal is only moderately distorted, so that the impulse response is compact and this value can be 2.0, then this constraint is equivalent to the second one, and both will give a better result than the first constraint. However, as will be seen in the second example at the end of this section, for more severely distorted signals, the differences between this and the first two constraints may be consid-

† Clark and colleagues [Cl1], [CH&D], [CK&H], [CL&M], and [Cl2] have described many hybrid detection schemes, but it is difficult to categorize and evaluate them.

erable. If the general shape of the attenuation distortion can be predicted, the calculation of the minimum distance of any given DIR can be greatly simplified, and this form of constraint may become as feasible as the second one; it would certainly be preferable to either of the first two.

4. Recognizing that the prefilter cannot be all-pass, and that its output noise must therefore be correlated, calculate the variance of the noise in the direction of each error sequence, and maximize the minimum ratio of the distance between the sequences and the length of the noise vector between them.†
Application of this form of constraint is so complicated that it is useful only as a method of calculating an upper bound on performance.

Structure of the Prefilter. Messerschmitt [Me2] showed that if the impulse response is to be shortened to, say, NS terms, then the optimum prefilter is a matched filter followed by a DSTDL that has the $(NS - 1)$ taps after the main tap fixed at zero. If the main tap, c_{NM}, is fixed at unity, the transfer function of this TDL can be written as a sum:

$$C(z) = C_1(z) + C_2(z) \tag{9.25a}$$

where

$$C_1(z) = \sum_{n=1}^{NL} c_n z^{-n+1} + z^{-NL} \tag{9.25b}$$

$$C_2(z) = \sum_{n=NM+NS}^{NTOT} c_n z^{1-n} \tag{9.25c}$$

$$NL = \text{the number of leading taps}$$
$$NS = \text{the number of terms in the SIR}$$
$$NT = \text{the number of trailing taps}$$
and
$$NTOT = NL + NS + NT$$

The functions C_1 and C_2 are sometimes called the *anticausal* and *causal* parts of C, but both of them are, of course, necessarily causal; we will eschew this terminology.

If we dispense with the matched filter, then, just as for the DFE case, the transfer function that would have been implemented by the combination of the matched filter and the TDL must now be implemented after the sampling. If, as before, the channel impulse response is defined as

$$H(z) = h_1 + h_2 z^{-1} + \cdots + h_M z^{-M+1}$$

then the transfer function of the MF/TDL combination is $z^{1-M} H(z^{-1}) C(z)$, and this could be approximated by a single DSTDL (now with no zero taps).

Since the prefilter can no longer be purely all-pass (it must transform the

---

† This is an inadequate explanation of a very complex process; the reader is referred to [Q&N] for details.

input amplitude response to that of an SIR), the best setting of its coefficients will, regardless of its structure, be somewhat dependent on the input noise level. Consequently, its performance will become worse than optimum when the noise level changes.

Nevertheless, if the channel impulse response has some leading terms greater than about 0.4 (relative to a main term of 1.0), then, just as for a DFE, it will be better to deal with these by using an all-pass, and then follow this AP with a TDL to deal with only the trailing terms beyond the range of the SIR. The form of this AP/SSTDL combination can be found by arranging the transfer function of the MF/TDL:

$$z^{1-M}H(z^{-1})\,C(z) = z^{1-M}\,H(z^{-1})\,C_1(z)\left[1 + \frac{C_2(z)}{C_1(z)}\right] \qquad (9.26)$$

The first factor can be well approximated by an all-pass, and the second (unstable) factor, by an SSTDL with taps 2 to NS fixed at zero. The output impulse response of the AP has only very small leading terms, so the coefficients of the SSTDL have very little effect on the terms of the SIR; therefore, it seems to be best to fix the first tap of the SSTDL at unity and learn the whole SIR, rather than to try to constrain the SIR in some way. Thus the transfer function of the complete prefilter is

$$F(z) = \left[\frac{c_1 + \cdots + c_{NL}z^{1-NL} + z^{-NL}}{1 + c_{NL}z^{-1} + \cdots + c_1 z^{-NL}}\right]$$

$$[1 + c_{NM+NS}z^{-1} + \cdots + c_{NTOT}z^{-NT}] \quad (9.27)$$

The AP should be implemented as shown in Fig. 9.11. Its output storage register (extended a little if $NT > NL$) can be used as the input register for the SSTDL, and the whole prefilter becomes as shown in Fig. 9.16. The output and the error are calculated from

$$v_1 = y_{NM} + \sum_{n=1}^{NL} c_n(y_n - v_{NL+2-n}) \qquad (9.28a)$$

$$w = v_1 + \sum_{n=NM+NS}^{NTOT} c_n v_{n-NL} \qquad (9.28b)$$

and $$\epsilon = w - \sum_{n=1}^{NS} \hat{a}_n hs_n \qquad (9.28c)$$

where the SIR is

$$HS(z) = hs_1 + hs_2 z^{-1} + \cdots hs_{NS}z^{-NS+1}$$

The output of the prefilter is input to the VA, and also to the updating algorithms for the taps and the terms of the SIR; a complete receiver is shown in

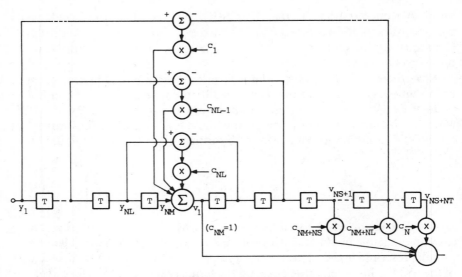

Figure 9.16 AP/SSTDL prefilter for MLSE.

Fig. 9.17. The received data $\hat{a}_n$ that enters the register of length NS may come from one of three sources — a local reference during training, a simple decision device (slicer), or the VA; the delay shown in the dotted box is used only to compensate for the delay through the VA.

Adaptation of the AP/SSTDL. The taps of the all-pass should be incremented according to (9.17), and just as for the DFE, a good argument can be made that this is truly a stochastic gradient update. However, the adaptation of the SSTDL taps is not so straightforward because they must follow those of the AP. (The two transfer functions are multiplicative rather than the simpler additive.) The simplest strategy is to deal with the trailing taps as though there were no all-pass preceding them and update them according to

$$c_n(k+1) = c_n(k) + \Delta_{ct}\,\epsilon(k)\,y_{n-NL}(k) \qquad \text{for } n = (NM + NS) \text{ to } N$$

Since the SSTDL taps operate only on single signals (rather than the difference of two), the step sizes recommended in (9.14) need not be reduced by the factor of $\sqrt{2}$.

Finally, the terms of the SIR should be updated as shown in (9.2) for the feedback terms of a DFE.

Ensured Convergence? As we said when discussing the AP/DFE, the fact that the AP is a recursive structure does not, contrary to some expectations, seem to result in instability or convergence to one of many local minima. This is because the poles of the transfer function are not independent, but are constrained to be the reflections (in the unit circle) of the zeros. However, for severely distorted channels (the ones for which MLSE is most likely to be used), the coupling between the AP and the SSTDL sometimes causes the MSE to diverge temporarily; that is, the convergence is not always monotonic.

More work is needed on the best way to ensure fast monotonic convergence; the results for the examples given in Section 9.3.1.3 indicate that the work will be well worthwhile.

9.3.1.1 Prefilters for QAM Systems.
Falconer and Magee [F&M2] pointed out that the implementation of MLSE for a QAM system is greatly simplified — the number of multiplications is approximately halved — if the SIR is made real. This cannot be achieved by the structure shown in Fig. 9.16; it can be appreciated that some imaginary taps would be needed in the gap in order to zero the imaginary part of the SIR. Therefore, a full DSTDL, constrained according to either [Me2] or [F&M1], would be needed.

However, forcing the SIR to be real means equalizing the input amplitude spectrum to be symmetrical about the carrier frequency. For the very skewed spectra often found on the PSTN, this may require a lot of amplitude equalization, and consequent noise enhancement. Careful calculation of the trade-offs in complexity and performance between short complex SIRs and longer real ones would be needed.

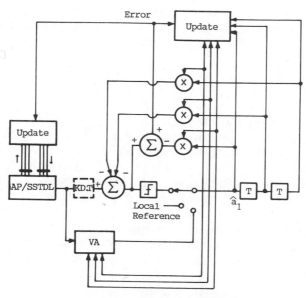

Figure 9.17 Complete MLSE receiver.

TABLE 9.4 Equivalent RMSE and Effective SNR after Transformation of Proakis's Five-Term Impulse Response to a Three-Term SIR (input SNR = 20 dB)

	Equivalent RMSE	Effective SNR (dB)
Input 5-term IR	0.1	15.0
3-term SIR from		
DSTDL trained at ∞ dB SNR	0.328	9.7
DSTDL trained at 40 dB SNR	0.317	10.0
DSTDL trained at 20 dB SNR	0.241	12.4
AP/SSTDL trained at ∞ dB SNR	0.235	12.6
AP/DFE trained at ∞ dB SNR	0.285	10.9

9.3.1.2 A Simple Academic Example for Familiarization.
Proakis [Pr2] considered a linear-phase channel with

$$H(z) = (1 + z^{-1} + z^{-2})^2 = 1 + 2z^{-1} + 3z^{-2} + 2z^{-3} + z^{-4}$$

which, when normalized to unit energy, is

$$0.2294 + 0.4588z^{-1} + 0.6882z^{-2} + 0.4588z^{-3} + 0.2294z^{-4}$$

This has severe attenuation distortion; the pulse is far from compact, and $d_{min} = 1.124$. The impulse response is already short, but let us consider the problem of transforming it to an SIR with just three terms.

Four different prefilters were designed: DSTDLs with input SNRs of ∞, 40, and 20 dB, and an AP/SSTDL with an SNR of ∞.† The equivalent RMSE when working with an SNR of 20 dB is defined by

$$\text{ERMSE} = \frac{2[\text{MSISI} + 0.01 \, \Sigma c^2]^{1/2}}{d_{min}}$$

where the summation of the squared tap coefficients is performed over all the taps of a DSTDL, but only over the (trailing) taps of the AP/SSTDL. This ERMSE was calculated for all four prefilters and, as an interesting comparison, for an AP/DFE as described in Section 9.1; the results are shown in Table 9.4.

It can be seen that the DSTDL trained at its operating noise level is almost as good as the AP/SSTDL, but the other two, trained in low and zero noise, are 2.6 and 2.9 dB worse. It should also be noted that there is a penalty of at least 2.4 dB to be paid for shortening the impulse response.

The DFE appears to be better than two of the DSTDLs, but this is misleading; the output impulse response has second and third terms that are bigger than

---

† AP/SSTDLs were not designed for any finite SNRs; when operating at 20 dB SNR, they would be better than the zero-noise one, but probably not significantly so.

the first, and error propagation would be severe; a DFE would not be practical for such a channel.

9.3.1.3 A Subscriber Loop IR at 160 kbit/s.

Richards [Ri2] calculated the impulse response of a worst-case subscriber loop (approximately 4 km of twisted-pair cable with two bridged taps) sampled at 160 kbit/s. The simulated channel was unrealistic because it did not contain a blocking transformer; however, the low-frequency portion of the spectrum that would be removed by a transformer is very narrow compared to 80 kHz, and if dealt with properly, it should not significantly degrade the noise and error performance.

Implementation of a VA at 160 kbit/s is probably beyond the present state of the DSP art, but such an extreme example provides a severe test of this new method of designing a prefilter and also establishes a benchmark of performance for other presently possible detection methods (see [B . . 2] and Section 10.3).

The amplitude response of the channel is shown in Fig. 9.18; its impulse response (probably minimum-phase) can be reasonably truncated to 0.35, 0.41, 1.0, 0.99, 0.93, 0.80, 0.70, 0.51, 0.43, 0.37, 0.32, 0.27, 0.23, 0.19, 0.16, 0.14, 0.12, 0.08, 0.06, 0.04, 0.03, 0.02, and 0.01. The pulse is extremely dispersed (i.e., far from compact), and if its energy r_0 is normalized to unity, then $d_{min} = 0.626$. Thus there is a 10.1 dB loss in the effective input SNR.

Since the pulse has very few leading echoes, an asymmetrical configuration of five leading and twelve trailing taps was chosen for both the DSTDL and AP/SSTDL prefilters. In order to keep the amount of processing in the VA in the realm of remote possibility, the SIR lengths were restricted to three and

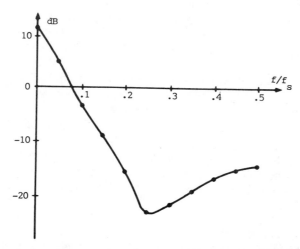

Figure 9.18 Amplitude response of a subscriber loop with bridged taps.

TABLE 9.5   Equivalent RMSE and Effective SNR After Transformation of a Subscriber loop Impulse Response to Four- and Three-term SIRs (input SNR = 26 dB)

	Training SNR (dB)	Equivalent RMSE	Effective SNR (dB)
Input 25-term IR		0.16	15.9
4-term SIR from			
DSTDL	∞	0.298	10.5
DSTDL	40	0.192	14.3
DSTDL	26	0.201	13.9
AP/SSTDL	∞	0.201	13.9
AP/SSTDL	40	0.201	13.9
AP/SSTDL	26	0.197	14.1
3-term SIR from			
DSTDL	∞	0.249	12.1
DSTDL	40	0.23	12.8
DSTDL	26	0.206	13.7
AP/SSTDL	∞	0.217	13.3
AP/SSTDL	26	0.210	13.6

four.† Both sets of prefilters were designed for input SNRs of ∞, 40, and 26 dB (about the minimum that could be tolerated on such a channel). The equivalent RMSE and effective SNR with an input SNR of 26 dB were calculated for each filter; the results are shown in Table 9.5.

The most important result is that for the four-term SIR, the DSTDL trained without noise is 3.4 dB worse when used with an SNR of 26 dB than the DSTDL trained at 26 dB, *but* the AP/SSTDL is only 0.2 dB worse. For the three-term SIR, the results are not so dramatic, but still significant.

A very strange, subsidiary result is that the DSTDL trained at 40 dB SNR appears to be better at 26 dB than the one trained at that level. Yet the algorithm is supposed to minimize the MSE at each particular noise level! A partial explanation of this paradox is that the DSTDLs were designed according to the simplest constraint, in which the first term of the SIR is fixed [Me2], but the RMSEs were calculated relative to the more meaningful $d_{min}/2$.‡ There is no apparent reason why the simple constraint should work better at the low noise level, and this may be just a fortunate coincidence. The only conclusion that can be readily drawn from this is that there may be substantial differences between the performances of DSTDL prefilters designed according to the various constraints.

The AP/SSTDL prefilters do not show this anomalous behavior; the differences between the noisy and noiseless RMSEs are less than 0.2 dB regardless of whether they are related to the first term of the SIR, $\sqrt{r_0}$, or $d_{min}/2$.

---

† The improvement in effective SNR from going to five terms would be only about 0.2 dB.
‡ The algorithm is doing what it is supposed to do because if the RMSEs relative to the first term are compared, the "noiseless" DSTDL is 2.7 dB worse than the noisy one.

The amplitude responses of the prefilters for the four-term SIR are shown in Fig. 9.19. The SSTDLs appear to be trying to match the high attenuation at the band edge, but the noiseless DSTDL seems to be equalizing it. This difference in approaches would become even more important if the noise were not white but were greater at high frequencies — as is the case in subscriber loops where the noise consists mainly of near-end crosstalk (see Section 11.3).

9.3.1.4 *Accelerating the Convergence of a VA Prefilter.* The convergence of the tap coefficients of the AP/SSTDL for the subscriber loop example was extremely slow, and indeed it would be for all the severely distorted channels for which a VA is typically needed. Application of any of the adaptive SSO techniques to the AP/SSTDL would be fraught with difficulties and uncertainties and is probably not worth trying. Two other approaches to fast adaptation have been suggested:

1. If, as for the subscriber loop, the general shape of the attenuation is predictable, then the fixed SSO method described in Section 8.4.1 might be used. Simulations have been run with the orthogonalizing algorithm — assuming a PR Class I response — applied only to the taps of the AP; the convergence was accelerated by a factor of about 3. Further possibilities for investigation are (1) extending the orthogonalization to the taps of the SSTDL also and (2) orthogonalizing for the generalized partial responses (impulse responses of . . . 0, 1, 1, 1, 1, 0 . . . , etc.) discussed in Section 10.3.

2. For HF radio signals that are subjected to severe, rapidly changing fades, any prefiltering that depends on a matrix inversion is probably impractical.

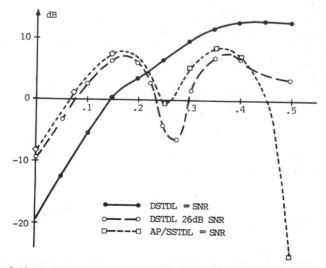

Figure 9.19 Amplitude responses of three MLSE prefilters for a subscriber loop.

Frenette and colleagues [FMPC] suggested biting the bullet and implementing a VA for the full length (nine terms) of the input impulse response—the rationale being, of course, that channel estimation can be much faster than any linear modification of the channel.

9.3.2 The Viterbi Algorithm

The VA was originally invented to decode convolutional codes [Vi1], [Fo1], and [Vi2]. In a seminal paper, Forney [Fo2] showed how it can also be used to perform only slightly suboptimum MLSE of distorted data signals, and it has been described many times since ([Ha2], [Un2], [HC&R], etc.). The description here is mainly based on that given by Hayes [Ha2] and summarized later in [HC&R].

The theoretical basis of all MLSE methods is that they estimate the set of transmitted data as that that maximizes the a priori probability of receiving the set of samples that was actually received. If the added noise is assumed to be white (uncorrelated, memoryless), this is equivalent to minimizing the amount of noise that would have to have been added to the transmitted and distorted signal in order to produce the received signal. This, in turn, is equivalent to minimizing the total squared distance between the received set of samples and the hypothesized transmitted and distorted set.

As usual, the received sample at time kT is

$$y_k = \sum_{m=1}^{M} a_{k-m+1} \, h_m + \mu_k \tag{9.29}$$

where the h_m are the samples of either the impulse response of the channel or the SIR of the channel plus prefilter as appropriate.† Then the estimates $\hat{a}_k$ of each of the K symbols of a transmitted message are chosen to minimize the total squared Euclidean distance between the samples of the received signal and all possible noiseless received (transmitted and distorted) signals,

$$\sum_{k=1}^{K+M-1} \left[y_k - \sum_{m=1}^{M} a_{k-m+1} h_m \right]^2 \tag{9.30}$$

This function includes a term Σy_k^2 that is the total power in the received signal and is not dependent on the choice of receive data; it can therefore be discarded, and the function to be minimized can be defined as

$$D_K = -2 \sum_{k=1}^{K+M-1} \left[y_k \sum_{m=1}^{M} a_{k-m+1} h_m \right] + \sum_{k=1}^{K+M-1} \left[\sum_{m=1}^{M} a_{k-m+1} h_m \right]^2 \tag{9.31}$$

---

† There is no suggestion here that a matched filter has been used, so there is no requirement that the h_m be symmetrical about some delayed time.

It is important to note that the aim here is to minimize D_K by choosing a set of a_i for $i = 1$ to K. Therefore, it is useful to rewrite (9.31) as

$$D_K = -2 \sum_{k=1}^{K+M-1} \left[y_k \sum_{i=1}^{K} a_i h_{k-i+1} \right] + \sum_{k=1}^{K+M-1} \sum_{i=1}^{M} \sum_{j=1}^{M} a_{k-i+1} a_{k-j+1} h_i h_j$$

and then interchange the orders of summation:

$$D_K = -2 \sum_{i=1}^{K} a_i \left[\sum_{k=1}^{K+M-1} y_k h_{k-i+1} \right] + \sum_{i=1}^{K} \sum_{j=1}^{K} a_i a_j \left[\sum_{k=1}^{K+M-1} h_{k-i+1} h_{k-j+1} \right] \tag{9.32}$$

Since there are only M terms in the impulse response, the range of k in the two summations in (9.32) can be reduced to $k = i$ to $(i + M - 1)$. Then it can be seen that the first bracketed term is the output at time iT (defined as z_i) of a noncausal matched filter with input y_i, and the second is r_{i-j},† one of the terms of the auto-correlation function of the impulse response. Therefore

$$D_K = -2 \sum_{i=1}^{K} a_i z_i + \sum_{i=1}^{K} \sum_{j=1}^{K} a_i a_j r_{i-j} \tag{9.33}$$

If the a_k are chosen from the set $\pm 1, \pm 3 \cdots \pm (L - 1)$, then a brute force approach to this minimization would require comparing L^K possible received sets.

On the other hand, at each symbol time the set of noiseless distorted signals comprises only L^M points, and the VA minimizes the final D_K by minimizing the intermediate, cumulative values of D_k (for $k = 1$ to K) and then invoking the principle of dynamic programming [Be1], [Dr], and [Ha2]—namely, that the optimal path (the one with the smallest cumulative value of D) to any point must include the optimal path to all previous points in that path.

These optimal paths can be constructed by first working backward from D_K to find the accumulation formula and then applying this formula from D_1 forward. Then D_K can be written as

$$D_K = D_{K-1} + \delta D_K \tag{9.34}$$

where D_{K-1} is the function defined in (9.33) with K replaced by $(K - 1)$ and, after a little algebraic manipulation,

$$\delta D_K = -2a_K z_K + a_K^2 r_0 + 2a_K \sum_{i=1}^{M-1} a_{K-i} r_i \tag{9.35}$$

---

† Note that $r_{i-j} = 0$ for $|i - j| \geq M$; this differs from the notation in [Ha2], where $m \, (=M - 1)$ is the number of extra (interfering) terms in the impulse response.

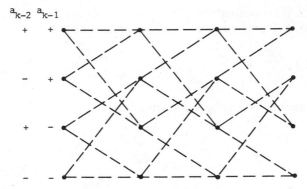

Figure 9.20 Trellis diagram showing possible transitions for $L = 2$ and $M = 3$.

This process can be continued backward from KT so that a general D_k (often called the *state metric* at time kT) is the sum of D_{k-1}, the previous state metric, and δD_k (often called the *transition metric*). It can be seen that δD_k depends on only the most recent output of the matched filter and $a_{k-M+1} \cdots a_k$.

The bookkeeping involved in constructing the optimal path is best illustrated by a trellis diagram. At each time kT a state vector $\{a_{k-M+1} \cdots a_k\}$, where $a_j = \pm 1 \cdots \pm(L-1)$, *could* be used to define L^M states, and each state would then uniquely define a possible level of the noiseless distorted signal. However, in practice it is more economical of storage and computation† to use a shorter vector, $s_k = \{a_{k-M+1} \cdots a_{k-1}\}$, to define just L^{M-1} states, so that each level is now defined by a pair of vectors, s_k and s_{k+1}. All the possible paths from state to state can then be drawn by connecting each point in the kth set to its L possible predecessors and successors.

With these definitions established, the use of the principles of dynamic programming in the VA can be best explained by a specific example. From here on a minimum of mathematics will be used, so that a designer may understand the general principles and then apply them in a specific medium (microprocessor program, dedicated DSP hardware, switched-capacitors, etc.) with the appropriate methods of defining and recording the variables and paths.

Consider a binary signal that is distorted so that the impulse response has three terms, h_1, h_2, and h_3 (viz. $L = 2$ and $M = 3$). The four possible states at each time can be ordered as shown in Fig. 9.20 and then connected by dashed lines to show all possible paths. The algorithm is then:

VA1. For each point (state) in the kth set, calculate the output of the matched filter as

$$z_k = h_3 y_k + h_2 y_{k-1} + h_1 y_{k-2} \tag{9.36}$$

† It also agrees with the notation developed in the theory of finite-state machines (see [Fo2]).

This filter is, of course, causal, and will introduce a minimum delay of $2T$ into the detection process.

VA2. Calculate two values of D_k as the sums of the D_{k-1} of the two tentatively connected points in the $(k-1)$th set and the δD_k associated with the two transitions. For the case of binary data, the second term on the RHS of (9.35), $a_k^2 r_0$, is a constant for all states and all paths, and so can be ignored. Then a factor of 2 can also be divided out and discarded, so that a simpler form of the transition metric given in [FMPC] can be used:

$$\delta D_k = a_k \left[-z_k + \sum_{i=1}^{M-1} a_{k-i} r_i \right] \tag{9.37}$$

Alternatively, if (9.37) is rearranged as

$$\delta D_k = a_k \sum_{i=1}^{M-1} a_{k-i} r_i - a_k z_k$$

it can be seen that the first term is constant for each of the eight possible transition paths; computation might be simplified by precalculating and storing these terms.

VA3. Select the smaller of each of the pairs of D_ks, and draw the appropriate link.

VA4. If the four paths that terminate in the points of the kth set all emanate from a single point in a previous set, it is certain that no subsequently received signals can change the estimate that that point (and all points on the path leading to it) lies on the optimal path. Therefore, at each symbol time, check whether the remaining four paths can all be traced back to a new (i.e., not previously identified) single point or series of points. These "merge" points and the path they build up define the best sequence of receive data.

The eight partially calculated transition metrics are shown in Fig. 9.21 for the particular case of $h_1 = 1.0$, $h_2 = 0.9$, and $h_3 = 0.5$, and the first 12 stages of the algorithm are illustrated in Fig. 9.22 with the correct path accentuated. Several points of comment and explanation may be helpful:

1. The paths were all started from the correct $(1,1)$ point with the assumption that a known training pattern had previously been transmitted. If this is not possible, the output of the slicer shown in Fig. 9.17 can be used, and any small errors in the starting point will be quickly cleared out.
2. At $4T$ the noise sample of -1.1 would have been large enough to cause an error in a DFE, with subsequent error propagation for three symbols.
3. At $4T$ all paths come from a single state $(1,1)$, at $2T$; therefore, a_0 could be decided unequivocally at that time.

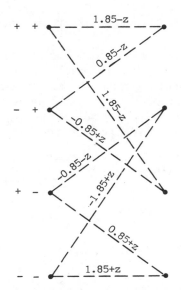

Figure 9.21 Partially calculated transition metrics for $H(z) = 1.0, 0.9, 0.5$.

4. At $6T$ all paths come from a single state at $3T$; therefore, a_1 could be decided then.
5. At $7T$ one can decide both a_2 and a_3.
6. Then one has to wait till $12T$, when a_4 to a_8 become available.
7. The "random" data and noise were chosen so as to keep the decision

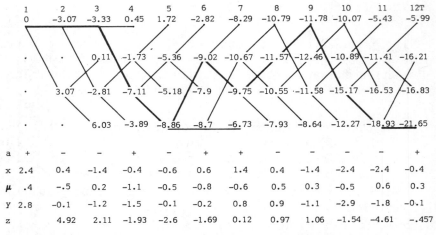

1	2	3	4	5	6	7	8	9	10	11	12T
0	−3.07	−3.33	0.45	1.72	−2.82	−8.29	−10.79	−11.78	−10.07	−5.43	−5.99
		0.11	−1.73	−5.36	−9.02	−10.67	−11.57	−12.46	−10.89	−11.41	−16.21
	3.07	−2.81	−7.11	−5.18	−7.9	−9.75	−10.55	−11.58	−15.17	−16.53	−16.83
		6.03	−3.89	−8.86	−8.7	−6.73	−7.93	−8.64	−12.27	−18.93	−21.65

a	+	−	−	+	−	+	+	−	−	−	−	+
x	2.4	0.4	−1.4	−0.4	−0.6	0.6	1.4	0.4	−1.4	−2.4	−2.4	−0.4
μ	.4	−.5	0.2	−1.1	−0.5	−0.8	−0.6	0.5	0.3	−0.5	0.6	0.3
y	2.8	−0.1	−1.2	−1.5	−0.1	−0.2	0.8	0.9	−1.1	−2.9	−1.8	−0.1
z		4.92	2.11	−1.93	−2.6	−1.69	0.12	0.97	1.06	−1.54	−4.61	−.457

Figure 9.22 Paths through the trellis for a particular sequence of data and noise.

delays of the trellis fairly short and therefore easily displayable. In practice, they might be considerably longer than the eight symbol periods shown.

It is apparent that the algorithm, defined this way, has a variable delay; at some symbol times no new data may be generated, and then there may be a burst of data. This could be handled by a long receive data buffer, but it is easier to use the trellis itself as the buffer and to fix the delay at some number KD of symbols. Then the last part of the algorithm might be restated as:

Amended VA4. At time k, trace the paths to all four points back to time $(k - KD)$. If they all pass through a single point at that time (i.e., they "merge"), choose a_{k-KD} accordingly; if they do not (i.e., there are "multiple survivors"), choose the path that begins, at time k, at the point with the smallest D_k and trace it back.

This procedure is useful during the development and simulation of the algorithm because it provides a measure of how often the performance is inhibited by the fixed delay, but in an actual modem there is no point in tracing back from all the points; it is sufficient to trace back only from the point with the smallest D_k.

Reemergence of the Matched Filter? It was said at the beginning of this section that we would not consider using a matched filter as an input filter for the VA and would be concerned only with the sampled, and probably shortened, impulse response. Yet a matched filter has reappeared, phoenix-like, in (9.32)! However, this filter is not performing any of the classical functions of an unsampled matched filter (delay equalization, noise minimization in the excess band); it is not protecting the integrity of the sampling process, nor is it ensuring that the samples form a set of "sufficient statistics." It is being used only as a computational device for processing the already sampled and shortened impulse response.

Choice of the Delay Through a VA. The amount of storage required for implementation of a VA is proportional to the chosen delay, so it is desirable to keep this as small as is feasible. The delay should be chosen so that the probability of multiple survivors is small, and the probability that the smallest D_k will lead back to the wrong survivor (one that would have been bypassed if a greater delay had been used), becomes insignificant.†

---

† Note that in the example in Fig. 9.22, even though one has to wait until $12T$ for there to be only one survivor at $6T$, with the algorithm that traces back from the smallest D_k, one has to wait only until $10T$.

The required delay depends on:

1. *The number of SIR terms M.*

2. *The magnitude of those terms.* For partial-response signals (see Section 9.3.3) and severely distorted signals for which the impulse response has been shortened by a prefilter, the ISI terms are comparable in magnitude to the main term, and many branches have equal or nearly equal signals assigned to them; consequently, the paths through the trellis take longer to merge, and for binary signals, a delay as large as $5M$ may be needed. Conversely, when using the full, unconcentrated, impulse response with small tails as in [FMPC], $3M$ will probably be adequate.

3. *The number of levels.* This dependence is because all paths tend to merge after the reception of a few outer levels — as they did in the, admittedly contrived, example in Fig. 9.22 — and these outer levels occur less frequently as the number of levels is increased. For multilevel signals, the previous delays should be increased to $8M$ and $5M$, respectively.

With these delays, the performance of the VA is usually within a few tenths of a decibel (SNR needed for a given error rate) of the optimum calculated from d_{min}.

Implementation of a VA. It is difficult to say anything general about this; everything will depend so much on the data speed and the technology used. Examples of microprocessor and dedicated DSP hardware implementations have been described in [C . . 1] and [R . . .] and in [G&S] and [FMPC], respectively.

Avoidance of Buildup in D_k. With the algorithm as stated so far the eight stored values of D_k will continue to grow and would eventually overflow (in the negative direction) any storage medium. This must be prevented by either periodically subtracting the smallest D_k from all the others and then resetting it to zero, or else subtracting the smallest transition metric from all the others before adding them to the stored D_ks; one example of this was described in [FMPC].

9.3.3 Maximum-Likelihood Detection of Partial-Response Signals

The partial-response signals we shall consider in this section include the well-known duobinary and modified duobinary and the most interesting examples of polybinary and polybipolar. The impulse responses of these are

Duobinary (PR Class I)	$1 + z^{-1}$
MDB (PR Class IV)	$1 - z^{-2}$
Polybinary	$1 + z^{-1} + z^{-2} + z^{-3}$
Polybipolar	$1 + z^{-1} - z^{-2} - z^{-3}$

and as with general ISI, the number of terms in the impulse response is designated by M. If the signal had L levels before p-r shaping (conveying $\log_2 L$ bits per symbol), the shaped signal will have $(ML - M + 1)$ separate levels, compared with L for a general M-term impulse response.

We have already established that we should certainly not detect these signals by linearly equalizing to a flat spectrum. Decision-feedback equalization of the channel is problematic, so the most common strategy is to linearly equalize just the channel and then slice the multilevel signal to recover the original (i.e., before precoding) data. As we have seen, this would incur an SNR penalty from the linear equalization and a further one of between 2.1 and 3.0 dB from the slicing; there is clearly a strong incentive to use some form of MLSE.

At the beginning of Section 9.3 we saw that, in order to reduce the computation required for MLSE to reasonable amounts, it is necessary to use a prefilter to shorten the input impulse response. Partial-response signals could be treated the same as full response signals; an SIR or DIR that took account of both channel and partial-response shaping could be found, and a conventional VA could be used.

However, there are two factors that favor a different approach: (1) the spectral weighting associated with partial-response shaping means that the SNR penalty incurred in linearly equalizing a *channel* is less than with a full-response signal, and (2) the integer values of the back-to-back partial impulse responses mean that much simpler (than a general purpose VA) algorithms based on the principles of error detection and correction (EDCA) can be used.

Error detection was described in [Le4] and [G&L], and then the full EDCA algorithm was developed more or less simultaneously in 1969 at Codex and GTE, Lenkurt, but published only in [Fo1]. Other approaches to simplified optimal detection of partial-response signals were described in [Fe2], [Fr1], and [Ko2]. To establish the basic principles, we will concentrate initially on duobinary ($M = 2$), in which a signal set $0, \ldots, (L - 1)$ is expanded to $0, \ldots, (ML - M + 1)$† and generalize to the others later.

9.3.3.1 EDCA for a Duobinary Signal.

At time kT the received signal y_k ($= x_k + \mu_k$) is sliced to yield an estimate x'_k of the transmit signal, and an apparent error amount $\epsilon_k = y_k - x'_k$ is calculated and "stored" (more details about the method of storing later). The $(1 + z^{-1})$ shaping that was applied to the b_k sequence in the transmitter is then inverted by calculating $b'_k = x'_k - b'_{k-1}$. If an error has been made because $\mu_k > 0.5$, then either immediately or at some later time, $(k + kd)T$, b'_{k+kd} will go outside the range 0 to $(L - 1)$; this constitutes the error detection. Upon detection of an error the stored ϵ_k are searched to find the largest estimated noise of the appropriate polarity, the assumption made that that was when the error occurred, and the estimate x'_k changed accordingly.

---

† In practice, bipolar signals will be used, but it is easier to explain the method with all positive signals.

TABLE 9.6 An Example of Duobinary Error Detection and Correction for L = 4

$t=0$	T	$2T$	$3T$	$4T$	$5T$	$6T$	$7T$	$8T$
a 0	3	2	1	3	1	3	0	1
b 0	3	3	2	1	0	3	1	0
x	3	6	5	3	1	3	4	1
μ	−0.4	0	0.7	0	0.4	0	0	0
y	2.6	6.0	5.7	3.0	1.4	3.0	4.0	1.0
x′	3	6	6	3	1	3	4	1
b′ 0	3	3	3	0	1	2	2	−1!
μ'	−0.4	0	−0.3	0	0.4	0	0	0
c	+1	−1	+1	−1	+1	−1	+1	−1
ϵ	−0.4	0	−0.3	0	+0.4	0	0	0
ϵ_{max}	?	?	?	?	+0.4	+0.4	+0.4	+0.4
k_{max}	?	?	?	?	0	1	2	3
ϵ_{min}	−0.4	−0.4	−0.3	−0.3	−0.3	−0.3	−0.3	−0.3
k_{min}	0	1	0	1	2	3	4	5

Details of the storing and searching will be understood better after we have examined the operation of a conventional VA. A sequence of data, transmit signals, noise samples, receive signals, and data and noise estimates—carefully contrived to illustrate the salient points of the error correction method—is shown in the top part of Table 9.6. The VA trellis, showing only the correct path and its nearest neighbor, is shown in Fig. 9.23. At $3T$ the noise sample of 0.7 causes a "raw" error, and it can be seen that the path with the temporarily smallest metric oscillates ±1 around the correct path until the latter reaches an appropriate limit at $8T$; then the previously favored path terminates, and the survivor can be traced back to the correct b_3.

The recoding also detects the error at $8T$ (by the b' going out of bounds to −1), but it is not obvious how to choose the "largest" of the estimated noise samples μ'; the suspects at T and $5T$ both have a larger magnitude than the real culprit. If, however, the noise at $5T$ had been −0.6 rather than +0.4 (so that

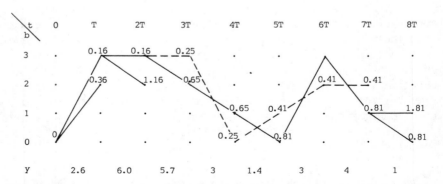

Figure 9.23 Paths through the trellis for a duobinary signal with three large noise samples.

$x'_5 = 2$), it could not have caused the *negative* error at $8T$. On the other hand, although the estimated noise at T is of the correct polarity, any error there would have been detected when the signal reached the 6.0 limit at $2T$.

Identifying the most likely source of a detected error therefore requires keeping track of two largest noise samples — those that could cause positive and negative errors — and resetting the appropriate record whenever an outer level is received. This can be done as shown in the bottom part of Table 9.6, and in the flow-chart of Fig. 9.24. A clock signal c is toggled each symbol time between

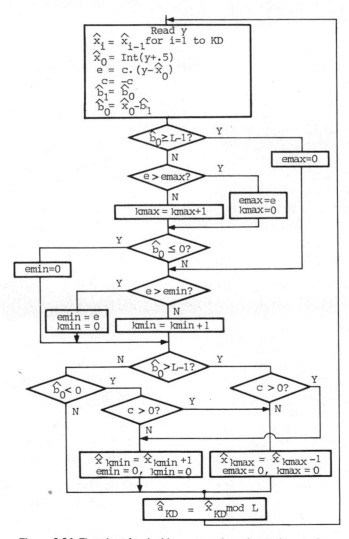

Figure 9.24 Flowchart for duobinary error detection and correction.

± 1 and used to multiply the noise estimates; the most positive and negative of the products, ϵ_{max} and ϵ_{min}, and their addresses, relative to the present symbol, k_{max} and k_{min}, are updated each symbol time and stored in four registers. Then, when a positive (negative) error is detected by b' going outside the upper (lower) bound, the value of x' that is stored k_{max} (k_{min}) symbols back can be corrected.

As with the VA, the delay through the basic algorithm would vary; it is best to fix it at some KD†, and store KD values of x_0 so that they have a chance to be corrected before being sliced and output.

Ways in Which the EDCA May Fail. The circumstances under which this simple algorithm may produce an error are:

1. An error in the raw data may not be detected within the span KD of the x storage register.

2. A second error may occur before the first is detected. If it is of the opposite "sign" (taking account of the oscillating effect), it will prevent detection of either error. If it is of the same sign, it could be detected and corrected by a more complicated algorithm that kept record of the two largest error signals and was prepared to circulate twice through the correction routine shown in Fig. 9.24 if the b_0 went out of bounds by 2; however, the improvement in SNR (<0.2 dB, since only half of the double errors could be corrected anyway) probably would not justify the complication.

3. An error may be detected, but the noise that caused it may be so large that another, "innocent," noise sample may appear larger than the culprit.

4. The "guilty" noise sample may be greater than twice the distance to the threshold and thereby appear to be of the wrong sign.

Forney showed that for any reasonable raw error rates causes 2 and 4 are much less significant than 1 and 3, and that, if they are discounted, the performance of the EDCA is the same as that of a VA.

Since detection occurs when an appropriate extreme value of b is detected, the probability of nondetection is $(1 - 1/L)^{KD}$. For the example of $L = 4$, in order to detect 99% of the errors, and thereby to have the potential of reducing the error rate by two orders of magnitude, KD should be greater than 16. Since there are only two terms in the impulse response, this is an extreme example of the dependence of the required detection delay on both the magnitude of the ISI and the number of levels of the signal, which was discussed in Section 9.3.2. However, it should be noted that the computation and storage requirements of the EDCA do not increase with the detection delay as do those of the VA.

Performance of the EDCA. All simulations of the EDCA have shown that its performance is within 0.2 dB SNR improvement of that of a VA with the same

---

† The required value of KD will be discussed shortly.

detection delay. However, it has sometimes been misleadingly reported that the improvement in SNR over conventional detection approaches 3 dB at low error rates; this is true only if the noise is added after the complete partial-response shaping. In practice, the shaping is split equally between transmitter and receiver, and the deterioration of a partial-response signal from a full-response signal — and therefore the possible subsequent improvement from any MLSE — is only 2.1 dB. The filtering of the receive signal colors the noise and probably further reduces the gain of a VA, EDCA, or any MLSE; no information is available about this effect.

A Name for this Algorithm. Variations of this algorithm are used when modified duobinary (see next section) is applied to high-speed magnetic recording [CLPW], but it has been called a "Viterbi algorithm"; this is a misnomer. It has none of the main characteristics of the VA; it does not calculate all the metrics at each sample time, it does not plot paths through a trellis, and as we have seen, the storage requirements do not depend on the chosen detection delay. It is best described as an Error Detection and Correction Algorithm (EDCA).

9.3.3.2 *Modification for Modified Duobinary (PR Class IV).* In Section 3.3.2 it was shown that MDB is equivalent to interleaved AMI. Therefore, the $(1 - z^{-2})$ shaping can be handled by considering the data stream as two interleaved streams at half the symbol rate, and using two parallel implementations of the flowchart in Fig. 9.24 *without* the oscillating clock addition.

9.3.3.3 *Extension to Polybinary.* With the polybinary shaping $(1 + z^{-1} + z^{-2} + z^{-3})$, $d_{min} = 2\sqrt{2(r_0 - r_1)} = 2\sqrt{2}$, and the minimum distance for symbol-by-symbol detection using decision feedback or multiple thresholds is 2. Therefore, only 3 dB of SNR can be retrieved by any MLSE,[†] and an EDCA need correct only all single and some double errors. The singles require only the simple algorithm of Section 9.3.3.1, but since a larger proportion of double errors can be corrected than in the duobinary case, it is more advantageous to keep a record of the two largest estimated noises of each sign and to perform the correction routine twice if necessary. This algorithm has not been simulated; it is included here only as a suggestion.

It should be noted that the polybinary pulse is far from compact, and even a perfect MLSE would still "waste" 3 dB of the signal power. For this reason the shaping is rarely used.

9.3.3.4 *Extension to Polybipolar.* With the polybipolar shaping $(1 + z^{-1} - z^{-2} - z^{-3})$, the impulse response is compact and $d_{min} = 2\sqrt{r_0} = 4$, but the minimum distance for symbol-by-symbol detection is again only 2; thus any MLSE must retrieve close to 6 dB of SNR, and, in particular, an EDCA

[†] The potential gain is 3 dB instead of 2.1 dB because this shaping is usually caused by a severely distorted channel, so that the noise is added — at least in theory — after the shaping.

must be capable of detecting and correcting all single and double, and even some triple errors.

The present estimate of the precoded data is given by $b_0 = x_0 - b_1 + b_2 + b_3$, and an error e ($= \pm 2$) in x_0 will propagate in the b values as $e, -e, 2e, -2e, 3e, \ldots$. As before, the error is detected when b_0 goes out of bounds (i.e., $< (1 - L)$ or $> (L - 1)$), and it can be shown that all double errors will eventually be detected if a third error of a particular type does not occur before detection. If the sizes and sources of the two largest noise estimates are stored, then on detection, the signs and positions of these presumed culprits can be decoded to correct both errors. Some triple errors can also be detected and corrected, but the algorithm and its limitations have not been worked out.

A lot of work is needed on this algorithm, but it might prove very worthwhile for high-speed transmission on subscriber loops, or high-speed, high-density magnetic recording.

Chapter 10

Coding Used for Forward-Acting Error Correction

This chapter can serve only as a very brief introduction to the subject of coding, particularly those types of coding that are most appropriately implemented in a modem; anything more would be beyond the scope of this book (and, indeed, beyond the capabilities of the author). For more thorough treatments of coding in general, the reader is referred to [V&O], [L&C2], or [C&C], and, for coding as it has been recently implemented in modems, to [Un4], [Un5], [F . . .], [Th], and [We3].

Coding can be used independently of the modem as shown in Fig. 10.1. The encoder accepts data at some bit rate f_{b_1} bit/s, introduces redundancy into the bit stream so that the decoder can detect and correct errors in the primary data, and outputs data at some greater rate f_{b_2} bit/s. The modems' transmitter and receiver do their best to deliver data at this rate to the decoder without knowing or caring that the original data was at f_{b_1} bit/s.

In general, this is inefficient in terms of reduction of error rate for a given amount of redundancy. It is better if the decoder makes only tentative ("soft") decisions about the data, and passes them, together with some estimate of their reliability, to the decoder. Since DCE/DTE interfaces seldom include any means for this transfer, the combined operations of demodulation and decoding must be performed in the modem.

Error-correcting codes may be either *block* or *trellis* codes. Forney and colleagues [F . . .] have argued that trellis codes can achieve everything that block codes can, and usually with less processing. This was debated by the CCITT working party during the discussions that led to the V.32 recommendation, and there was a strong consensus of agreement. We will therefore be concerned only with trellis codes. These have also been called *convolutional*

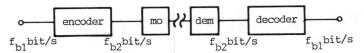

Figure 10.1 Encoder and decoder separate from a modem.

codes, but strictly speaking, that term should be reserved for *linear* trellis codes,† and we shall have to consider some nonlinear codes.

The term "error correcting" is slightly misleading; it implies that there are three stages of detection in the receiver—a first stage of detection, checking of this "crude" data for errors, and correction to produce the "refined" data. In practice, all three stages are combined, and there is no correction per se; perhaps the method would be better called "coding for forward-acting error minimization."

Spectrum of the Encoded Signal. Partial-response precoding and mapping is a form of trellis coding that is used especially to shape the spectrum of the transmitted signal by introducing correlation between successive samples. As we describe trellis coding in this chapter, it may seem that it also introduces correlation and must therefore shape the output spectrum. It is certainly possible to devise trellis codes of the form used here that do this, but the constraints that we put on the trellis in order to minimize the probability of error when the signal is contaminated by AWGN also have the effect of zeroing all terms of the auto-correlation except r_0.‡ Therefore, the spectrum of the signal that is input to the shaping filters is flat, and the spectrum of the transmitted signal is determined only by those filters.

10.1 CODING GAIN

For band-limited channels, our main interest in this book, coding is most often used when it is necessary to transmit several bits per symbol. Then the coding is most conveniently implemented after the serial-to-parallel converter as shown in Fig. 10.2. Past and present values of the M inputs are used to generate M_1 outputs; this is defined as a *rate M/M_1* encoder. The outputs are then mapped into 2^{M_1} points in signal space. Very significant improvements in the effective SNR (up to 6 dB of *coding gain*) can be achieved with just $M_1 = M + 1$, so this is the only case we will consider.

This doubling of the number of points increases the power in one- and two-dimensional signal spaces by amounts that asymptotically—and rapidly

---

† A linear code is one in which the coder outputs are modulo-2 sums (exclusive-ORs) of combinations of the present and previous inputs and/or the previous outputs. A nonlinear code uses other logical operations (NANDs and NORs).

‡ The theoretical proof of this is beyond the scope of this book, but it can be easily checked arithmetically for any of the codes that are commonly used.

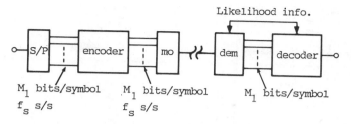

Figure 10.2 Coding incorporated into a multilevel modem.

—approach 6 and 3 dB, respectively. If all constellations are such that the minimum distance between points is 2.0, the gain in effective SNR from coding and modulation is defined by

$$\text{Gain} = [10\log\left(\frac{d_{\min}^2}{4}\right) - \Delta P]\ \text{dB} \tag{10.1}$$

where $d_{\min}$ is the minimum distance between sequences of the encoded signal and ΔP is approximately 6.0 or 3.0 as appropriate. Coding gains of 3–6 dB are usually aimed for and achieved.

10.2 DESIGN OF THE ENCODER

The design of the encoder and the mapping proceeds in three steps that were originally described by Ungerboeck [Un4]. We will consider them first for one-dimensional signals in order to establish the basic principles, and then consider the two-dimensional case.

10.2.1 One-Dimensional Signals

The design procedure is best illustrated by a simple example: 2 bits per symbol, which, without coding, would be mapped as shown in Fig. 10.3a, are encoded in a rate 2/3 encoder and then mapped as shown in Fig. 10.3b. The signal points are labeled binarily† in order to define the interface between the encoder and the mapping. For the moment each number refers to a single point; later they will refer to subsets of points.

Step 1: Signal Set Partitioning. The expanded set of 2^{M+1} points is partitioned sequentially into two sets of four points, four sets of two, and so on as shown in Fig. 10.3c. It can be seen that with each step the minimum distance

---

† The mapping of 111 into +7.0 follows our convention of translating the logic variables 1 and 0 into positive and negative physical signals, respectively.

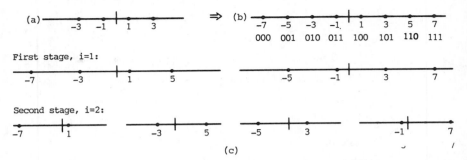

Figure 10.3 Expanding, mapping, and partitioning a (4/8)PAM: *(a)* 2PAM constellation; *(b)* 2PAM constellation; *(c)* partitioning.

between any point and the others in its subset is doubled, so that with the subscript denoting the level of partitioning, $d_i = 2^{i+1}$.

The number of partitioning steps should be equal to the number — M_c, say — of coded output data lines from the encoder (see step 2). This number must be greater than one, since one partitioning step merely compensates for the extra power used in expanding the signal set, but it need not be as great as $(M + 1)$; the smallest subsets contain 2^{M+1-M_c} points.

Step 2: Design of the Trellis. The encoder will consist of L storage elements,† which store some or all of the input signals, and logic operators, which generate the $(M + 1)$ outputs. If $M_c < (M + 1)$, then some of the input data lines are passed through uncoded. The stored inputs $s_1, s_2, \ldots, s_L$ define the *state* of the encoder, and the transition from one state to the next defines an output and thence a signal point.

The general form of the encoder can be described by a trellis diagram in which there are 2^L states (nodes) at each symbol time and 2^M possible transitions (branches) from each state to a next. For our example, the three simplest encoders and associated trellises are shown in Figs. 10.4–10.6. It can be seen from Figs. 10.4 and 10.6 that if one of the inputs is not stored, *parallel transitions* — two or more transitions between the same two nodes — will occur. The form of the eight-state trellis with no parallel transitions (Fig. 10.5) that was given in [We3] is preferred to that in [Un4] because it is derived from a more natural labeling of the stored inputs, and also because we will refer mainly to Wei in the study of rotationally invariant codes in Section 10.4.

The encoding rules are defined by associating each of the 2^{L+M} transitions with one of the 2^{M+1} signal points according to the set partitioning and Ungerboeck's rules:

---

† Some writers have used "v" for the number of storage elements, but we will stick to our FOR-TRAN convention (fixed integers are $I, J, \ldots,$ or N).

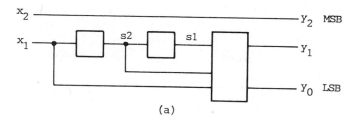

(a)

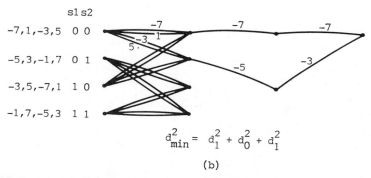

$$d^2_{min} = d^2_1 + d^2_0 + d^2_1$$

(b)

Figure 10.4 Four-state encoder (rate 2/3) for a one-dimensional signal: *(a)* general form of encoder; *(b)* trellis with two minimally separated paths.

U1. Each of the points (eight in this example) should occur 2^{L-1} times "with a fair amount of regularity and symmetry."

U2. Parallel transitions should be assigned points from the same smallest subset. With at least two partitioning steps, this ensures that the minimum distance associated with the uncoded bits — and therefore with single error-events — is greater than that associated with the coded bits and multiple error-events. The improvement in performance at high SNR from making it much greater is small — with only a 2 dB difference in the minimum distances the overall error rate will be dominated by that in the coded bits. Therefore, since the complexity of the Viterbi decoder increases with the number of inputs that are encoded, it is generally agreed that for a one-dimensional signal, two stages of partitioning are sufficient for all but the most complex codes.

U3. All transitions that start from the same node or finish at the same node should be assigned points from the same one of the first two subsets. This ensures that for all multiple error-events (paths through the trellis that diverge and then remerge after two or more symbol periods) $d_{min} > 2d_1$, and the coding gain is at least 3 dB.

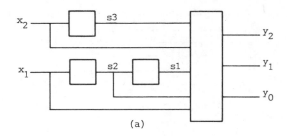

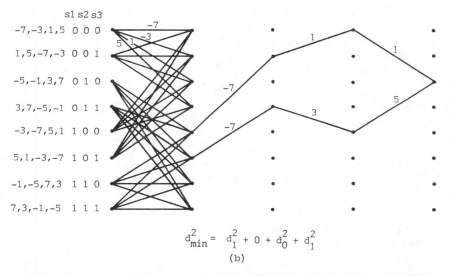

$$d_{min}^2 = d_1^2 + 0 + d_0^2 + d_1^2$$

(b)

Figure 10.5 Eight-state encoder (rate 2/3) for one- or two-dimensional signals: *(a)* general form of encoder; *(b)* trellis with two minimally separated paths.

It should be noted that the parallel transitions considered in U2 are a special case of the transitions in U3.

These rules are necessary, but not sufficient, to uniquely define the best trellis and code for a given number of states. Some diligent searching — preferably computer-aided for L greater than 3 — through all possible trellises and sequences may be needed. This was done by Ungerboeck, and the point assignments shown in the left columns of the figures are adaptations of those in [Un4].

Examples of two minimally separated paths and the calculation of the resultant d_{min} are shown in Figs. 10.4–10.6. The coding gains are given in Table 10.1; Ungerboeck showed that the gain could be increased to 5 dB by

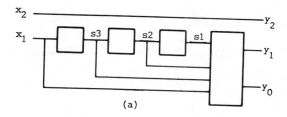

(a)

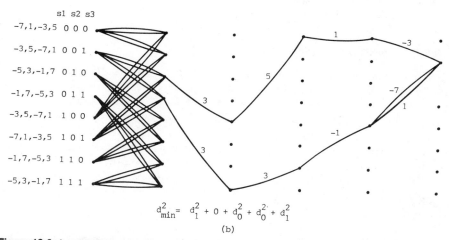

$$d^2_{min} = d^2_1 + 0 + d^2_0 + d^{2\cdot}_0 + d^2_1$$

(b)

Figure 10.6 Another eight-state encoder for one- or two-dimensional signals: *(a)* general form of encoder; *(b)* trellis with two minimally separated paths.

using five states in the encoder, or almost to the theoretical limit of 6 dB by using seven states. The gain of the second eight-state encoder (with parallel transitions) is greater than that of the first; with more states the first form — with both inputs being coded — would probably be better.

Step 3: Design of the Encoder. The encoder can now be designed — without any knowledge of the theory of convolutional coding — by writing the truth

TABLE 10.1   Minimum Distances, Coding Gains, and Encoder Outputs for 2/3 Encoders

Figure	L	d^2_{min}	Coding Gain (dB)	y_0	y_1	y_2
10.4	2	36	3.3	s_2	$x_1 \oplus s_1$	x_2
10.5	3	36	3.3	s_2	$x_2 \oplus s_1$	$x_1 \oplus s_3$
10.6	3	40	3.8	s_2	$x_1 \oplus s_1 \oplus s_3$	x_2

table implied by the assigned transitions; one example of this, for the trellis and encoder in Fig. 10.5, is shown in Table 10.2. The resultant encoders are defined in Table 10.1; it can be seen that they are all linear.

Encoders designed this way are called *nonsystematic*. Greater knowledge of coding theory is needed to design equivalent *systematic* encoders in which the M inputs are passed directly to the output, and the $(M + 1)$th output, which is a type of parity-check bit, is generated from those inputs plus feedback of the previous values of the $(M + 1)$th output itself. Wei [We3] also described some nonsystematic encoders that use feedback. The relative advantages of systematic and nonsystematic encoders with or without feedback are controversial; the reader is referred to [F . . .], [C&C], [We3], and [AT&T] for much more detailed discussion of this.

10.2.2 Two-Dimensional Signals

The basic encoding method for one-dimensional signals could be applied to QAM by treating each dimension separately; then, for example, a 4 × 4 (16-pt) QAM would be expanded to 8 × 8 (64-pt) and encoded at a 4/6 rate. However, this would be inefficient because it neglects the opportunity to introduce constraints between the signals in the two dimensions.

A better method is to only double the number of points in the signal constellation, and then partition and encode jointly in the two dimensions. Thus, for example, a 4 × 4 QAM set would be expanded to a 32-pt (preferably the cross or "chipped diamond") and encoded at a 4/5 rate. The design steps given in Section 10.2.1 need be modified only very slightly.

Step 1: Signal Set Partitioning. The smallest set that can be partitioned is the eight-phase constellation.† The partitioning is shown in Fig. 10.7; it can be seen that it proceeds exactly as for one-dimensional 8PAM with the line of Fig. 10.3 bent to form a circle. If the signal power is normalized to 2.0 in order to facilitate comparison with the original 2 × 2, then $d_0 = 1.08$, $d_1 = 2.0$, and $d_2 = 2.83$.

Larger constellations based on a square lattice can all be partitioned by using a two-dimensional equivalent of the one-dimensional procedure. An infinite lattice can be divided into two sets as shown in Fig. 10.8a; the minimum *squared* distance between points is doubled. The points of each subset now lie on a sparser square lattice rotated by 45°; they can therefore be further subdivided as shown in Fig. 10.8b, with a further doubling of the minimum squared distance. The original expansion of the signal set increased the power by approximately 3 dB, and each stage of partitioning increases the distance between nearest neighbors within the set (the distance that determines the probability of single-error events) by 3 dB. Just three

---

† The more efficient star constellations shown in Figs. 4.4b and 4.4c, which are 1.6 and 0.9 dB better, respectively, to start with, are not suitable for partitioning.

TABLE 10.2 Truth Table for the L = 3, Rate 2/3 Encoder of Fig. 10.5

x_2	0	1	0	1	0	1	0	1	0	1	0	1	0	1	0	1	0	1	0	1	0	1	0	1	0	1	0	1	0	1	0	1
x_1	0	0	1	1	0	0	1	1	0	0	1	1	0	0	1	1	0	0	1	1	0	0	1	1	0	0	1	1	0	0	1	1
s_3	0	0	0	0	1	1	1	1	0	0	0	0	1	1	1	1	0	0	0	0	1	1	1	1	0	0	0	0	1	1	1	1
s_2	0	0	0	0	0	0	0	0	1	1	1	1	1	1	1	1	0	0	0	0	0	0	0	0	1	1	1	1	1	1	1	1
s_1	0	0	0	0	0	0	0	0	0	0	0	0	0	0	0	0	1	1	1	1	1	1	1	1	1	1	1	1	1	1	1	1
Point	0	2	4	6	4	6	2	0	6	4	6	4	2	0	4	6	2	0	6	4	7	5	7	5	3	1	3	1	5	7	5	3	1
y_2	0	0	1	1	1	1	0	0	1	1	1	1	0	0	1	1	0	0	1	1	1	1	1	1	0	0	0	0	1	1	1	0
y_1	0	1	0	1	0	1	1	0	1	0	1	0	1	0	0	1	1	0	1	0	1	0	1	0	1	0	1	0	0	1	0	1
y_0	0	0	0	0	0	0	0	0	0	0	0	0	0	0	0	0	1	1	1	1	1	1	1	1	1	1	1	1	1	0	0	1

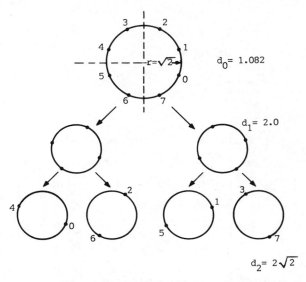

Figure 10.7 Partitioning of an eight-phase constellation.

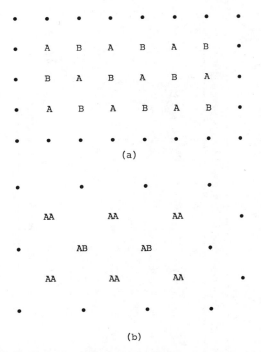

Figure 10.8 Partitioning an infinite square lattice: *(a)* first stage — partitioning into As and Bs; *(b)* second stage — partitioning As into AAs and ABs.

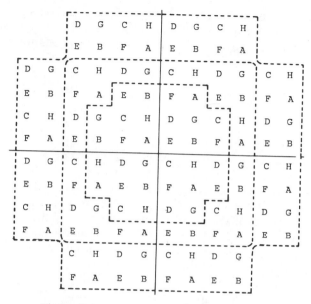

Figure 10.9 Partitioning a 128CR/64SQ/32CR.

stages of partitioning would therefore allow a coding gain for *parallel transitions* of 6 dB. Since the gain for multiple-error events will typically be about 2 dB less than this, there is usually little point in partitioning further.

A 128CR constellation (used to implement a rate 6/7 code) partitioned into eight subsets is shown in Fig. 10.9, together with the smaller 64SQ and 32CR constellations (used for 5/6 and 4/5 codes, respectively).† The alphabetic labeling conforms with [We3]; it is derived from the simple partitioning:

[A,B,C,D,E,F,G,H]	minimum distance within set(s) = 2.0		
[A,B,C,D] [E,F,G,H]	.... = $2\sqrt{2}$		
[A,B] [C,D] [E,F] [G,H]	.... = 4.0		
[A] [B] [C] [D] [E] [F] [G] [H]	.... = $4\sqrt{2}$		

Binary labeling of the subsets is also necessary in order to define the interface between the encoder and the mapping, but, unlike the one-dimen-

---

† In practice, as we saw in Chapter 4, in order to maintain a constant signal power with a simple integer grid, the 128CR and 32CR constellations are usually rotated by 45° to what we have called the chipped diamond constellations; however, the partitioning is more easily understood with the square grid.

sional case, there is no obvious ordering of the sets. One labeling was defined by the CCITT in the V.32 and V.33 recommendations, but unfortunately the translation between it and the alphabetic labeling does not seem to follow any rules:

000	001	010	011	100	101	110	111
A	H	B	E	C	F	D	G

Each subset contains 2^{M-2} points, so $(M-2)$ of the inputs — the more significant bits — will not be encoded, and each branch of the trellis will comprise 2^{M-2} parallel transitions. The assignment of these MSB bits to the points in each subset is not very critical because it has only a small effect on the multiplier of the error function and therefore a very small effect on the Bit Error Rate (BER). The simplest way to do it — and the most economical of storage if a modem has to be capable of several speeds (e.g., a V.33) was suggested by Ungerboeck [IBM1] and [IBM2]: number the two points in each subset of a 16SQ, and then "grow" the constellation to 32CR, 64SQ, and 128CR in stages by each time adding a leading zero to the points in the inner core and using a leading one for all the points in the added outer ring.

The schemes used in V.32 and V.33 are not as elegant as this, but they do have a certain regularity. The 32CR used for V.32 has two shapes of subsets as shown in Fig. 10.10 — the T-shaped A, E, C, and G, and the square B, F, D, and H. The points in A and B are numbered according to the conventional rotational Gray code, and then the other subsets are numbered by rotating A or B by 90°, 180°, and 270°. The interested reader is referred to the V.33 recommendation for more details.

Step 2: Design of the Trellis. This proceeds just as before, except that for all constellations other than the eight-phase, testing prospective trellises for

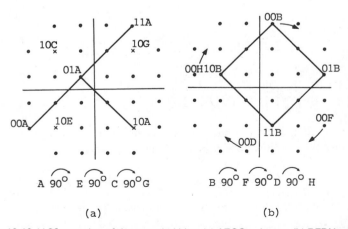

(a)   (b)

Figure 10.10 V.32 mapping of the uncoded bits: *(a)* AECG subsets; *(b)* BFDH subsets.

their minimum distances must be a little more strict. Each point now has four nearest neighbors instead of two, and all four transitions must be checked.

If the three simple trellises shown in Figs. 10.4–10.6 are used for two-dimensional signals, the coding gains are as given in Table 10.3. It can be seen that the eight-state trellis with parallel transitions is no better than the four-state one; this is because $d_{\min}$ for both is now limited by d_2, and the errors are dominated by single-error events.

Step 3: Design of the Encoder. This follows the same rules as for the one-dimensional case, with the same choices to be made between systematic and nonsystematic, feed-forward, and feedback structures.

Step 4: Mapping. In the one-dimensional case this was a trivial operation; the outputs were just binarily weighted, and treated as though the signal had not been coded. For the two-dimensional signal the binary labels must be used as addresses in two look-up tables to find the in-phase and quadrature components.

10.2.2.1 90° Rotational Invariance.

All common constellations — whether used for coded or uncoded signals — look the same when rotated by any multiple of 90°. Consequently, a conventional carrier-recovery circuit or algorithm in the receiver can adjust the carrier phase only modulo $\pi/2$, and differential phase modulation must be used in the transmitter so that the receiver need be concerned only with phase changes. Therefore, the rules developed so far for design of the trellis and encoder must be made more restrictive in order to accommodate this differential encoding.

Ungerboeck, [IBM1], proved that a linear encoder cannot satisfy these rules, but Wei [We3] found several eight-state nonlinear encoders that do. The trellis diagram and the systematic encoder with feedback that were adopted in V.32 are shown in Fig. 10.11. The differential encoder is different from any used in previous recommendations, which all used a Gray code; in this one the quadrants are numbered anticlockwise, and the two-bit input *binarily* defines a change of phase (viz., $00 = 0°$, $01 = 90°$ $10 = 180°$, and $11 = 270°$), so that

$$\text{New quadrant} = \text{old quadrant} + \text{input,} \quad \text{Mod 4}$$

TABLE 10.3 Coding Gains of Rate 2/3 Codes for Two-Dimensional Signals

Figure	L	Trellis Configuration	Gain (dB)	
			8φ/4QAM	32CR/16QAM
10.4	2	Parallel transition	3.0[a]	3.0
10.5	3	No parallel transition	3.6	4.0[b]
10.6	3	Parallel transition	3.0	3.0

[a] This has been used in an experimental 64 kbit/s satellite modem [UH&A].
[b] This is the basis for the rotationally invariant codes used in V.32 and V.33.

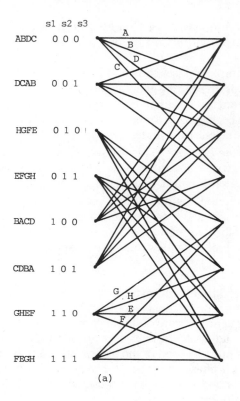

(a)

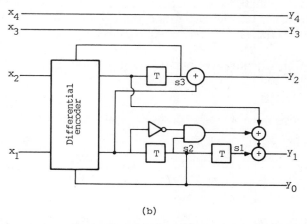

(b)

Figure 10.11 V.32 90° rotationally invariant encoder: *(a)* trellis diagram; *(b)* encoder.

The squared minimum distance of this trellis and code is

$$d_{\min}^2 = d_1^2 + d_0^2 + d_1^2 \qquad (10.2)$$

and the coding gain is 4.0 dB, which is as good as could be achieved by a eight-state linear encoder without the rotational invariance requirement. However, Ungerboeck has suggested that the eight-state nonlinear code may be a special case; with 16 states, a linear code can achieve 4.8 dB gain, but a nonlinear one that attains this has not been found. The theory of these codes is still very much in its infancy; an explanation of it will not be attempted in this book.

10.3 DECODING WITH THE VITERBI ALGORITHM

The version of the Viterbi algorithm described in Section 9.5 can deal with ISI, and in this section we will consider a simpler version for use with trellis coded signals. Unfortunately, very little information is available about how to deal with both ISI and trellis coding. Therefore, in this section we will assume that the received signal has been *linearly* adaptively equalized, and that the impulse response of the "channel" is indeed an impulse. Later, in Section 10.3.2, we will speculate a little about what forms of ISI might be manageable.

Just as when dealing with ISI, the VA is used to find the optimum path through the trellis. As before, this is defined as the sequence of estimated transmit signals that would have needed the least noise added to them in order to produce the sequence of received signals.

If the trellis included any parallel transitions, the algorithm can be greatly simplified by splitting it into two stages. In the first stage the receiver must choose the one point out of each subset that is closest to the received sample and store the uncoded bits thereby associated with each subset. This is equivalent to choosing just one out of the 2^{M-2} branches (four in a V.32) that constitute each parallel transition. Then each transition in the trellis can be labeled with a single nominal receive signal, and the second step—the VA itself—need be concerned only with a trellis that has single transitions; the description of the algorithm can therefore ignore the possibility of parallel transitions.

At each symbol time kT, a transition metric, $\delta D_{k,j}$, is calculated for each of the 2^{M_c} possible transitions; this is defined as the squared distance† of the received signal y_k from the jth possible *noiseless* received signal x_j; that is,

$$\delta D_{k,j} = (y_k - x_j)^2$$

This contains the term y_k^2, which is the energy in the received sample, and is the

---

† For simplicity we will show this only for the one-dimensional case; it can be easily extended to two dimensions.

same for all transitions. For some means of implementation it may be simpler to subtract this from all the $\delta D_{k,j}$ to give

$$\delta D'_{k,j} = x_j (x_j - 2y_k) \tag{10.3}$$

A multiplication is still required, but now it is by one of a finite set of x_j, and (10.3) could, in a switched-capacitor implementation, for example, be achieved by a differencing amplifier with appropriately ratioed input capacitors.

Then 2^L potential new state metrics $D_{k,i}$ for each state (node) are calculated by

$$D_{k,i} = D_{k-1,ij} + \delta D_{k,j} = D_{k-1,ij} + \delta D'_{k,j}$$

as appropriate, where ij is the subscript of the node at time $(k-1)T$ that is connected to the ith node at time kT by the jth transition. The smallest of these new state metrics is then selected, the corresponding branch — known as the "survivor" — marked as part of the optimal path leading to that node, and all other branches leading to that node discarded.

As when dealing with ISI, the optimum detection method for the kth transmit data symbol is to wait until, at some later time $(k + kd)T$, the survivors to the 2^L nodes all emanate from — that is, can be traced back to — a single node at time kT. The "detection delay" kdT, is variable, and implementation of this method would require a very long elastic buffer. It is easier to fix the delay — at KD symbols for instance — and then trace back the path that produces the lowest state metric at time $(k + KD)T$.

When dealing with ISI, a delay of five times the number of ISI terms is adequate, but for trellis codes Thapar [Th] has shown that a delay of as much as eight times the constraint length (i.e., $KD = 8L$) may be needed to achieve the last few tenths of a decibels of coding gain; then the probability of making a nonoptimum decision — one that would have been different if a longer delay had been used — becomes much smaller than the probability that the optimum decision is wrong, and therefore the effect on the average error rate is small.

An example of a data sequence encoded by the trellis in Fig. 10.5b and then corrupted by just one large noise sample at time T† is shown in Fig. 10.12. It is instructive to note that only at $9T$ does the accumulated metric of the correct path become the smallest and can a path be traced back to the correct state (011) at time 0. The fact that one could then decide all the data up to $9T$ is a result of using just one noise sample; the buffer would not usually need to be as elastic as that!

It can also be seen that at $9T$ seven of the eight states still lead back to the wrong state at 0; the paths would probably not merge until considerably later.

---

† In practice, with a high SNR, one large noise sample that would have caused an uncoded error would typically be followed by many smaller samples, but this simplification is used for the sake of clarity. With noise on all the samples, the paths would probably take even longer to merge.

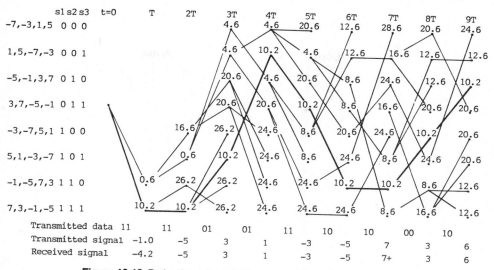

Figure 10.12 Paths through trellis for eight-state 4PAM/8PAM code.

The fixed decoding delay does not need to be sufficient to give a high probability of merging — only that the survivor with the smallest metric can be traced back correctly.

10.3.1 Carrier Recovery for a Two-Dimensional Encoded Signal

In an uncoded eight-phase or QAM system the error rate seen by a decision-directed carrier recovery circuit or algorithm is the same as the final data error rate; it is necessarily low, and, once the carrier has converged, the effect of errors on operation of the loop is negligible. In a coded system this may not be the case; the distance from any point to the nearest decision threshold has been decreased, and the "raw" error rate is high — maybe as high as 10^{-2} for the noise levels and final error rates that concern us here.

It has been suggested [AT&T] that partially delayed, and therefore more reliable, decisions from the VA† should be used to derive the phase error; we will consider using both raw and delayed data.

Estimate of Phase Error from Raw Decisions. In this case the $\mathscr{S}$ curve (see Section 6.4.3) will contain more segments (typically from two to four times more) than in the uncoded case; this is illustrated in Figs. 10.13 and 10.14 for the cases of expansion from 4QAM to 8ϕ and from 16SQ to 32CR. This affects the carrier recovery in two ways:

---

† One of the reported advantages of systematic encoding is that the intermediate decisions of the VA are more accessible.

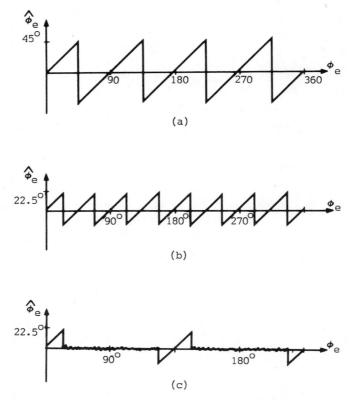

Figure 10.13 $\mathscr{S}$ curves for 4QAM/8ϕ expansion: *(a)* 4QAM; *(b)* 8ϕ using raw decisions; *(c)* 8ϕ using delayed decisions.

1. In the converged state, when the phase error is in the principal segment of the $\mathscr{S}$ curve, centered around $\phi_e = 0$ or any other phase offset from which the decoder is able to extract the correct data, the combination of noise and closer thresholds will reduce the slope of the $\mathscr{S}$ curve (estimated phase error divided by actual phase error), and thereby reduce the gain of the loop and the input SNR of the loop filter. If these reductions are taken account of in the design of the loop, the effects on the jitter of the recovered carrier should be small.

2. The greater number of segments means that the receiver is more sensitive to phase hits (steps) on the channel. With small steps ($<18°$ for a 32CR), the discontinuities considerably reduce the gain of the loop, and slow down reconvergence. With larger steps, and with a very low noise level, the carrier loop might stay locked at a strong false zero for a long time; therefore, if (a) the $\mathscr{S}$ curve has a strong false zero, (b) phase steps are a likely phenomenon in the transmission medium, and (c) long periods of low noise are likely, then some special methods of detecting loss of lock, and "kicking" the recovered carrier may be necessary.

The curve for 8ϕ in Fig. 10.13b is the same in all segments, and a basic carrier recovery algorithm is unable to distinguish between them; a carrier could remain locked for ever at any of the zeros at integer multiples of 45°. This problem must be solved by (1) using one of the codes described by Oerder [Oe] that have 45° rotational invariance, (2) detecting and kicking, or (3) using more reliable, delayed decisions.

The curve for 32CR shown in Fig. 10.14b shows only three weak false zeros; special action would probably not be necessary.

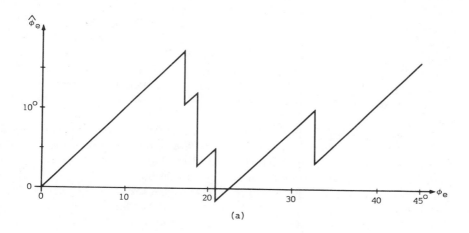

(a)

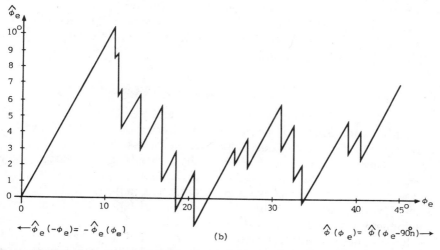

(b)

Figure 10.14 $\mathscr{S}$ curves for 16QAM/32CR expansion: *(a)* 16QAM; *(b)* 32CR using raw decisions.

Estimate of Phase Error from Delayed Decisions. The only case for which an analysis has been published is the simplest 4QAM to 8ϕ (see [Un4]). The $\mathscr{S}$ curve in Fig. 10.13c is for the eight-state encoded 8ϕ that has only 180° rotational invariance; no other rotations produce possible coded sequences. Consequently, with ϕ_e outside the ranges $0 \pm 22.5°$ or $180° \pm 22.5°$ the errors appear random, and all estimates of ϕ_e have zero mean. If the phase is displaced out of one of the principal segments, it will wander around, buffeted by random noise, until it falls, by chance, into one of those segments.

This is different from the situation when using raw decisions—then the carrier would never reacquire lock unless kicked—but the number of errors made while reacquiring will still be large. The number would probably be acceptable if the data were transmitted in blocks, with the opportunity for retransmission; if the bit error rate is more important than the block error rate, the 45° rotational invariant codes are the only solution.

For SQ to CR and CR to SQ expansions and encoding, the situation is much more complicated. What follows is only informed speculation, because no analysis of the problems is available; the use of the word "may" should be noted.

By using delayed, and more reliable, decisions, some of the earlier discontinuities in the $\mathscr{S}$ curve may be eliminated, and the gain of the loop may remain nearly constant over a wider range of phase offsets. For larger-phase offsets, the estimate of the phase error would be small—or perhaps even zero—and the weak false zeros of the $\mathscr{S}$ curve would probably be eliminated altogether; the carrier may reacquire lock via a random walk. Whether this happens faster or slower than when using raw decisions would depend on the noise level and the exact amount of the phase step.

Tracking Frequency Offset or Discrete Components of Phase Jitter. To do this, the open-loop gain must be very high (preferably infinite) at d.c. and/or the frequencies of the ringing current or the a.c. power supply. It is very difficult to keep such a loop stable if it contains a lot of delay, so use of the raw decisions is probably the only feasible strategy.

10.3.2 Performance of Trellis Coding with a VA Decoder

Performance with Gaussian Noise. The codes and the detection algorithm are designed to minimize the probability of a sequence error at high SNR, and the simplest comparison of the codes is made on the basis of the ratio of d_{min} to the power in the signal. However, the BER further depends on the average number of nearest neighbors to a path, and the average number of symbol errors in those nearest paths; as Ungerboeck has pointed out, these, in turn, depend on the bit labeling of the branches of the trellis and the realization of the encoder as systematic or nonsystematic, with or without feedback.

All results of simulations and tests that have been published have shown that, for BERs of 10^{-5} or less, the needed SNR is less than 0.5 dB (usually 0.2 to

TABLE 10.4   Changes in Peak-to-Average Power Ratio Involved in Trellis Coding

Uncoded		Coded			
Constellation	P_p/P_{av}	Constellation	P_p/P_{av}	Coded/Uncoded	Modem
4PAM	1.8	8PAM	2.33	1.3	
4QAM	1.0	8ϕ	1.0	1.0	
16QAM	1.8	32CR	1.7	0.94	V.32
32CR	1.7	64QAM	2.33	1.37	V.33
64QAM	2.33	128CR	2.07	0.89	V.33

0.3 dB) greater than would be calculated by comparing coded and uncoded systems on the basis of d_{min} alone. That is, the practical coding gain is within a few tenths of a decibel of the theoretical.

Performance with Nonlinear Distortion. Second- and third-harmonic distortions act like "noise" that is correlated with the signal, and a VA provides no advantage in dealing with them. If the expansion of the signal set, which is needed for coding, increases the ratio of peak to average signal level, the sensitivity to nonlinear distortion will be increased, and the coding gain may be reduced.

The levels of distortion typically encountered on the PSTN do not by themselves cause errors, but only reduce the noise margin; as pointed out by Thapar, the correlated noise component is small compared to the random component, and, trellis encoding and VA decoding still provide nearly the theoretical decibel improvement in tolerated noise level.

With higher levels of harmonic distortion, the correlated noise component becomes significant, and the change in the peak-to-average power ratio that is involved in coding becomes more important. The changes are shown in Table 10.4 for all the useful expansions. It is interesting to note that the cross constellations are more "efficient" in this respect, because they eliminate the corner (high-power) signal points; expansion from a square to a cross should reduce the sensitivity to nonlinear distortion.

With high levels of distortion, such as might be encountered on a satellite channel, multilevel signals are probably not usable, and the expansion from 4QAM (or SQAM or whatever variation was used) to eight-phase is probably the only useful one.

Performance with Phase Jitter. If any of the phase jitter introduced by the channel can be tracked by the carrier recovery circuit or algorithm,† it can be ignored in this present discussion; we will consider only untracked jitter.

---

† We must bear in mind that tracking phase jitter may be less effective when coding is used.

If successive samples of the "noise" are strongly correlated, the VA cannot deal with them. For voice-band signals, even the maximum frequency of phase jitter that is usually considered, 120 Hz, is low compared to the symbol rates of all single-band modulation systems, and the VA is weakened; for jitter at 20 Hz and lower, it is powerless. However, this must not be interpreted as meaning that encoding is useless on channels that have low-frequency phase jitter; even for jitters as high as 20° ptp, the major part of the noise will still be uncorrelated, and encoding will be useful.

Performance over ADPCM Systems. These systems are designed for voice signals so as to take advantage of the fact that successive samples at an 8 kHz rate are strongly correlated. With data signals that are, for the most part, uncorrelated they introduce various forms of distortion. Because of their planned proliferation, all new modulation methods must be judged by their performance over both linear analog channel and these somewhat nonlinear digital channels.

Tests done on the systematic feedback eight-state code adopted for V.32 and V.33, and reported in [AT&T], show that it always maintains its 3.5 dB coding gain, and sometimes will achieve an acceptable block error rate in conditions (linear distortion and three codecs in tandem) where an uncoded modem cannot.

10.3.3 The Viterbi Algorithm for both ISI and Trellis Coding

For channels with a moderate amount of attenuation distortion, the various receiver strategies can be ranked in ascending order of performance (descending order of error rate with AWGN): (1) no trellis coding and linear equalization; (2) no trellis coding and decision feedback equalization; (3) no trellis coding and MLSE (VA) to deal with the ISI; (4) trellis coding and linear equalization; and (5) trellis coding with MLSE to deal with both ISI and coding. (There is one combination missing from this list: decision feedback equalization seems to be incompatible with Viterbi decoding of trellis codes because of the delay through the latter.)

If there is a large amount of attenuation distortion, the noise enhancement that accompanies linear equalization may completely eliminate the SNR advantage gained by using trellis coding, and the order may become (1), (2), (4), (3), (5), or even (1), (4), (2), (3), (5). In all cases, (5) would be much better than all the rest, but unfortunately an MLSE method that can deal with both ISI and trellis coding has not been developed; A. P. Clark (private correspondence) has called it the most difficult problem that remains to be solved in data transmission.

A simpler problem should be MLSE for trellis coding and very simple, controlled ISI such as is used in PR Classes I and IV. For systems that use partial-response shaping to allow tight band-limiting without very sharp filters (e.g., a 60–108 kHz groupband modem, or a FDX 4800 bit/s voice-band modem), or to match the amplitude characteristics of the channel (e.g., high-speed magnetic recording systems), the ability to add trellis coding would be very attractive. Wolf and Ungerboeck [W&U] have described some codes, but the coding gains were not as great as might be expected. Pizzi and Wilson [P&W2] described the application of convolutional coding to continuous phase modulation, which also involves controlled ISI.

Chapter 11

Duplex Operation

The title of this chapter needs some explanation. As defined in Chapter 1, "full-duplex" means simultaneous transmission and reception (data transfer) at the same "full" speed. Although most of this chapter will be concerned with this type of transfer, most of the problems arise whenever any form of duplex operation (symmetrical or not) is attempted, and so we will also discuss the more general asymmetrical case. We shall find that sometimes the problems are easier if the speeds are equal—particularly in echo canceling—and sometimes, on the other hand, extreme asymmetry provides a useful flexibility in configuring the modem—particularly in FDM.

The Need for Full-Duplex Capability. When discussing duplex operation we must be careful to distinguish between duplex data capability and actual communication. An end-to-end connection may sometimes include a link on which "data" is traveling at full speed in both directions,† but full-duplex communication between users is rare. High-speed transfer of information in one direction is usually accompanied by only low-speed transfer in the other for purposes of interrupts, acknowledgments, error control, and so on.

The interface between data terminal (DTE) and modem (DCE) defined by EIA-232-D and V.24 (CCITT) includes lines for primary and secondary (low-speed) data and control, but these are generally considered inconvenient to use. The appeal of the original FDX 1200 bit/s modem (the VA 3400) was not that users often needed full-duplex communication at 1200 bit/s, but simply that terminals and computer front ends could "turn the line around" without

---

† Packet data systems are examples of this but are beyond the scope of this book.

switching data and control leads. However, as data speeds increase, the proportional extra cost of providing the full-speed reverse channel increases, and the asymmetrical, or split-speed, modem may come into its own again.

FDM, EC, or TCM? Since duplex operation on four wires is trivial (the signals are spatially separated), we will concern ourselves only with two-wire operation, and consider other methods of separation. In Frequency-Division Multiplexing (FDM), the two signals use different, partial, frequency bands, and they are separated by fixed filters in the transmitter and receiver. When using Echo Canceling (EC), the signals use the same, full, band, and each receiver adaptively learns the characteristics of the reflection path from its own transmitter, so that it can subtract the echoes of its known transmit signal from the receive signal. In Time-Compression Multiplexing (TCM), the two modems operate half-duplex; each transmits and receives alternately — with a guard time between to allow the previous signal to traverse and decay — and the modems are made to appear full-duplex to their users by buffering in both transmitter and receiver. The efficiency of TCM decreases as the propagation time of the medium increases, and the method should be used only for high-speed data on the subscriber loops; we will pass it over for now, and only consider it briefly in Section 11.3.

A symmetrical (FDX) modem realized by FDM is just a special case of an asymmetrical (duplex) one, but we will concentrate on it initially and then discuss the more general case in Section 11.4.

In order to compare FDM and EC, we should first calculate the theoretical performance of each, assuming perfect removal of the reflected transmit signal by one means or the other. In a voice band from about 300 to 3200 Hz, the maximum practical symbol rate is 2400 s/s, and with perfect filtering in an FDM system, the maximum symbol rate in each channel is 1200 s/s. If we use the SNR (measured in the full 300–3400 Hz band) needed to achieve an error rate of 10^{-5} in a basic full-band 4QAM modem as the norm,† then we can calculate the SNR needed by both methods at several speeds for the same error rate. Each FDMed signal uses half the bandwidth and thus needs 3 dB less SNR; however, it must modulate twice as many bits per symbol, and as we saw in Chapter 4, each increase of one in the number of bits per symbol increases the required SNR by approximately 3 dB.

In practice, perfect filtering is impossible, and so far quite large excess bandwidths have been used to allow adequate separation. The voltage spectra of 212, V.22, and V.22 bis modems are shown in Fig. 11.1; the symbol rate is only 600 s/s, and they could use up to 100% of excess bandwidth.

The SNRs (in decibels relative to a wide-band 4QAM EC system) for EC, and FDM at 1200 and 600 s/s are shown in Table 11.1.

---

† Note that now we are using the notion of the "SNR of a modem" — that needed to achieve a certain error rate. Normally high SNRs are "good"; here a high SNR is "bad."

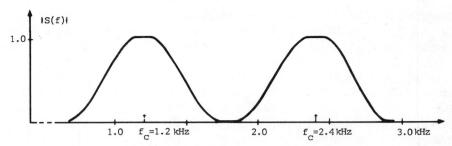

Figure 11.1 Voltage spectra for V.22 and V.22 bis.

It can be seen that at 1200 s/s the SNR penalty of FDM would become prohibitive for speeds over 4800 bit/s, and at the presently used 600 s/s only 2400 bit/s is feasible.

11.1 FDM

The communication between two FDX modems over a two-wire line is illustrated in Fig. 11.2. The leakage of the transmitted signal through the 4W/2W "hybrid" is caused by the mismatch between the characteristic impedance of the hybrid and the input impedance of the line. This is sometimes called *near-end crosstalk* (NEXT), but that term is rather misleading; some of the crosstalk is caused by reflections from the far-end hybrid, and some from discontinuities at intermediate points. This distinction is not important for FDM — we can just say that the reflection is caused by an impedance mismatch — but it becomes very important for EC.

Band Assignment. It is generally expected that FDX modems that are designed to operate on the PSTN should be able to communicate with any other modem of the same type. Therefore, they must be able to operate in either mode — transmit in the low band and receive in the high or vice versa — and in order to allow this, there must be a convention to define which modem does what. For modems that use automatic dialing and/or answering, the rule is that the *calling* (dialing or "originating") modem transmits in the low band.

TABLE 11.1   Extra SNR (in Decibels Relative to a Wide-Band 4QAM System) Needed by EC and FDM Methods at Various Speeds

f_b	EC (2400 s/s)	FDM (1200 s/s)	FDM (600 s/s)
1200	—	−6	−3
2400	−3	−3	−6
4800	0	3	15
9600	6	15	36

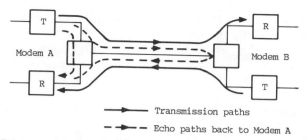

Figure 11.2 Echoes from near and remote 2W/4W hybrids (R = receiver; T = transmitter).

[Acoustically coupled (a-c) modems transmit to the line via a speaker into the hand-set mouthpiece and receive via a microphone from the hand-set speaker. Therefore they do not need any special switches or connectors, and have the great advantage of portability and flexibility. However, the carbon microphone of a hand-set is a very nonlinear device, and an a-c modem that transmits in the low band would see its receive signal grossly contaminated by second-harmonic distortion of its own transmit signal. Since all a-c modems operate in the calling mode, there was an attempt with the introduction of the VA3400 in the early 1970s to reverse the convention, and have the calling modem transmit in the high band from which harmonic leakage into the low band is much less serious. Unfortunately, the operating companies and PTTs, which have the final say on standards, do not like acoustic coupling, and the adverse convention survived.]

11.1.1 Development of Specifications for the Hybrid and the Filters

A V.22 bis calling modem is much more susceptible to Adjacent-Channel Interference (ACI) than an answering modem because the receive signals in the high band are more attenuated by the line, so the signal : interference ratio is much lower. Since there is seldom any advantage in using different specifications for the two modes, we will concentrate on the more difficult, calling mode.

The attenuation of ACI at any frequency can be calculated by adding the attenuations provided by filters, AT and AR, and hybrid, AH. The ACI power integrated over the whole band (from 600 to 3000 Hz) must then be compared to the power of the received signal.† This latter varies across the band, but its attenuation can be approximated by that at 2400 Hz, the carrier frequency. From Figs. 1.7 – 1.9 a 99 percentile line has:

Attenuation at 1000 Hz	27 dB
70% of the distortion from 1000 – 3000 Hz	13 dB
Total attenuation at 2400 Hz	40 dB

† Note that the critical number is a ratio of powers; the actual transmit level does not enter into this calculation, since both modems transmit at the same, maximum allowable level.

The suggestions in Table 5.1 were that all the impairments in a V.22 bis receiver should degrade the SNR by no more than 3 dB (i.e., the effective "noise" should be no more than double the input Gaussian noise), and that ACI should account for no more than 0.5 dB of that degradation. A signal:total noise ratio of at least 19 dB is necessary for an acceptable error rate, so a signal:ACI ratio of 31 dB is desirable.

If the attenuation of the filters and hybrid must be specified by a single number, then 71 dB should be prescribed. However, as we shall see in Section 11.1.3, it is more efficient to consider the aggregate interference throughout the whole stopband of a filter, and 71 dB — less whatever can be provided by the hybrid — is an "average" specification.

11.1.2 Design of the Hybrid

The simplest way of connecting both transmitter and receiver to an unbalanced line with input impedance Z_L — and at the same time terminating the line in its

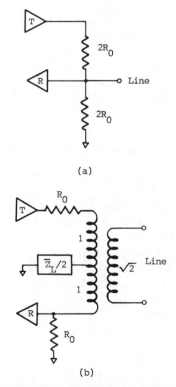

(a)

(b)

Figure 11.3 2W/4W converters: *(a)* direct connection; *(b)* passive center-tapped "hybrid" transformer; *(c)* active-*RC* hybrid; *(d)* compromise line impedance emulator; *(e)* splitting of line emulator to change FLR.

nominal impedance, R_0—is shown in Fig. 11.3a; with this arrangement the full transmit signal is seen at the input to the receiver. A conventional hybrid transformer, in which an impedance $\overline{Z}_L$ is used to match some average line, is shown in Fig. 11.3b. A simple active-RC implementation that does not need a center-tapped transformer is shown in Fig. 11.3c.

The *trans-hybrid transfer function* (FTH) is defined as the ratio of the transfer function from transmitter to receiver (FTR), to the product of those from transmitter to line (FTL), and from line to receiver (FLR). That is,

$$\text{FTH} = \frac{\text{FTR}}{\text{FTL} \times \text{FLR}} = \frac{2R_0(Z_L - \overline{Z}_L)}{(R_0 + Z_L)(R_0 + \overline{Z}_L)} \tag{11.1}$$

In the special case that $Z_L = R_0$, then $\text{FTH} = (Z_L - \overline{Z}_L)/(Z_L + \overline{Z}_L)$, the more familiar reflection coefficient.

As shown in Fig. 1.14, the input impedance of a telephone line varies widely. Since the loss of the local loop increases with loop length, it is most important to match $\overline{Z}_L$ to the average input impedance of long loops, which is marked in the figures. This can be done with acceptable accuracy by a simple circuit of the form shown in Fig. 11.3d. When using the active-RC circuit, the transfer function from line to receiver can be adjusted slightly—without affecting the im-

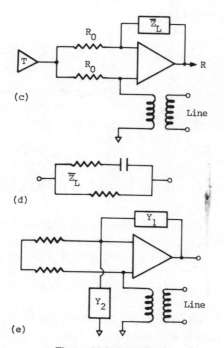

Figure 11.3 Continued

pedance match — by any variation of the form shown in Fig. 11.3e, in which $(Y_1 + Y_2) = \overline{Y}_L = 1/\overline{Z}_L$.

With such a hybrid, a crude estimate based on Fig. 1.14 is that for 90% of all long loops, the trans-hybrid loss can be made greater than 12 dB from 600 to 3000 Hz.

11.1.3   Specification of the Filters

Since the signals are supposed to be in separate bands, the ACI comes about in two ways: a reflected in-band transmit signal leaking through the receive filter, and a reflected out-of-band transmit signal passing straight through the filter. Thus the total ACI power is given by

$$\text{ACI} = \int_{600}^{3000} |\text{FTH}(f)\,\text{FL}(f)\,\text{FH}(f)|^2 \, df \tag{11.2}$$

where $\text{FL}(f)$ and $\text{FH}(f)$ denote the transfer functions of the low-band and high-band filters, without specifying whether they be calling or answering.

It is very difficult to design both filters or sets of filters (usually consisting of both low-pass and bandpass) together, so it is assumed that for FDX modems the overall system is nearly optimized when the two halves of the integral — from 600 to 1800 Hz and from 1800 to 3000 Hz — each contribute half of the allowable ACI. If FL and FH are normalized so that

$$\int_{600}^{1800} |\text{FL}(f)|^2 \, df = \int_{1800}^{3000} |\text{FH}(f)|^2 \, df = 1$$

then the stopband specifications for each can be expressed as

$$\text{AL} > 71 + 3 + 10 \log |\text{FTH}|^2 \tag{11.3}$$

where

$$\text{AL} = -10 \log \int_{600}^{1800} |\text{FL}(f)\,\text{FH}(f)|^2 \, df \tag{11.4}$$

and AH is similarly defined in the band from 1800 to 3000 Hz.

If we assume that the matching impedance in the hybrid is the best compromise, and the line impedance is a worst case, this means that

$$\text{AL} \simeq \text{AH} > 71 + 3 - 12 = 62 \text{ dB}$$

When each filter (or set) is designed separately from the other, the response of the other — particularly in the crossover region from 1500 to 2100 Hz — must be taken into account. This may require two iterations for the first set, using an initial guess at what the second set will provide (e.g., about 28 dB at 1800 Hz), and then refining the design using the actual attentuation achieved by the second.

All efficient filter designs have less attenuation close to the passband than far out in the stopband(s), so it is to be expected that the ACI power spectrum of a V.22 bis will have a peak at the "crossover" frequency, 1800 Hz, as shown in Fig. 11.4.

11.1.4 Design of the Filters

A computer program to design sets of filters is described and listed in Appendix I; in this section we will discuss the general configuration of the filters.

The basic decision is between baseband (low-pass) and passband filtering; factors to be considered are as follows:

1. Doing all the filtering in the baseband makes it very easy to change bands (by merely changing the frequency of the carrier input to each modulator) and bandwidths (by changing the clock frequency in sampled-data filters). This is not so important for FDX, where only two fixed bands are needed, but it might be very valuable in an asymmetrical modem that requires great flexibility of configuration (see Section 11.4).

2. Analog implementation is simpler because Q factors are much lower in LPFs.

3. Two baseband LPFs perform the same filtering—and are of the same *total* degree—as a symmetrical BPF; since the BPFs need high attenuation in only one stopband, the baseband approach results in overdesigned filters.

4. Baseband removal of ACI in the receiver is feasible only if we use free-running demodulation to keep the delay through the LPF out of the carrier loop—and then only if the demodulator is extremely linear. The quadrature switched-capacitor modulator described in [Bi5], implemented with balanced operational amplifers, would be one solution.

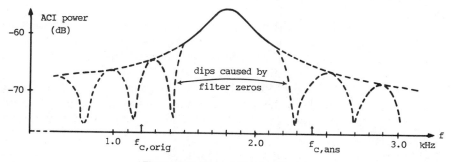

Figure 11.4 Overall filtering of ACI.

5. Direct-current offset in the receive LPFs would be a severe problem because nearly all ACI must be removed before the low-level receive signal can be amplified. Switched-capacitor technology with balanced operational amplifiers may be equal to the task.

The usual decision after consideration of these factors for an FDX modem is to split the filtering between passband and baseband and either (1) do only as much filtering in the baseband as is necessary to prevent aliasing on modulation in the transmitter and to remove extraneous products of demodulation in the receiver or (2) do only as much filtering in the passband as is needed to reduce the ACI to about the same level as the lowest receive signal, so that the AGC can work at almost full gain.

A point that needs to be stressed here is that each LPF/BPF combination must be designed as a set, with the square-root-of-a-raised-cosine shaping in the passband being distributed between the two filters as needed to reduce the sharpness of cut-off. One filter should *not* be designed according to some cherished procedure (e.g., [Pe]), because then the other may have to be "perfect" (i.e., have a flat passband); this would be very wasteful.

11.2 ECHO CANCELING

Early echo cancelers were independent of a modem; they were adaptive 4W/2W networks. The simplest versions adjusted the impedance $\overline{Z}_L$ in Fig. 11.3b so as to match the line impedance according to some criterion.

If a more precise match is needed,† it is achieved with a TDL, as suggested in [K&W] and shown in Fig. 11.5. The tap coefficients are usually adapted according to the SG LMS algorithm by correlating each sample of the canceler's error signal r'_k, which includes the wanted receive signal, with the signals x_n, present at the taps. The speed of convergence depends on the spread of the eigenvalues of the auto-correlation matrix of the transmitted signal. For data signals, which generally strive to use the whole frequency band equally, this spread is small — in contrast to the case of a TDL used for equalization, where the input signal may be severely distorted — and the convergence of such a canceler is fairly well behaved and predictable.

The major disadvantage of this method is that the input signal to the TDL must be quantified very finely (to at least 12 bits of resolution), and all multiplications by the tap coefficients must be of two full-length words.

Data-Driven Echo Cancelers. Since the inputs that generate the transmit signal are defined by very few bits (the transmit data *after* it has been subjected to any encoding), it is much more efficient to correlate the error with this data than

---

† We have changed to the present tense because, although the method is now seldom used for data signals, it is still the only practical one for voice echo cancelers.

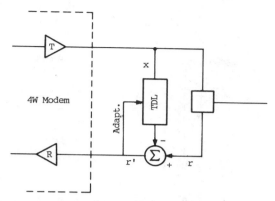

Figure 11.5 Echo canceler implemented separately from modem.

with the transmit signal. These more efficient *data-driven* methods were originally described in [Mu3] and [We2]; they are the subject of the rest of this chapter.

The basic principle of a data-driven echo canceler is the same as that of a DFE, except that an EC is simpler because (1) it does not have to make any decisions about the data whose echoes are to be subtracted (they are the known transmit data) and (2) there are no "leading echoes" that must be dealt with in some other way. However, although the basic principles are simplified, implementation of an EC is much more difficult. The main problems are that the level of the echo is usually much higher than that of the signal, and the delay of part of the echo may be as great as 550 ms.

The modifications that must be made to Fig. 11.5 depend on whether (1) the near and far data sources are synchronized,† (2) the data signals are baseband or modulated into a passband, and (3) symbol-rate or Nyquist rate canceling and receive processing is used. We will begin the discussion by considering one very special case. This may seem strange, but it will allow us to begin studying the basic problems of adaptation, full-duplex operation, and required arithmetic precision; the other problems of timing recovery and demodulation will be discussed in subsequent sections.

11.2.1 A Baseband, Synchronized, Symbol-Rate EC

Suppose that, by some means, receiver and canceler can be made to operate at the same clock frequency and phase. (See the discussion at the end of this section as to how this phase is decided.)

---

† Synchronization is usual for high-speed FDX communication on the subscriber loop but is not generally possible on the voice-band PSTN.

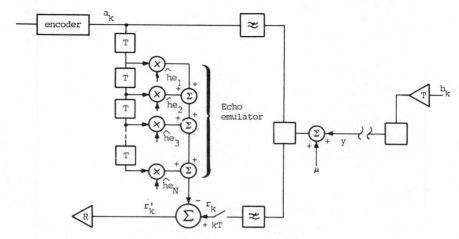

Figure 11.6 Data-driven, baseband EC.

In a basic EC, such as is shown in Fig. 11.6, the received signal at time kT is

$$r_k = x_k + \sum_{n=1}^{\infty} a_{k-n+1} \, he_n + \mu_k \tag{11.5}$$

where $x_k \, (= x(kT))$ is the desired signal from the far end, the he_n are samples (at the symbol rate, and at an arbitrary but constant phase) of the echo path impulse response, the a_k are the assumedly uncorrelated transmit data, and μ_k is a sample of noise with variance σ^2. The output of the canceler is

$$r'_k = x_k + \sum_{n=1}^{N} a_{k-n+1} \, (he_n - \widehat{he}_{n,k}) + \sum_{n=N+1}^{\infty} a_{k-n+1} \, he_n + \mu_k \tag{11.6}$$

where the $\widehat{he}_{n,k}$, the tap coefficients, are the estimates at time kT of the he_n. If the a_k are normalized so that the average power of the transmit data $\langle a_k^2 \rangle$ is unity, then — ignoring any ISI in x_k, which obviously the EC can do nothing about — the MSE at time kT has three components:

$$\text{MSE} = \sum_{n=1}^{N} (he_n - \widehat{he}_{n,k})^2 + \sum_{n=N+1}^{\infty} he_n^2 + \sigma^2 \tag{11.7}$$

Clearly the MSE is minimized if, after convergence, $he_n = \widehat{he}_n$.

When considering adaptive equalizers we recognized that successive samples of the received signal might be fairly strongly correlated, and that this weakened the theory of steepest descent to an LMS solution; nevertheless, the conclusion was still that it is best to update with every sample. For ECs, the input samples are much less correlated, and the conclusion is even easier to accept. The stochastic update equations for the tap coefficients are

$$\widehat{he}_{n,k+1} = \widehat{he}_{n,k} + \Delta\, a_{k-n}\, r'_k \qquad \text{for } n = 1 \text{ to } N \tag{11.8}$$

There are no problems here with eigenvalue spread,† and the step size for fastest convergence was shown in [Mu2] to be $\Delta_{\text{opt}} = 1/N$.

Half- and Full-Duplex Adaptation. Even though the samples x_k in (11.6) are the "desired" signal, they act like noise for the adaptation according to (11.8). Therefore, whenever possible, initial training of the EC is done in a half-duplex mode — that is, with the far transmitter silent. When the adaptation has fully converged Mueller showed that with a small step-size the MSE of (11.7) is given approximately by

$$\text{MSE} \approx \frac{N\Delta\sigma^2}{2 - N\Delta} \qquad \text{(HDX operation)}$$

$$= N\Delta\sigma^2 \qquad \text{if } \Delta = \Delta_{\text{opt}} = \frac{1}{N} \tag{11.9a}$$

When full-duplex transmission ("double-talking") begins, the power of the received samples, defined as P_r, generally increases the tap jitter, and

$$\text{MSE} \approx \frac{N\Delta(P_r + \sigma^2)}{2 - N\Delta} \qquad \text{(FDX operation)} \tag{11.9b}$$

Clearly, if the step size were maintained at $1/N$, the MSE would be intolerable — at approximately the same level as the received signal! There are three possible strategies to reduce the MSE: reduce the step size, average the update quantities, $a_{k-n}\, r'_k$, before updating the taps, or use an *adaptive reference*.

Reducing the Step Size. In order to reduce the MSE to, say, 30 dB below the received signal, the step size must be reduced to $\Delta = 1/(500\,N)$. An echo emulator with 100 taps is fairly typical, so this would require $\Delta \approx 2 \cdot 10^{-5}$. As we saw when discussing equalizers, reducing the step-size below a certain amount when using finite-precision arithmetic may be self-defeating; the jitter may increase. The jitter can only be reduced to near its theoretical level if each increment of a tap can be defined fairly precisely. For this example, this would require a tap precision of about 20 bits.

In practice, as reported by Weinstein, correlation between successive samples of the $y_k\,(= x_k + \mu_k)$ caused by amplitude distortion of the received signal

† As pointed out by Mueller, there is no matrix inversion involved — only an estimation.

may further reduce the filtering effect by about 6 dB, so that 22 bits would be needed for the coefficients.

Averaging Before Updating. Each update quantity $a_{k-n}\, r'_k$ can be accumulated for K_{av} symbol periods before updating the taps. If Δ were maintained at its training value of $1/N$, then $K_{av} = 500$ would be needed. The disadvantage of the method is the extra storage required for the cumulative updates. The method would be attractive only if it made the difference between using double and single precision storage for the coefficients—very unlikely for the example considered.

A Combination of Reduced Step Size and Averaging. If averaging is useful, then using $K_{av} = N$ (and therefore, for the example, $\Delta = 1/500$), and updating one coefficient per symbol period—rather than updating them all after K_{av} periods—would have the advantage of spreading the computation load evenly.

Adaptive Reference. The most effective way of reducing the tap jitter is to subtract an estimate of the x_k from the r'_k in (11.6) before correlating with the transmitted data. If there were no ISI in the received signal, the x_k would be equal to the far-end transmitted data, so that the output of the receiver could be fed back and subtracted from a suitably delayed r'_k.

To deal with ISI, the canceler must learn the impulse reponse of the channel (h_m for $m = 1$ to M). The learning algorithm is the same as in a DFE, but there should be no prefilter to make the impulse response "causal"; both leading and trailing echoes are learned by delaying the receive data enough that all terms, including the main one, look like trailing echoes. If the x_k are defined as usual by

$$x_k = \sum_{m=1}^{M} b_{k-m+1}\, h_m \qquad (11.10)$$

then h_m can be learned in the usual way by subtracting the effects of the impulse response learned so far ($h_{m,k}$) to form an error signal,

$$\epsilon_k = r'_k - \sum_{m=1}^{M} b_{k-m+1}\, h_{m,k} \qquad (11.11)$$

and then correlating this with the received data†:

$$\hat{h}_{m,k+1} = \hat{h}_{m,k} + \Delta_b\, \epsilon_k\, b_{k-m+1} \qquad \text{for } m = 1 \text{ to } M \qquad (11.12)$$

---

† During training—or at least the early stages of it—the received signal is hopelessly contaminated by the echo, and decision-directed learning is impossible; synchronized stored reference data must be used for both (11.10) and (11.11).

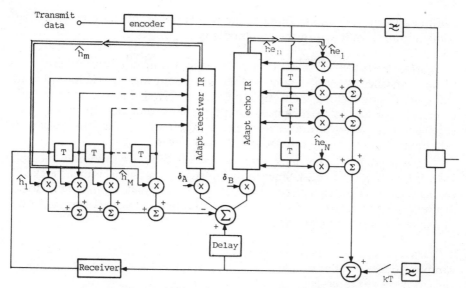

Figure 11.7 Baseband synchronized EC with adaptive reference.

The error signal given by (11.11) is then also used in (11.8) instead of r'_k to learn the $\widehat{he}_n$. A complete EC is shown in Fig. 11.7.

This method was first described by Falconer [Fa2]; he showed that the step size and the FDX convergence rate can be greatly increased over the previous *zero-reference* methods. This has three potential advantages:

1. It could reduce the total training time for both modems by using just one block of FDX transmission rather than separate blocks of HDX transmission from each end. However, half-duplex training is fairly generally accepted and is allowed for in all CCITT hand-shake protocols; this advantage is seldom significant.

2. The canceler can follow faster variations of the echo path. This is seldom important for the near echo, which is caused mostly by the local subscriber loop, and therefore changes only very slowly, but it may be very important for far echoes that have phase jitter or frequency offset (see Section 11.2.3).

3. The precision required of the taps is much less; this may be very important in some implementations.

Deciding the Sampling Phase. In a synchronized system one modem, the master, provides the clock, and the other, the slave, must lock its transmitter to its recovered receive clock. If the attenuation distortion of the channel is such that the received signal uses no excess bandwidth, and there is therefore no danger of aliasing distortion from sampling at the wrong phase, then the best receiver configuration is different in the two modems:

1. *Master Modem.* Cancellation of the echo, the most important task, can be accomplished most accurately — that is, without any degradation due to clock jitter — by locking the sampling of the received signal to the transmit clock. This phase can then be used — with perhaps some penalty in length of equalizer required — for the received signal. Alternatively, the samples can be interpolated at any phase decided on in the receiver.

2. *Slave Modem.* The clock must be recovered by a symbol-rate method [M&M], [ATMH], or [Li2]. Since the reflected transmit signal is then "locked" to this clock, the echo canceler automatically follows any slow jitter of the clock. However, as discussed in [Me4], because of delay in the clock recovery and echo paths, fast jitter cannot be tracked.

It should be noted that if the receive signal uses any excess bandwidth, the sampling phase must be decided by the receiver, and master and slave configurations become the same.

Positioning of the A/D Converter. In order to digitalize the canceler and receiver of either Fig. 11.6 or 11.7, the first thought would be to insert an ADC immediately after the sampler. However, this would mean that the ADC would have to have a dynamic range from the largest echo signal to the precision needed to process the smallest receive signal — often an impossible demand on the art of ADC design. Two more practical configurations adapted from [F&M3] are shown in Figs. 11.8a and 11.8b.

The arrangement in Fig. 11.8b is a development of that discussed in Section 5.3.2.3. It further reduces the needed dynamic range of the ADC by adjusting the gain before conversion. In order to control the gain of the EC loop, any signal gain added by the ADC (compressor) must be removed in the feedback path by a matched expandor.

11.2.2   Unsynchronized Systems with Excess Bandwidth

The sequence of design decisions may depend on whether an adaptive reference is used. If one is not, then:

1. As we have seen, the step size will depend on the signal : echo ratio and will probably range from very small to minute. It is likely that a receiver-timed canceler would not be able to keep up with the cyclic changing of the transmitter-timed samples of the echo impulse response. Therefore, the initial sampling must be at the transmit rate.

2. This means that this sampling must be at the Nyquist rate of the signal, in order that subsequently, after echo cancellation, either (*a*) the signal can

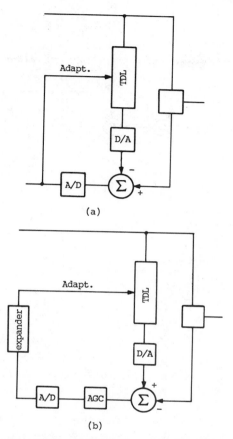

Figure 11.8 Positioning of the ADC in an EC: *(a)* after echo subtraction; *(b)* after echo subtraction and AGC.

be interpolated to produce samples of the right phase for symbol-rate receiver processing or (*b*) an FSE can be used.

Use of an adaptive reference would allow a larger step size and probably enable the canceler to follow a precessing receive clock. However, implementation of one in an unsynchronized canceler is much more complicated because, as discussed in [Fa2], the transmit data must be buffered, and the learned receive impulse response "slipped" by one symbol period occasionally as the two clocks precess. The consensus seems to be that an adaptive reference is impractical in an unsynchronized system—just when it is most needed!

11.2.2.1 Nyquist-Rate (Fractionally Spaced) Echo Emulators. If the sampling shown in Figs. 11.6 and 11.7—or any variation thereof involving some

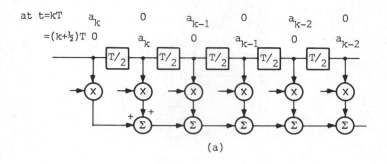

(a)

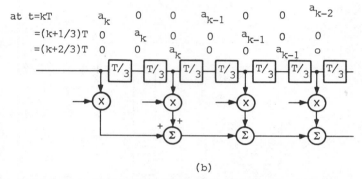

(b)

Figure 11.9 Fractionally spaced echo emulators: *(a) T/2; (b) 2T/3.*

configuration of ADC and DAC—is performed at a simple multiple of the symbol rate ($f_{samp} = Lf_s/M$, where L and M are small integers), the data input to the TDL can be considered as impulses at the symbol rate; consequently, at each time kMT/L only one Lth of the taps contribute to the output and can be updated. Emulators for the common cases of $f_{samp}/f_s = 2/1$ and $3/2$ are shown in Figs. 11.9a and 11.9b.

It can be seen that the number of calculations per tap per second is the same as for symbol-rate sampling.† However, the programming of a digital signal processor may be considerably more complicated—particularly in the 3/2 case—and the increases in code and total execution time will depend very much on the flexibility of the processor's instruction set and memory indexing, and on the programmer's ingenuity.

Also, it must be noted that we were comparing the numbers of calculation per second *per tap*. Although, as pointed out in [Q&F], a $T/2$ equalizer can sometimes do a better job than a symbol-rate equalizer *with the same number of*

---

† This is also true for fractionally spaced equalizers, but the two structures and algorithms are in a sense complementary. In an FSE all the taps contribute to the output each time, but the output need be calculated at only one (M/L)th of the sampling times.

taps (i.e., with twice the time span), this is not true for echo cancelers. For them the operation is subtraction of impulse responses, not multiplication; an EC is only effective over its time span. A $T/2$-spaced EC will need twice as many taps and twice as many calculations per second as a T-spaced EC. Nevertheless, the choice between symbol and fractional spacing does not now depend on a subtle assessment of the relative performance of equalizers, but on a much more basic fact; if the system is asychronous, and an adaptive reference is not used, then a fractionally spaced echo emulator is essential.

11.2.3 Cancelers for Unsynchronized Passband Modems

The most important modems of this type are high-speed voice-band modems used on the PSTN—typified by the V.32 recommendation. We will discuss their specification and design in some detail.

11.2.3.1 Attenuation Specifications. The echoes to be canceled must be considered in two groups: *near echoes,* delayed by less than some amount T_D (typically about 25 ms), which have traversed only a short or medium-length terrestrial path; and *far echoes,* delayed by more than T_D, which have traversed a long four-wire path including, perhaps, two satellite ups-and-downs, as shown in Fig. 1.4. The conditions that result in large near and far echoes are quite different, and, indeed, mutually exclusive, so we can calculate a specification for each based on the worst case, and assuming that each is the sole contributor to the echo.

Near Echoes. Near echoes can be caused by reflections from anywhere in a short or medium-length connection, and the impulse responses from these reflections may overlap. The calculation of required echo reduction is similar to that in Section 11.1.1 for FDM but is less demanding because the received signal is centered around a carrier frequency of 1.8 instead of 2.4 kHz:

Attenuation of received signal at 1000 Hz	27 dB
40% of distortion from 1000 to 3000 Hz:	7.5 dB
Attenuation of received signal at 1800 Hz:	34.5 dB
Average attenuation of signal in 600 to 3000 Hz band	33 dB

The 16QAM constellation of a V.32 *without* convolutional coding would require an effective SNR of about 20 dB for an error rate of 10^{-5}. If the echo could be canceled perfectly, then with a noise enhancement figure of 4 dB, and 2 dB degradation from all other impairments, as suggested in Table 5.1, one would expect to need an SNR of only about 26 dB. However, the few performance figures for V.32 modems that have been published so far [AT&T] suggest that an input SNR of about 28 dB would be needed without coding. Presumably the "other impairments"—particularly uncanceled echo—were much larger than expected.

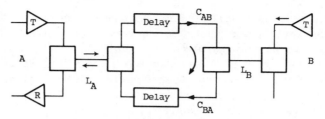

Figure 11.10 Calculating signal : far-end echo ratio.

This suggests that a reasonable requirement would be that uncanceled echo should not degrade an input SNR of 28 dB by more than 2 dB; that is, the signal : remanent echo ratio should be at least 30 dB. This means that hybrid and canceler must accomplish about 63 dB of echo reduction.†

Far Echoes. Far echoes are caused by reflections from one of the hybrids at the far end of a long four-wire circuit. Figure 11.10 shows a long (e.g., satellite) connection, with the losses of the loops (L_A and L_B) and four-wire carrier (C_{AB} and C_{BA}) segments. The Signal : Far-end Echo Ratio (SFER) is:

$$\text{SFER} = L_A + C_{AB} + \text{THL}_B + C_{BA} + L_A - (L_B + C_{BA} + L_A)$$
$$= L_A + C_{AB} + \text{THL}_B - L_B$$

It can be seen that the worst case arises when the losses of the near-end loop, outgoing carrier, and far-end hybrid are low, and the loss of the far-end loop is high. The appropriate extrema at 1000 Hz (see, e.g., [Fr4]) are:

Minimum near-end local loop loss	2 dB
Minimum four-wire loss‡	4 dB
Minimum trans-hybrid loss	6 dB
Maximum far-end local loop loss	(8 dB)
Worst-case SFER	4 dB

In this case the modem hybrid is no help—it obviously cannot match the delay of the long link; therefore, for a signal : remanent echo ratio of 29 dB, the canceler must reduce the far-end echo by 25 dB.

11.2.3.2 *Design of Unsynchronized Passband ECs.* Just as the adaptive equalization of a QAM signal can be performed in the passband or baseband, so

---

† This may be too undemanding a specification. Perhaps eventually echo cancelers will be able to do almost as well as FDM filters, and provide about 60 dB of echo reduction, so that in combination with the hybrid they can keep the degradation of SNR from uncanceled echo to less than 0.5 dB.
‡ The transmission losses of the two 2W/4W hybrids (typically about 3.5 dB each) are always included in the stated four-wire loss.

can echo cancellation; in both cases the differences in theory and implementation between the two bands are slight.

Baseband cancellation is the simplest in concept; a basic configuration is shown in Fig. 11.11*a*. The composite receive signal is demodulated to baseband,† and an error signal is formed just as in Section 11.2.2.1, except that now all quantities are complex. The echo-path emulator coefficients $\widehat{he}_n$ are updated by the complex extension of (11.8); that is,

$$\widehat{he}_{n,k+1} = \widehat{he}_{n,k} + \Delta b^*_{n-k}\,\epsilon_k \tag{11.13}$$

where ϵ_k represents either the net error after subtraction of the learned receive signal, or r'_k if an adaptive reference is not used.

The emulated echo can be modulated as shown in Fig. 11.11*b*, and the *analytic* (i.e., complex) passband echo subtracted from the analytic received signal. The complex error must then be remodulated before being used to update the taps. It can be seen that this is mathematically equivalent to the baseband method; however, its implementation is more complicated and there is little to recommend it.

A true passband emulated echo can be formed as in any conventional QAM modulation (see Section 4.2) by taking the real part of the modulator output. This is then subtracted from the real received signal, as shown in Fig. 11.11*c*. Weinstein [We2] showed that by discarding the imaginary part of the error signal, the adaptation is simplified — we shall see later by how much — but the gradient of the error with respect to the tap coefficients is halved, and the convergence time, for the same residual jitter, is approximately doubled.

The other relative merits of Figs. 11.11*a* and 11.11*c* can be assessed only if sampling rates and amounts of computation and storage are taken into account. It is easiest to illustrate this for a particular case, so let us consider a V.32 ($f_b = 9.6$ kbit/s, $f_s = 2.4$ kHz, and $f_c = 1.8$ kHz). Baseband and passband sampling rates of 3.6 and 7.2 kHz, respectively, are sufficient to characterize the signals ($\alpha \simeq 0.125$, so the highest baseband and passband frequencies are 1.35 and 3.15 kHz). However, to simplify the band-limiting filters, 4.8 and 9.6 kHz are often used, and we will assume these here. Since the time span of the emulator must remain more or less constant regardless of the configuration or the sampling rate, we will use the number of multiplications per millisecond for one millisecond of time span (mult/ms²) as the critical number for any considered realization.

Even though one of the multiplicands (the transmit data) used in calculating the output and tap update of the emulator has only a very few bits, so many more multiplications are needed for the emulator than for all the other operations combined, that its computation requirements dominate; in comparing methods, we can ignore the other operations.

---

† The demodulation is done using the transmit carrier; this is "free-running" for the receive signal, and an adaptive rotator is incorporated into the equalizer.

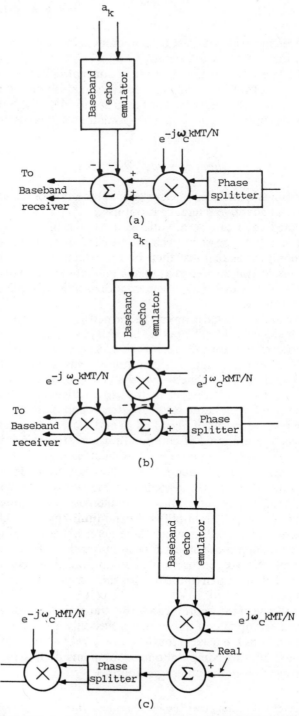

Figure 11.11 Subtraction of emulated echo: *(a)* in baseband; *(b)* in complex passband; *(c)* in real passband.

Baseband Cancellation. The baseband emulator shown in Fig. 11.11a operates at $f_{samp} = 4.8$ kHz and needs four multiplications each for output calculation and tap update. The critical number is therefore $(8 \times 2.4 \times 4.8) \simeq 92$ mult/ms².

Passband Cancellation. The passband emulator in Fig. 11.11c operates at $f_{samp} = 9.6$ kHz. Since only the real output is needed, it would seem that it should be possible to calculate the output with only two multiplications per sample. However, the multiplication by the carrier intervenes, and no such algorithm has been found. Therefore, with four multiplications for the output and only two for updating, the critical number is $(6 \times 2.4 \times 9.6) \simeq 138$ mult/ms².

Passband Cancellation with Interpolation. The previous number can be reduced by operating the emulator at 4.8 kHz and interpolating its output halfway between samples (i.e., to 9.6 kHz)† before multiplying by samples of the the carrier. Then the critical number becomes $(6 \times 2.4 \times 4.8) \simeq 69$ mult/ms².

Relative Merits of Baseband and Passband Canceling. The main problem with the baseband method illustrated in Fig. 11.11a is that the phase splitter operates on the full received signal (with high-level echo), and therefore must be much more precise than for the passband method in Fig. 11.11c. It must suppress the upper sideband that results from demodulation by more than 60 dB so that the sideband does not fold over into the band when the sampling rate is reduced to 4.8 kHz. Its design and implementation are formidable tasks; the design is discussed in Appendix I; implementation with switched capacitors and balanced operational amplifiers is probably within the capability of present, but rather esoteric, switched-capacitor technology.

Another way of implementing the baseband method — not necessarily simpler, but more conventional and less worrying — is to demodulate the signal and then use two low-pass filters to remove the upper sideband. This approach is very similar to the baseband approach to filtering for FDM, and the only problem is the linearity of the demodulator. However, it constitutes overdesign in the sense that the high attenuation of the echo is not really necessary.

If the doubling of the convergence time that results from using only the real error is acceptable, the passband method with interpolation in the baseband is preferable.

11.2.3.3 Separate Cancelers for Near and Far Echoes.

As we have seen, once a signal has entered the four-wire part of a connection, there is no further mechanism for echo until the 4W/2W hybrid at the far end. Consequently, on connections with an end-to-end delay of greater than about 8 ms, the near and far echoes are separated by a quiet period. For long terrestrial and satellite

---

† A symmetrical four-tap TDL might be used for this; the interpolation error is about −60 dB. More study is needed to determine whether this would be good enough.

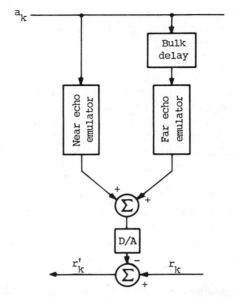

Figure 11.12 Separation of two echo emulators by a bulk delay.

connections, it would be prohibitively expensive and noisy to use adaptive taps throughout that span. Therefore, a *bulk delay* is used as shown in Fig. 11.12. The echo delay is measured during the initial hand-shaking by the phase-reversal method described in Recommendation V.32. This actually measures two end-to-end delays, so it must be reduced by an estimate of the extra delay through the far-end two-wire link, the far modem's receive and transmit filters, and the two-wire link again; a typical value would be 3 ms.

To prevent overlap of the near and far echo emulators on short and medium connections, the bulk delay must not be made less than the span of the near emulator. It might be thought that some of the far emulator could be disabled in these cases; however, since the "far" echoes on these connections are usually at a higher level than on long connections,† and double echoes (twice around the four-wire loop) may even be significant, it is worthwhile to leave both emulators fully operative.

Tracking of Frequency Offset in Far Echoes. All short- and medium-length four-wire links, and some satellite links, route both directions of transmission through the same, SSB, carrier system. Consequently, any frequency shift that is introduced into one two-wire path will be canceled, as far as the echo is concerned, by an opposite shift in the reverse path. However, some long con-

---

† Long-delayed echoes are more annoying to talkers than short ones, so the four-wire paths are more carefully engineered to reduce echoes.

nections (e.g., transatlantic) are routed via a satellite in one direction and a submarine cable in the other, and there may be a net frequency shift of the echo of as much as 1 Hz. Since the 99 percentile of the SFER is only about 7 dB,† it is clear that a continual rotation of this echo must be tracked. On the other hand, it is probably not necessary to track phase jitter, which has a maximum of only about 4° peak.

The complex taps of an echo emulator can adapt to any carrier phase (rotation) of the echo, but only very slowly. Even if an adaptive reference is used, the step size cannot be made large enough to track anything but the slowest phase jitter, or the most trivial frequency offset. Therefore, just as in adaptive equalizers, it is advisable to use a separate circuit or algorithm whose sole purpose is to adjust the phase of the far-echo emulator.

A baseband implementation of an adaptive echo rotator is shown in Fig. 11.13; the near and far echo emulators have N_1 and N_2 taps, $\widehat{he}_n$, with a bulk delay of $N_D MT/L$ between them, and it is assumed for the sake of simplicity that there are no significant echoes outside these ranges. The output of the second subtractor is

$$
r'_k = x_k + \sum_{n=1}^{N_1} a_{k-n}\,(he_n - \widehat{he}_{n,k})
$$

$$
+ \sum_{n=N_1+N_2-1}^{N_1+N_D+N_2} a_{k-n}\,(he_n\,e^{j\theta_k} - \widehat{he}_{n,k}\,e^{j\hat{\theta}_k}) + \mu_k \tag{11.14}
$$

where θ_k and $\hat{\theta}_k$ are the actual and estimated rotations of the far echo, respectively. If no adaptive reference is to be used, then, as shown by Weinstein, the squared modulus of the error term, $|r'_k|^2$, can be minimized with respect to $\hat{\theta}_k$ by the stochastic gradient method; the estimate of the gradient is

$$
\frac{\partial |r'_k|^2}{\partial \hat{\theta}_k} = r'_k \frac{\partial r'^*_k}{\partial \hat{\theta}_k} + r'^*_k \frac{\partial r'_k}{\partial \hat{\theta}_k} = -2\mathrm{Im}[r'_k\,v^*_k\,e^{-j\hat{\theta}_k}] \tag{11.15}
$$

where $v_k\,e^{j\hat{\theta}_k}$ is the rotated output of the far-echo emulator TDL, and the asterisks indicate complex conjugates.

If an adaptive reference is used, r'_k in (11.14) should be replaced by ϵ_k as given by a complex version of (11.11). In either case the gradient is estimated from the imaginary part of the correlation between the error and the output of the rotator. It can be seen that this estimate of the gradient of the error is analogous to the phase error in a conventional PLL, and, indeed, Weinstein showed that the expectation of the gradient is proportional to $\sin(\theta_k - \hat{\theta}_k)$.

Just as in a conventional PLL, a frequency offset can be tracked with no residual phase error by using a second order loop in which the loop filter shown

---

† Slightly larger than the number calculated in Section 11.2.3.1 because these long connections typically have more loss in their four-wire paths.

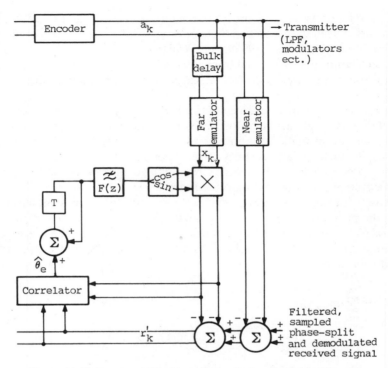

Figure 11.13 Baseband EC with adaptive rotation of emulated far echo.

in Fig. 11.13 has a pole at d.c.; that is, $F(z) = \gamma z/(z - 1)$. The gain γ and hence the closed-loop bandwidth are made very small in order to filter out the "noise" of the far-end signal. Then the new phase is given by

$$\theta_{k+1} = 2\theta_k - \theta_{k-1} + 2\gamma \text{Im}[r'_k\, v_k^*\, e^{-j\hat{\theta}_k}] \qquad (11.16)$$

It appears that this derivation is equivalent to that given in [We2].

Use of the Real Error Signal. Just as when updating the emulator taps, a passband simplification of the baseband algorithm can be made by using only the real part of the error signal, and correlating this with the imaginary part of the output of the rotator:

$$\theta_{k+1} = 2\theta_k - \theta_{k-1} + 2\gamma\, \text{Re}[r'_k]\, \text{Im}[v_k\, e^{j\hat{\theta}_\kappa}] \qquad (11.17)$$

However, the savings in computation are not now as great as before (25%, because six multiplications per tap per sample were needed compared to eight), because the full complex output of the far-echo emulator must be calculated; taking account of the relative lengths of the near and far emulators, the computational saving would now probably be about 15%.

11.3 DUPLEX DATA TRANSFER ON THE WIDE-BAND SUBSCRIBER LOOP

The present need for high-speed data transfer on the subscriber loop is for access to the Integrated Services Digital Network (ISDN). The standard that has been established is a combination of two 64 kbit/s "B" channels for digitized voice or high-speed data, and one 16 kbit/s "D" channel for data and/or signaling — requiring a total of 144 kbit/s full-duplex operation on a Digital Subscriber Loop (DSL). Although, as we have discussed previously, full-duplex communication between two users is rare, the requirement for FDX capability here is justified because it is anticipated that one modem may serve several terminals connected by a four-wire system within a customer's premises [M&G2] asymmetrical capability only would greatly complicate the protocol and reduce the throughput.

The specification and design of modems for the DSL is a very controversial subject at present (spring 1987); there have been special sections in three journals devoted to it,† and the CCITT is discussing an international standard. This section can serve as only a very brief introduction to the subject — describing some of the fundamental issues, and attempting to interpret some of the conclusions. The reader is referred first to one of the recent surveys [Me3], [GS&B] or [Me5], and then to the more detailed papers cited therein.

11.3.1 Multiplexing Methods

Of the four possible methods for FDX operation, EC, FDM, TCM, and ECBM, only the last two now need any explanation. Time–compression multiplexing, or *ping-pong,* is merely half-duplex transmission with periodic line turn-around, and buffering of input and output data to make the line appear full-duplex. It can be efficient on a subscriber loop because propagation delays are very short, and little or no guard time is needed between one modem seeing the end of a message and beginning its own — between a "ping" and a "pong". The data rate in one direction can be as little as 2.1 times the rate seen by the composite users.

This "excess time" factor is much less than the excess bandwidth that would be needed by the filters in an FDM system to achieve adequate channel separation. Furthermore, TCM is simpler to implement than FDM. It is clear, therefore, that FDM must be discarded as a candidate for the multiplexing method.

Echo-Canceling Burst Mode (ECBM) (see [BM&T] and [FT&M]) is a hybrid of EC and TCM. Each modem transmits in bursts at somewhat higher than the user's required rate; the bursts from the two ends mostly overlap but are staggered as shown in Fig. 11.14. During the time that B is silent, A trains its echo canceler, and B acquires its receive timing, and then vice versa. These

---

† The November 1981 and September 1982 issues of the *IEEE Transactions on Communications* and the November 1986 issue of the *IEEE Journal on Selected Areas in Communications.*

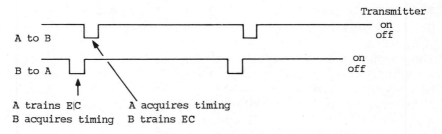

Figure 11.14 Timing of transmitters in EC burst mode.

short periods of half-duplex transmission greatly simplify the design of the appropriate circuits with only a fairly small penalty in bandwidth, loss, and crosstalk. However, as circuits and/or algorithms for EC adaptation and timing recovery improve [ATMH], [TH&M], and [SD&M], the attraction of this hybrid scheme will diminish.

Comparison Between Echo Canceling and Ping-Pong. A ping-pong system will have to use slightly more than twice the bandwidth of an EC system. If the channel had no attenuation distortion, this would mean a reduction of the SNR by slightly more than 3 dB. Since the channel has much more distortion at high frequencies (those that only the ping-pong system uses), the deterioration is certainly greater than this, but the actual figure would depend strongly on the signal format(s) chosen.

One main source of interference with a low-level receive signal is coupling from high-level transmit signals in other twisted-pair cables in the same cable sheath, as shown in Fig. 11.15; this is called *near-end crosstalk,* (NEXT). Since the coupling that causes NEXT is mainly capacitive, it increases with frequency (it is usually insignificant in the 300 to 3400 Hz voice band); therefore, it is much more potent in the wider bandwidth ping-pong system.

Some national PTTs (e.g., the Japanese) will synchronize transmission on their subscriber loops, so that all out-stations transmit simultaneously and there can be no NEXT. The elimination of this worst form of interference will usually—always, on long loops—more than compensate for the use of the wider bandwidth and the resultant higher noise level; ping-pong becomes the preferred method [Y&T] on the basis of both performance *and* cost.

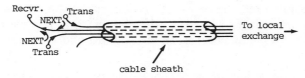

Figure 11.15 Sources of NEXT in a multiconductor cable.

On the other hand, for cases where synchronization is not possible, it has been shown [AM&S], [AC&N] that EC systems have a greater range because of their lesser bandwidth and consequent lower attenuation and NEXT. However, it must be pointed out that most calculations of performance of EC modems have been based on the assumption that the echo can be canceled perfectly. In practice, this may be far from true, and careful studies would be needed to determine whether the remanent echo is comparable to the extra NEXT.

11.3.2 Signal Formats (Line Codes)

Many different line codes have been proposed [B . . 1], [Sc], [F&W], and [B . . 2], and many different conclusions have been reached. The CCITT will probably have agreed on one by the time this book is published, but the decision is likely to have been influenced as much by political factors as by technical ones. However, because subscriber loops never cross national boundaries, one international standard is not as important here† as it is for voice-band data. Perhaps PTTs and operating companies will eventually buy DSL systems that best suit their needs and conform to their own independent technical judgment.

In considering line codes it soon becomes apparent that the choice should depend very strongly on the length of the loop. The subscriber loop is very different from other low-pass and bandpass channels in that its attenuation distortion, considered as the rate of change of attenuation with frequency, is greatest at low frequencies. Therefore, for short loops (less than about 3 km) on which the signal is not attenuated so much that NEXT becomes serious, it is best to use a code such as AMI that has the maximum of its power spectrum at $f_s/2$, where the attenuation *distortion* is not so great; then only a very simple equalizer — or perhaps no equalizer at all — is needed, and, particularly for synchronized ping-pong systems, a very inexpensive implementation is possible.

On the other hand, for longer loops (≤ 7 km), the more intuitive strategy — to put the signal energy where the line attenuation is least, regardless of the distortion — becomes the best one. A low-pass spectrum (perhaps duobinary) comes first to mind, but the low-frequency cut-off caused by transformer coupling results in an impulse response with a very long tail; d.c. restoral techniques [B&D], which are a type of DFE, become very cumbersome.

Of all the codes that produce spectra with zeros at d.c. and $f_s/2$ and are skewed toward d.c., the most promising is polybipolar [Le2], discussed in Section 3.3.4. It has been claimed that five level signals are optimum in a NEXT-dominated channel, and polybipolar has the further advantage that most of the noise immunity apparently lost by using more levels can be retrieved by the error correction techniques described in Section 9.3.3.

† The format of the data must, of course, conform to ISDN standards, but we are talking here only about the means by which that data is delivered to the local office — the *physical layer* as it is called in ISDN parlance.

11.3.3 Receiver Methods

Once the multiplexing method and line code have been chosen, many of the decisions about receiver design follow from what has been discussed elsewhere in this book. Let us assume echo canceling and polybipolar shaping:

1. Since no excess bandwidth is used, symbol-rate sampling can be used for both the canceler and equalizer, with the sampling phases found as described in Section 11.2.1.
2. In order to allow error correction after equalization, linear adaptive equalization must be used. The enhancement of noise and NEXT will be less than with other signal formats because the equalizer is trying to restore only the $(z + 1) \cdot (z - 1)^2$ spectrum shown in Fig. 3.6.

However, there are still some problems:

1. Equalization of the distortion caused by bridged taps is much more effectively done with a DFE, which appears to be incompatible with error correction. Perhaps this problem will be solved or some simplified form of MLSE, as discussed in Section 9.4, will become feasible.
2. The $(z + 1) \cdot (z - 1)^2$ shaping allows several variations in how the shaping is divided between transmitter and receiver. Following the suggestions in [Fo1], [N&P], and [B . . 2], it would seem that $(z + 1)$ should be in the transmitter, and the difference between $(z - 1)^2$ and the line shaping achieved with a fixed and/or adaptive equalizer. However, the high-frequency content of the transmit signal may cause problems of coupling (RFI) into other twisted pairs in the cable sheath.

11.3.4 A Speculation

The subcriber loop might be used for other types of high-speed data than FDX access to ISDN—for example, user-to-user communication via just the local exchange. Then, if this communication follows the usual pattern, the required data speeds will be high in one direction and low in the other for a while, and then vice versa. It would seem that these needs could be met much more efficiently with ping-pong than with echo canceling (essentially a symmetrical method since both signals use the same, full bandwidth), because the data could be buffered, and then the duration of the pings and pongs allotted to fit the demand. Consequently, the throughput of a ping-pong system using *the same transmission rate* as an EC system would be only slightly less than that of the EC system. The cost of the modems would probably be about half that of those using echo canceling.

11.4   AN ASYMMETRICAL DUPLEX VOICE-BAND MODEM FOR 9600/600 BIT/S

Most of the requirements for high-speed user-to-user data communications on the PSTN can be met with a very asymmetrical speed capability; accordingly, the CCITT is now (spring 1987) considering proposals for a 9600 bit/s modem with a slow-speed, reverse channel. Proposals so far have been limited to just 300 bit/s for this channel, but many users think that at least 600 bit/s are needed for efficient operation of most error correcting protocols.

The following suggestions may come too late to influence the discussions, but at least they can serve as an illustration of the application of many of the ideas in this book. Some of them have not been subjected to peer scrutiny, or tested—beyond some preliminary computer simulation—and are offered mainly as food for thought.

11.4.1   Multiplexing Method

Because of the great difference in speed and bandwidth required for the two directions, the performance advantages of echo canceling shown in Table 11.1 do not apply; the symbol rate of the main 9600 bit/s channel must be 2400 s/s whether EC or FDM is used. Moreover, since filtering is a much more effective way of removing a reflected transmit signal than echo canceling, the performance of an FDM system will be better. Since FDM is also much less expensive to implement, it is the clear choice.

11.4.2   Assignment of Bands

The traditional band assignment—going all the way back to the Bell 203 in the mid 1960s—has been with the slow-speed channel at the lower end; this has been continued in one proposal to the CCITT [CDSI]. However, this assignment should now be reversed for two reasons:

1. The relative effects on the throughput of errors in the two directions have not been analyzed; they would depend very much upon the error correction protocol used (see Section 12.2). Lacking such an analysis, it is reasonable to strive for equal error rates in the two directions. Because of the high-frequency roll-off of the voice channel, the SNR at the lower end is considerably higher than at the upper end. This means that the upper end should be less loaded (in terms of bits per symbol). This principle is followed scrupulously in the multichannel OMQAM modem discussed in Section 4.5. It can be followed here in a much less precise way by putting the low-speed channel at the upper end and adjusting its number of bits per symbol to compensate for the lower SNR. There is no flexibility for the high-speed channel; it must use 4 bits per symbol, and so should be placed as low in the overall band as possible. The calculation of the

loading for the low-speed channel goes as follows: The carrier frequencies for the two channels will be approximately 1600 and 3000 Hz, and interpolation from Fig. 1.9 shows that the 90 percentile difference in attenuation between these frequencies is 10.5 dB. Since most of the noise in a voice-band signal is added either at the receiving local office or before, it is also filtered by the receiving subscriber loop. A reasonable estimate of the 90 percentile SNR difference between channels centered around 1600 and 3000 Hz would therefore be 6 dB. Table 4.2 shows that the SNR penalty in going from 4QAM to 16QAM is 7 dB, so it is clear that the loading of the upper channel should be only 2 bits per symbol.

2. Much greater flexibility in choice or carrier frequencies and bit rates can be achieved by doing all the filtering for band separation in the baseband. Also, great simplicity of implementation can be achieved by using simple square-wave modulators.† These two desiderata are unattainable if the low-speed channel is put at the lower end (f_c around 400 Hz), because sidebands of harmonics of the carrier would fall in the high-speed band.

11.4.3   Various Speed Mixes

It might sometimes be both desirable and possible to increase the low-speed to 1200 bit/s. Also, the option of 4800/4800 bit/s might be useful; in this case we could not afford the luxury of using only 2 bits per symbol in the upper band.

11.4.4   Signal Format

Some voice channels—particularly those in Europe—cut off very sharply below 300 Hz, and the usable band extends from 300 to about 3200 Hz (sometimes as high as 3400 Hz). For the 9600/600 bit/s configuration, the total symbol rate is (9600/4 + 600/2) = 2700 s/s, and full-response signal formats could be allowed excess bandwidths of only about 10%; this would require impossibly steep filters.

On the other hand partial-response shaping makes the filters quite realizable. For 9600/600 bit/s, the total band should extend from about 400 to 3100 Hz with carrier frequencies around 600 and 2950 Hz.‡ The 9600/1200 bit/s combination would be feasible only if a band from 300 to 3300 Hz could be used; then the carrier frequencies should be 1500 and 3000 Hz. FDX 4800/4800 bit/s should be placed in the band from 400 to 2800 Hz.

Initially, simple partial-response error correction (as described in Section 9.3.3) could be used, but better performance using trellis coding as described in [W&U] should be investigated.

---

† See [Bi3–Bi5] for switched-capacitor implementation of these.

‡ With the implementation envisaged, there would be no advantage in having the ratio of carrier frequency and symbol rate equal to the ratio of small integers; the exact carrier frequencies would be chosen for easy division down from some high-frequency crystal oscillator.

11.4.5 Implementation

The essence of this modem would be simplicity and low cost. These could be achieved with a mixture of switched-capacitor implementation of the filters, modulator, demodulator, and A/D converter, and DSP for the equalizer and error corrector. Band separation would be done in the baseband, and symbol-rate switching (mainly between high and low speeds, but perhaps also between two low speeds) would be achieved by changing the switched-capacitor clock frequency. The unsolved problems are (1) timing recovery from a complex baseband duobinary signal that is not locked in phase and (2) d.c. offset in the receive low-pass filter; if this latter could be kept low enough—by the use of balanced compensated operational amplifiers—that it did not impinge on the dynamic range of the ADC, the remaining offset could be compensated for digitally.

11.4.6 A Postscript (January 1988)

Two things have happened in the past nine months that will affect the development of asymmetrical modems and the agreement upon a standard for them: (a) the price of V.32 modems has fallen considerably, and it may soon fall so low that the development of a more logical (because it would be cheaper in the long run)—but nevertheless redundant—9600/600 bit/s modem will not be attractive, and (b) the CCITT has shown interest in a PSTN version of a V.33 with a slow reverse channel, to be used primarily for the ultrareliable transfer of facsimile.

It is unlikely that the partial-response approach proposed in Section 11.4.4 could be extended to 14.4 kbit/s. On the other hand, it seems very unlikely that the voice band as defined by some "international-composite" attenuation specification would have enough room outside the 400–3200 Hz main channel for an FDM 600 bit/s reverse channel. It will be an interesting problem!

Chapter 12

The Ancillary Functions Needed to Make a Full Data Set

12.1 ASYNCHRONOUS-TO-SYNCHRONOUS CONVERTERS

Everywhere else in this book we have assumed that the data that is presented to a modem is accompanied by a clock signal at frequency f_b, and that the data changes only on one edge of that clock.† However, as noted in Section 1.2.5, a considerable amount of data is presented in character format without a clock and must be converted to what we will call *synchronized data* — to distinguish it from originally synchronous data — before it can be accepted by the "synchronous" modem. A typical character with K bits, including *start* (a zero) and *stop* (a one), is shown in Fig. 12.1. Values of $K = 8$, 9, 10, and 11 are used, but 10 — consisting of a 7-bit ASCII character with one parity bit — is by far the most common and will be assumed in all examples in this chapter.

For this form of data, two rates are important: the *intracharacter* asynchronous bit rate f_a and the *character* rate f_{ch}. The bit rate is often imprecise, but modems and receiving terminals (RDTEs)‡ seldom attempt to adapt to it. The data is detected by noting the negative edge of the start bit, and then sampling the data signal at $t = (2k - 1)T_{a,r}/2$, for $k = 1$ to K, where $T_{a,r}$ is the inverse of the f_a of the receiving device. After they have recognized the stop bit, they are ready to resynchronize on another start bit. It can be seen that if all operations were performed perfectly, with no jitter on any clocks, the new start

---

† EIA Standard 232-D specifies that the data passed across the DTE/DCE interface should change on the positive-going (p-g) edge of the clock.
‡ The term "DTE" will be used generally to denote all sources and destinations of data, including multiplexers.

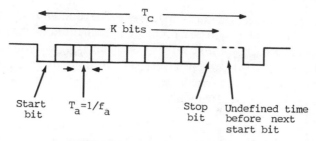

Figure 12.1 Character-formatted data.

bit could begin immediately after the stop bit was sampled; the absolute maximum transferable character rate would be

$$f_{c,\max} = \frac{f_{a,\min}}{K - \frac{1}{2}}$$

(12.1)

It should be noted that it makes no difference whether (1) all the bits are of the same shortened duration and the sampling slowly slips back as the character progresses; (2) all but the stop bit are nominal, and the stop bit is only half-duration; or (3) any combination in between exists.

With a precise receiving device, this would allow 5% "overspeed." In practice, however, to allow for jitter on all clocks, only half of this amount is usually permitted. Coincidentally (?), it has been observed that the data rate of some DTEs can vary about $\pm 2.5\%$ from nominal.

There are two basic ways of synchronizing this data.

12.1.1 Synchronized Data Rate Greater than the Maximum Asynchronous Rate

If the (accurately controlled) synchronous data rate $f_b > f_{a,\max}$, the basic algorithm for the asynchronous to synchronous converter (ASC) can be described very simply. Sample the state of the synchronous clock at the beginning (negative edge) of each start bit. When the sampled value changes from, say, zero to one, this means that the incoming data has slipped relative to the synchronized data. Then the start bit in the synchronized data must be delayed by inserting one full extra stop bit at the end of the previous character; this will not worry the RDTE at all. This method is simple to implement in the transmitter and requires no processing in the receiver; the synchronized, character-formatted data with occasional extra stop bits can be passed directly to the RDTE.

This method has one great advantage, simplicity, but two potential disadvantages:

1. The synchronous data rate is no longer a nice round number such as 1200 or 2400 bit/s; this might be a problem if the method of implementing the modem relied on a simple ratio of clock and carrier frequencies.
2. Modems that must accept either synchronous or asynchronous data would have to have the ability to change their synchronous data rate (and perhaps their carrier frequencies)

This method was used in the first FDX 1200 bit/s modem, Vadic's VA3400, but was rejected for 212, V.22, V.22 bis, and V.32 modems in favor of the following more complicated method.

12.1.2   Synchronized Data Rate Equal to Nominal Asynchronous Rate

The basic algorithm for the ASC is a development of the previous one. The state of the synchronous clock sampled at the beginning of the start bit may now change in either direction; therefore, it must be defined by two bits (i.e., samples of the f_b and $2f_b$ clock signals considered as the MSB and LSB, respectively) as shown in Fig. 12.2. If it changes from 01 to 10 (or 10 to 01) in response to slow (or fast) input data, the converter must add (or delete) a stop bit accordingly. To deal with a possible 2.5% overspeed, the converter must be allowed to delete up to one stop bit from every four characters.

With this type of Asynchronous-to-Synchronous (A/S) conversion there must also be a receive S/A Converter (SAC) to restore deleted stop-bits. The implementation of this would be simplest if it passed data to the RDTE at the maximum possible character rate, so that it only ever had to insert stop bits. However, CCITT recommendations specify that the bit rate be the accurate f_b; that is, that the input and output bit rates of the SAC be the same. The SAC must therefore (1) start a counter on the beginning of each character; (2) count K bits, and if the Kth bit is not a one (i.e., if the stop bit has been deleted), insert a

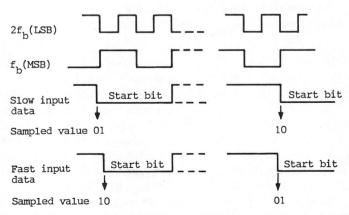

Figure 12.2 Sampling $2f_b$ and f_b clocks to detect slipping of asynchronous data

shortened stop bit of duration $(N-1)T_b/N$; and (3) shorten the stop bits of the $(N-1)$ subsequent characters by the same $1/N$ amount, so that it can get back into character synchronization with its receive data. The CCITT recommends that N be 8 for normal operation and 4 in an overspeed mode.

Overspeed and Inaccuracy in the DTEs' Bit Rates. An ability to delete stop bits is provided to allow for inaccuracy in the TDTE's $f_{a,T}$, but it must be recognized that the critical quantity is the *difference* between the bit rate of the TDTE and the sampling rate of the RDTE; that is, $(f_{a,T} - f_{a,R})$. The interposing of a synchronous modem, and the particular A/S algorithm (i.e., fast or nominal f_b) that is used *make no difference* to the total mismatch that can be handled. With jitter free clocks the maximum proportional mismatch is 5%, but 2.5% is generally specified. If a multiplexer occasionally generates bursts of data at $1.025 f_{a,nom}$, then the RDTE's $f_{a,R}$ must be exact. The CCITT's normal mode of operation, allowing for 1.25% of excess speed (no more than one stop bit in eight deleted), is probably intended to distribute the tolerance equally between TDTE and RDTE; however, it is not easy to understand why the overspeed mode should not be used all the time.

The Problem of Jittery Data from the TDTE. If the data presented to an ASC is jittery but its average rate is close to nominal, it is quite possible that the simple algorithm described so far would alternately insert and delete stop bits. The inserted ones would pass unnoticed through the receiver's SAC, but the deleted ones (every other character) would violate the rule and could not be replaced.

Therefore, the ASC must filter out the jitter in some way. A simple — and probably adequate — way is to store the pairs of bits that represent the states of the clock at the beginning of four successive start bits and only insert or delete in these cases:

Oldest			Present	
01	01	10	10	confirmed slow data; insert stop bit
10	10	01	01	confirmed fast data; delete stop bit

The Effects of Errors in Start and Stop Bits. If a simple asynchronous modem — typically using Frequency Shift-Keying (FSK) modulation — makes an error in a start bit, an RDTE would not start its character count until it received a zero and so would lose character synchronization. Depending on the data, it might make many character errors before it regained sync.

If a stop bit is incorrect, most RDTEs would interpret that character correctly, but then wait until they received a one/zero transition before restarting their character count; the character-error multiplication would be similar to that for incorrect start bits.

Interposing two synchronous modems can reduce the character-error rate significantly. There are four situations that must be examined:

1. *Start Bit Wrong, First Data Bit a Zero.* The ASC is one bit late in its character count, thinks that the next stop bit is missing, inserts one, and is immediately back in character synchronization; there would be only one character error.

2. *Start Bit Wrong, First Data Bit a One.* The ASC is more than one bit late in its count and will make several character errors as it regains synchronization.

3. *Stop Bit Wrong, Last Data Bit of Next Character a One.* The ASC assumes the stop bit has been deleted, inserts one, becomes one bit early in its count, thinks that the next character has a double stop bit, ignores it, and is immediately back in synchronization.

4. *Stop Bit Wrong, Last Data Bit of Next Character a Zero.* The ASC inserts a stop bit, and then "worries" that the next character apparently also has a missing stop bit; the algorithm should realise that this cannot happen, change the bit to a one, then observe and ignore an apparent double stop bit, and be immediately back in character synchronization.

It appears that a properly contrived ASC algorithm can reduce the occurrence of multiple character errors to one quarter of what it would be without the converters.

12.2 ERROR CORRECTION BY RETRANSMISSION

This can be only the briefest of introductions to a vast subject that is mostly beyond the scope of this book. The reader is referred to the documents submitted to TR-30.1 under project number PN-2012 for more details. The "error-free" transfer of data from modem A to modem B requires feedback — using either duplex transmission or ping-pong — from B to A. The transmission from B to A can, in some protocols, also be used for true data transfer, but we will be concerned here only with the feedback needed for error correction. This feedback is the responsibility of the modem alone; the DTEs are concerned with only the primary data.

We will first describe a generic method and then consider some of the variations of it being discussed by the CCITT for a probable recommendation, V.erc.†

1. Modem A assembles the data into blocks and appends extra characters for framing, control, and addressing.

2. It also generates a Frame-Check Sequence (FCS) — a generalization of

---

† During the early stages of consideration, proposed recommendations are given only temporary letters — often mnemonic.

what is also known as a Block-Check Character (BCC)—by dividing, modulo 2, the data sequence by a Cyclic Redundancy Check (CRC) polynomial, and appends the FCS to the block.

3. Modem B checks each block for integrity; if there are no errors, it passes the block to the RDTE and sends back an acknowledgment (ACK) to A.† If there are errors—in block k, say—it witholds the block from the RDTE, and sends back k, the address of the block, and a negative acknowledgment (NAK). Its action then depends on the retransmission strategy used.

4. On receiving a NAK, A has two possible strategies: for *(a)* "Go Back N," it retransmits the block k, and all subsequent blocks, or for *(b)* "Selective Retransmit," it sends only the block k, and then continues the message where it was interrupted.

5. With the "Go Back N" method, modem B discards the erroneous block k and all subsequent blocks until it receives block k again. With "Selective Retransmission," it discards block k, and then stores blocks $(k + 1)$ to $(k + N)$—assuming that they are judged to be error-free—until it receives block k again; it then passes k to the RDTE, followed by the stored blocks.

The most important versions of this basic algorithm are Link Access Procedures (principally LAPD), and Microcom Network Protocol (MNP). Both operate according to the rules of High-level Data-Link Control (HDLC), which works with eight-bit bytes (octets) throughout. LAPD as used for ISDN is defined in CCITT Recommendations Q.920 and 921. Modifications and refinements of the generic algorithm include:

1. The block, or frame, size is negotiable between the modems during their hand-shaking; LAPD suggests a maximum of 128 bytes (approximately 1000 bits).

2. In order to conform to HDLC, explicit NAKs are not used; modem B waits until it receives another block without errors, and then acknowledges that and requests retransmission of the previous block.

3. Neither LAPD nor MNP uses selective retransmission; the small increases in throughput at typical block error rates of 10^{-5} that it would bring do not seem to justify the extra complications.‡

4. With HDX modems, turning the line around in order to acknowledge every block may drastically reduce the throughput. Conversely, if the reverse channel of a duplex modem is fast enough to allow the acknowledgment of

---

† With an "agressive acknowledgment" policy—usually implemented only with duplex modems —each block is acknowledged as it is received; otherwise, B may wait for several blocks.
‡ For efficiency, the extra memory needed to store blocks $(k + 1)$ to $(k + N)$ in the receiver would have to be shared with the transmitter.

every block, then this will increase the throughput, because it will minimize the number of blocks that must be retransmitted after an error. If, for example, the forward block and the acknowledge signal contained 128 and 8 bytes, respectively, this would require one-sixteenth of the forward speed—600 bit/s for 9600 bit/s.

5. The two modems will usually agree on a *window size,* which is the maximum amount of time that a transmitter will be prepared to wait for an acknowledgment; this is usually determined by the amount of storage the modems have available for the transmitter. Since this allocation generally does not affect throughput, there is seldom any advantage in trying to reduce it adaptively.

Error Correction with Synchronous Data. The appending of framing and error checking characters and the occasional need for retransmission would require that the modem be able, from time to time, to refuse to accept data from the TDTE. This refusal can be signaled by turning off "Clear to Send" (Circuit 106) (see Section 12.4 for a definition of this interface lead), but that is often not acceptable, and consequently error control for synchronous transmission is usually considered to be the responsibility of the DTEs. The only effect that this has on the modem is that sometimes the system designer may prohibit scrambling and/or differential phase encoding within the modem in order to reduce the correlation between errors.

Error Correction with A/S Conversion. The octets required by HDLC are created by stripping the start and stop bits from a 10-bit ASCII character, error control is performed by the two modems, and then the "asynchronous" characters are reconstructed from the nearly error free synchronous data. Although the original character-parity bits are now essentially extraneous, they are treated as data. The 20% reduction in the number of bits resulting from the stripping is greater than the increase due to the addition of framing and error control bits for all practical block lengths. Therefore, with enough buffering capability—preferably switchable between transmitter and receiver—a pair of modems can provide nearly error free character-formatted transfer at the nominal "bit" rate.

Other character formats are handled as follows:

1. Nine data bits cannot be dealt with.
2. When there are 8 data bits plus a parity bit the parity bit is stripped, and then replaced when the character is reconstructed.
3. All parity bits appended to characters with 7 or fewer bits are treated as data.
4. All characters with fewer than 8 bits (including parity) are padded out to 8 by adding zeros at the end.

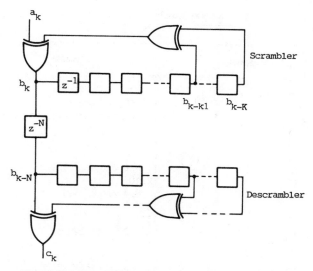

Figure 12.3 Scrambler and self-synchronizing descrambler.

12.3 SCRAMBLING

As we have seen, adaptive equalizers and echo cancelers rely on the "random-ness" of the data.† There are many ways in which quasi-randomness can be achieved, but one is particularly useful, and almost universally used, because the inverse process—the descrambling—is self-synchronizing. We will con-sider only this method.

The two processes are illustrated in Fig. 12.3 for the simple case where scrambler and descrambler have only two taps. It can be seen that if the trans-mission path is only a delay (i.e., there are no errors), then, after an initial loading period, the contents of the second shift register are a delayed version of those of the first. Thus

$$b_k = a_k \oplus b_{k-k_1} \oplus b_{k-K} \tag{12.2}$$

and
$$\begin{aligned} c_k &= b_{k-N} \oplus b_{k-N-k_1} \oplus b_{k-N-K} \\ &= (a_{k-N} \oplus b_{k-N-k_1} \oplus b_{k-N-K}) \oplus b_{k-N-k_1} \oplus b_{k-N-K} \\ &= a_{k-N} \end{aligned} \tag{12.3}$$

In the terminology of coding theory, the scrambler divides the input data sequence, modulo 2, by the generator polynomial $(1 + z^{1-k_1} + z^{1-K})$, and the descrambler multiplies the scrambled sequence by the same polynomial. In the

---

† Carrier and timing recovery operations do also, but in a different way; they typically work better with some steady data patterns, and not at all with others.

TABLE 12.1   Simplest Sets of Scrambler Taps

K	k_1, k_2, etc.
3	2
4	3
5	3
6	5
7	6 (V.27), and also 4
8	6, 5, 4, and several other sets
9	5
10	7
11	9
12	11, 8, 6, and also 10, 9, 3
13	12, 10, 9
14	13, 8, 4, and several other sets
15	14
16	15, 13, 4
17	14 (V.22 and V.22 bis)
18	11
19	18, 17, 14
20	17
21	19
22	21
23	18 (V.29, V.32[a], and V.33)
24	23, 22, 17
25	22

[a] This is for the calling modem; the answering modem uses the time inverse.

terminology of digital filters, the scrambler has an all-pole transfer function, and the descrambler, an all-zero one.†

If the input to a scrambler is some simple repeated pattern, the output is also repetitive. In order to maximize the period of these output patterns, the generator polynomials are usually chosen to be *irreducible* and *primitive* [P&W1]. Then, if the input is held at zero, the first exclusive-OR becomes transparent, and a pseudo-random maximum-length ($2^K - 1$) sequence is generated.

It is preferable to implement these scramblers with the minimum number of feedback taps; there is always a tap at the end of the shift register (at the Kth position), and most can be realized with just one more tap. Table 12.1 lists the simplest configurations for $K = 3$ to 25‡; another complementary set whose sequences would be the time inverse of those of the listed set can be obtained by using $k_1' = K - k_1$, $k_2' = K - k_2$, and so on.

---

† The complementary arrangement with all-zero scrambler and all-pole descrambler would work properly only if both registers were initially loaded with the same data; it would not be self-synchronizing.

‡ The longest scrambler recommended so far by the CCITT has $K = 23$.

Hang-Up. If the input of these scramblers were held at zero and the shift register somehow entered the all-zeros state, it would stay there and the output would be a steady zero. Similarly, if the input were held at one, then the all-ones state would be self-perpetuating. Several of the CCITT recommendations have included requirements that a transmitter test for these and other simple repeated output patterns, and reset the scrambler if they occur.

Error Multiplication in a Descrambler. It can be seen from Fig. 12.3 that if a transmission error occurs at time k, it will invert the descrambler output at times k, $(k + k_1)$, and $(k + K)$; that is, the bit error rate in this case will be multiplied by 3. In general, it will be multiplied by the number of nonzero coefficients in the generator polynomial.†

If the modems use differential phase encoding and coherent detection, then, as we have seen, any quadrant errors will be decoded as pairs of errors separated by M, $(M - 1)$, or $(M - 2)$ correct bits with probabilities $\frac{1}{4}$, $\frac{1}{2}$, and $\frac{1}{4}$, respectively, where M is the number of bits encoded per symbol. These error patterns at the input to the descrambler can cause some puzzling effects at the output. For example, for both a V.22 ($M = 2$) and a V.22 bis ($M = 4$), 25% of the errors would be separated by two correct bits, and when they reached the 14th and 17th taps, they would cancel each other. Thus 75% of the output errors would occur in groups of six, and 25%, in groups of four.

12.4 INTERFACE TO A DTE

The connection of a modem to a DTE has traditionally been made by 25-pin connectors and cable defined by EIA standard 232-D and CCITT Recommendation V.24. There are currently two opposing efforts to change this—to a 37-pin connector defined by EIA-449, and to a six-pin "mini" connector. The expansion to 37 pins was intended to allow a much more sophisticated control of a modem, but concurrently with its development the range of functions that a modem can perform on its own initiative without that control (e.g., signal quality assessment and error correction) has been greatly widened.

Most of the important aspects of the DTE/modem interface can be understood by studying EIA-232-D, and we will restrict our discussion to that. The CCITT circuit numbers, the connector pin numbers, the EIA alphabetic labels, and descriptions of all the interface signals are shown in Table 12.2. Before beginning the design of any particular modem it is essential to study the detailed sequencing of these signals that is required for calling, answering, line conditioning (e.g., disabling echo suppressors), and receiver training; these are prescribed in the appropriate modem manuals, proposals, and recommendations, and there would be no value in giving details here. The following notes and

---

† This is another reason for keeping the number of taps to a minimum.

TABLE 12.2   EIA-232D and V.24 DTE/DCE Interface

V.24 Circuit No.	Pin No.	EIA Label	Description — Abbreviation	Direction
101	1	AA	Chassis ground	DTE to modem
102	7	AB	Signal ground — SG	
103	2	BA	Transmit data — TD	DTE to modem
104	3	BB	Receive data — RD	Modem to DTE
105	4	CA	Request to send — RS	DTE to modem
106	5	CB	Clear to send — CS	Modem to DTE
107	6	CC	Data set ready — DSR	Modem to DTE
108/1	20	CD	Connect data set to line	DTE to modem
108/2	20	CD	Data terminal ready — DTR	DTE to modem
109	8	CF	Carrier detect (received line signal detect)	Modem to DTE
110	21[a]	CG	Data signal quality detect	Modem to DTE
111	23	CH	Data signaling rate selector	DTE to modem
112	23(12)[b]	CI	Data signaling rate selector	Modem to DTE
113	24	DA	Transmit clock	DTE to modem
114	15	DB	Transmit clock	Modem to DTE
115	17	DD	Receive clock	Modem to DTE
118	14	SBA	Secondary transmit data	DTE to modem
119	16	SBB	Secondary receive data	Modem to DTE
120	19	SCA	Secondary request to send	DTE to modem
121	13	SCB	Secondary clear to send	Modem to DTE
122	12	SCF	Secondary carrier detect	Modem to DTE
125	22	CE	Ring (calling) indicator	Modem to DTE
140	21	RL	Remote loop-back	DTE to modem
140	21	RL	Remote loop-back	DTE to modem
141	18	LL	Local loop-back	DTE to modem
142	25	TM	Test mode	Modem to DTE
	9		+12 V for modem testing	DTE to modem
	10		−12 V for modem testing	DTE to modem
	11		Unassigned	

[a] Circuit 110 (CG) is now seldom used, and pin 21 is generally used for circuit 140 (RL).
[b] If SCF is used, only one "Data Signal Rate Select" can be used; pin 23 is assigned to either circuit 111 (CI) or 112 (CH).

those in Section 12.5 can serve only as an introduction and a companion guide to such a study.

104. "Receive data" *must* be clamped to one ("mark") whenever 109 is off, but it is also desirable to clamp it immediately an internal signal called "Signal Power Detect" (see circuit 109 below) goes low. This protects against spurious data during the time that a modem is deciding whether the receive signal has truly finished.

105. "Request to send" plays a different role for half-duplex and full-duplex modems. In the former it controls the turning on and off of the transmitter for line turn-around. In the latter it is largely superfluous, and can be left on all the

time; the transmitters of calling and answering modems will be turned on at the appropriate times in the hand-shaking sequence.

106. "Clear to send" being on means that the transmitter either assumes or has been told that the far receiver has turned on its carrier detect† and is ready to accept data from the TDTE.

107. "Data set ready" is a controversial, inconsistent, and not very useful signal. Usually it means that a modem has finished the calling and answering sequence defined by V.25, but it is used differently in V.32.

108. Signal 108/1 is used for direct (usually manual) control of the modem connection to the line. Signal 108/2 is used for automatic calling and answering; for example, a modem tells a DTE that it has received a ringing signal by turning 125 on, and the DTE shows willingness to accept the call by turning on 108/2. The call is terminated (the line is "dropped", the telephone/modem goes "on-hook") by turning off either version of 108.

109. "Received line signal detect" being on means that (1) a signal within the specified amplitude range has been received; (2) all receiver training has been completed, so that the received signal has been decoded and found to meet all requirements for that modem; and therefore (3) the received data line, 104, has been unclamped, and all subsequent data on it will have come from the TDTE. The turning on of 109 usually corresponds with the finishing of all fast training modes in the receiver—adaptive equalizer step sizes are set to their final small value, and so forth. Many DTEs will terminate a call by turning off 108/2 if 109 goes low for more than a few milliseconds. However, temporary line fades may cause the receive signal to fall below the acceptable level for longer than this, or it may even be necessary to turn off the transmitter (e.g., to end a loop test—see Section 12.5). Therefore, an internal signal, called "Signal Power Detect" or "Energy Detect," that responds only to level should be used‡; when it goes low, 104 should be clamped and a timer started; if SPD does not turn on within a specified time, 109 should be turned off.

111, 112. In their normal mode of operation (i.e., static "On" or "Off"), these signals can only select one of two speeds. Their use with multispeed modems has not been agreed on. Both are called data rate "selector," but in case of conflict between the two signals—either DTE or modem "selecting" a higher speed than the other is capable of—the lower speed must prevail; the signal specifying the higher speed becomes only an "indicator."

113, 114. The usual direction for the transmit clock is from modem to DTE, but some systems—particularly those involving multiplexers—require that all clocks be slaved to one DTE. In both cases the DTE should change the data on

---

† Even though, in most high-speed modems, there is seldom any pure carrier, the old name for this signal persists.

‡ SPD is also useful for controlling fast resets of timing recovery PLLs, that are used to prevent hang-up (see Section 7.1).

the positive edge of the clock,† and the modem should sample it on the negative edge.

118–122. These signals perform the same functions for the reverse channel as 103–106 and 109 perform for the main channel. If the reverse channel is used only for error correction, it is not available to the DTE, and 121 and 122 should be held off and 119 clamped to one.

125. When a modem equipped for automatic answering receives a ringing signal, it will not respond (go off-hook) if 108/2 is off. The usual sequence is that the modem turns on 125, and the DTE decides whether to accept the call.

12.5 CONNECTION ESTABLISHMENT THROUGH HAND-SHAKING

Each modem shall perform — and help the other modem to perform — some or all of the operations:

1. Send answer tone and disable network echo suppressors and/or cancelers.
2. Identify itself, and recognize other members of the same modem family.
3. Train its echo canceler — usually in a half-duplex mode.
4. Train its AGC and timing and carrier recovery "circuits."
5. Train its adaptive equalizer.
6. Agree with the other modem, via an exchange of rate signals, on a maximum data rate of which the two are capable.
7. Refine its adaptive equalizer at the agreed-on rate.

12.6 LOOP-BACK TESTING

Useful diagnostic information can often be obtained by looping signals back to their source, and comparing the original data and the recovered data. Of the four loops defined by the CCITT in Recommendation V.54 and shown in Fig. 12.4,‡ only two are appropriate for modems; these are "Local Analog Loop-Back" (loop 3) and "Remote Digital Loop-Back" (loop 2), which is possible only with FDX (two-wire or four-wire) modems.

Details of the test conditions and interface signal sequences for these loops are given in V.54 and the appropriate modem recommendation. The following additional notes may be helpful.

---

† Note that the interface signals are nominally ±6 V, with positive (negative) control and data signals indicating an "On" ("Off") and a zero (one), respectively. These signals are usually inverted inside the modem.

‡ The numbering went awry!

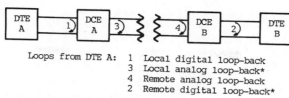

Loops from DTE A: 1 Local digital loop-back
                   3 Local analog loop-back*
                   4 Remote analog loop-back
                   2 Remote digital loop-back*

* Most important for modem testing

Figure 12.4 Diagnostic loops as defined by V.54.

12.6.1 Loop 3: Local Analog Loop-Back

This is very simple for half-duplex modems. Provided transmitter and receiver do not share any circuitry, the output of the former can just be connected directly to the input of the latter. For FDX modems, it is not quite so simple; the procedure depends on whether they use FDM or EC.

FDM. The transmitter and receiver must be switched into the same band. If, as is often the case, there is one high-band and one low-band BPF, one of these must be bypassed. For a full, rigorous test, loop-back should therefore be performed in both modes.

EC. If the transmitter were merely looped into the receiver, the received signal would be fully correlated with the data, and the echo canceler should subtract the entire signal and leave nothing for the receiver.† A first thought might be to disable the canceler, but since it is such a major part of the receiver, that would result in a very incomplete test. The merits of other solutions depend on the implementation technology:

1. With some transmitters, it may be possible to add in a small amount of signal scrambled in the other mode; this would then pass through the canceler.

2. The output of the transmitter could be attenuated and delayed by one symbol period, and then added to the net receive signal after it has been correlated with the transmit data. Since the data sequences via the looped-back and subsidiary paths are the same, the delay through the two paths would have to be known precisely so that it could be verified that the decoded data had the correct phase.

3. An echo-canceling modem will usually require a period of half-duplex transmission to train its canceler. An analog loop-back test could pause after this part and check that the remanent echo was low enough; if the data input to the canceler's TDL were then clamped, the canceler taps should be driven to zero, and the receiver could be checked.

---

† This is the same effect as in FDM; the receiver cannot tell the difference between a looped-back signal and a reflected signal and will remove both regardless of whether there be any genuine receive signal.

12.6.2   Loop 2: Remote Digital Loop-Back

After two modems have completed hand-shaking, either can initiate a loop 2 test by sending a data pattern as defined in the particular recommendation. Unfortunately, there is a small (but significant) chance that this pattern will occur during normal operation and stop communication by putting the modems into loop-back; some embarrassing field retrofits have been needed to cure this problem. A safer way of initiating the test is to briefly squelch the transmitter first, and then restart with the pattern.

If errors are detected in the returned data, the conventional test will not indicate whether they occurred from A to B or from B to A. This can be remedied by modem B checking the data and partially correcting any errors; for example, it could correct the second bit error that occurs because of coherent detection; single errors would be recognized by modem A as having occurred in the outgoing path.

In order to test as many functions as possible in both modems, it is recommended that the data be looped right at B's DTE interface. It should be recognized, however, that if asynchronous data is input to A at a rate less than f_b (the more usual situation), the data delivered to B's ASC will have full stop bits and will be synchronized with the f_b clock; consequently, the converter will not be truly tested.

Appendix I

Some Useful Computer Programs

The programs listed at the end of this appendix are written in FORTRAN 80 and have been used on an Apple IIe computer. Only small modifications should be needed to make them suitable for any FORTRAN compiler. All possible combinations of inputs have not been checked, and there may still be a few small bugs—particularly in FILPROX. However, Murphy's law applied to programming states that the removal of the last bug immediately triggers obsolescence!

A1.1 FILPROX: A FILTER APPROXIMATION PROGRAM

This program is designed to solve many of the filter approximation problems that conventional programs (FILSYN, McClellan and Parks, etc.) cannot; for example, it can design Infinite Impulse Response (IIR) filters with linear phase, complex filters, combinations of s-plane and z-plane filters, and sets of low-pass and/or band-pass filters with modulators interposed. The capabilities will become clear as you work through the detailed operating instructions.

The basic method is to analyze, calculate some error measurement, and iterate to minimize that error. Because of this simple approach, the program can design anything that it can analyze, but it is rather slow and requires a fair amount of user knowledge and assistance. This write-up is intended only to provide enough information for the use of the program; the theory belongs in a book on filter design.

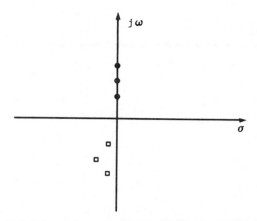

Figure A1.1 Poles and zeroes of a complex filter for LSB generation.

Complex Filters. In themselves, complex filters constitute a large subject that has not, unfortunately, had a comprehensive treatment. The reader is referred to [L&B], [S&S3], and [SSLB] for the best discussions of them. The following brief description is intended only to stimulate the reader.

The simplest type of complex filter is a 90° phase shifting network, whose purpose, when combined with a modulator (frequency shifter), is to pass one sideband and attenuate the other; the attenuation at other frequencies is not important. The poles and zeros for a lower sideband modulation are as shown in Fig. A1.1. As with most filters for data transmission, such a filter would be most conveniently realized as the tandem connection of biquadratic sections. In the sampled (z) domain, this means that

$$F(z) = \frac{N(z)}{D(z)} = \prod \frac{N_i(z)}{D_i(z)}$$

The input to the first section is real, but thereafter each section has a complex input and output, and complex coefficients in numerator and denominator.†
With a digital realization this poses no conceptual problems, but with an analog (e.g., switched-capacitor) realization it can become quite complicated.

It is most efficient to design the receive filter and phase splitter together (so that neither by itself has to have a flat passband). The combination would then be realized by real sections followed by complex sections.

---

† The familiar 90° phase-splitting networks, which are two separate chains of first-order networks (often all-pass), are special (less efficient and more sensitive) cases of the more general complex networks.

A1.1.1 Initial Inputs

It is first necessary to define the frequencies at which the analysis and optimization should take place. For a real filter, all the frequencies should be positive. For a complex filter (always assumed to be a bandpass), there are two possibilities — lower stopband negative, and passband and upper stopband positive, or lower stopband and passband negative, and upper stopband positive. In each case, of course, the stopband with frequencies of the same sign as the passband will be created by the real part of the filter (what is sometimes called *common-mode attenuation*).

"Number of Freqs in LSB,PB,USB".† For low-pass filters, the number of frequencies in the lower stopband should be zero. Generally speaking, the number of frequencies at which the filter is analyzed should be about twice the number of singularities in that band (zeros in the stopbands, poles in the passband). Usually, the frequencies should be evenly spaced because this will result in the lowest mean-squared error, but a result closer to equal ripple can be induced (if you insist!) by reducing the spacing of the frequencies near the edge(s) of the passband.

An input of 0,0,0 should be used if you have already entered all the frequencies and want to rerun the program with different stopband weights or type of fit.

"Simple Form of Input?"‡ The simple form is evenly spaced frequencies in all bands (not necessarily the same spacing in the different bands) with the input magnitude equal to unity everywhere.

"Fstart, Fstep for LSB,PB,USB (B.P.) or PB,USB (L.P.)?" For the simple form of input, the lower and upper frequency limits of the lower stopband (LSB), passband (PB), and upper stopband (USB) are defined as f_1, f_2, f_3, f_4, f_5, and f_6. The input would be f_1, $(f_2 - f_1)/\text{NLSB}$, f_3, $(f_4 - f_3)/\text{NPB}$, f_5, and $(f_6 - f_5)/\text{NUSB}$.

"Type Freq, Input Amp, Wanted Amp." This program does not use weighted errors; the effect is achieved by using input and wanted magnitudes. An illustrative example is a filter for a receiver in which the spectral shaping is to be split evenly between transmitter and receiver. The "input" amplitudes would be the square root of the desired overall shaping, and the "wanted" would be the overall shaping.

The input in the stopband(s) is a scaled representation of the amplitude spectrum of any interfering signal (often more than just white noise); the level of this signal relative to the passband signal is input later by the stopband weights.

---

† Prompts from the computer are shown in quotation marks; many of the prompts will, of course, depend on previous inputs.
‡ All such simple questions should be answered by 1 = yes or 0 = no.

"Mistakes? (0 for None, N to Change Line N)." This prompt gives you a chance to look at what you have typed and correct it if necessary.

"Amp Fit (1) or Amp and Phase (2)?" Amplitude fit is just what it says; the program does not concern itself with phase and seeks to minimize the sum of the squares of the differences between the output and wanted amplitudes. (The wanted amplitude in a stopband is, of course, zero.)

For an amplitude and phase fit, the program assumes that the response should be real in the passband and calculates the passband contribution to the error as the sum of the squares of the imaginary part and of the error in the real part.

"Band-Edge Timing (1), or LMS Phase Fit (2)?" An amplitude and phase fit is generally used for data transmission; then band-edge timing is the best to use because it is faster for the program to calculate, and it minimizes the error that will be produced in a real data system.

If you are just trying to design a linear phase filter, and have no knowledge of its intended use, an LMS phase fit should be used.

"Line Number(s) of the Band-Edge Frequency(s)." The band-edge frequencies are those at which the overall response should be 0.5. The program expects one line number for a low-pass and two for a band-pass.

"Stopband Weight(s)." These weights are measures of the ratio of amplitude of the interfering signals or noise to that of the wanted signal; this is how system specifications should be written. If somebody insists that you design a filter with a specified maximum ripple in the passband (dBPB) and stopband (dBSB), then assuming that you have estimated the degree of the filter correctly, the stopband weight can be calculated from

$$\text{Wt} \simeq \frac{\text{dBPB}}{20.10^{\text{dBSB}/20}}$$

"Number of Factors in s-Plane, z-Plane (0,0 If Already Stored)?" This program does not estimate the degree of the filter required. You specify a degree, and eventually it calculates the best it can do; if that is too good or not good enough, you must reduce or increase the degree accordingly.

It is okay to combine s-plane and z-plane factors. An example might be a digital filter with a continuous interpolating filter on its output. It would not be a good idea to design the digital filter to the overall specification, and then require that the smoothing filter be perfect. The ability to design the combination will allow one to use a lower sampling rate in the digital filter part.

Note that a "factor" consists of two independent parameters (e.g., the real and imaginary parts of the singularity in the s plane), so that it is either a single complex singularity or a complex conjugate pair.

Note also that since this program deals with singularities of the transfer function, you can have more zeros than poles (part of the filter must be FIR) or more poles than zeros (the other zeros are at infinity).

A 0,0 input is used if you have changed the set of frequencies or the stopband weights and want to start with the factors that had been used before.

"Sampling Frequency?" The listed program assumes that all the z-plane factors use the same sampling frequency, but this could be easily changed.

"Type First Guesses at Factors." For s-plane singularities, the format of the input is

$$\text{Real part, imaginary part, exponent, C/CC/M}$$

For the z-plane, it is

$$(\text{Radius} - 1.0), \text{ angle in radians, exponent, C/CC/M}$$

The following points are noted:

1. This unusual form for the z plane input was chosen because it most closely resembles the s plane; the first parameter being zero means a singularity at a real frequency in either plane.
2. The real part or $(R - 1)$ must be negative for poles.
3. The imaginary part may be positive or negative.
4. The exponent is $+1$ for zeros and -1 for poles.
5. C/CC/M $= 1$ if the pole or zero is singular (for complex filters)
   $= 2$ if it is to be combined with its complex conjugate; do not enter the complex conjugate
   $= 3$ if it is in the baseband and is to be modulated to the passband (the computer will then ask for the modulation frequency; the effect is shown in Fig. A1.1)
   $= -1, -2$ or -3 means the same as 1, 2 or 3, but the singularity is fixed (i.e., not to be adjusted during the iteration; an example where this is needed is a zero at $z = -1$)

Making the first guess is the difficult part of the design process. For a pure amplitude fit, familiarity with conventional maximally flat and equal ripple solutions will help you make a reasonable first guess.

For an amplitude and phase fit, the distribution of the poles is similar to that for an amplitude fit, but the spread of the Q values is less (i.e., the high Q values are reduced and the low ones are increased). The stopband zeros are easy to guess at, but the passband zeros in the right half-plane (outside the unit circle) which "equalize"† the phase are more difficult; the problem is to know how many of them to use.

For a typical problem, the number should be approximately half the number of stopband zeros; if the permitted passband error is small, the number of passband zeros should be increased; if the stopband weight is large (i.e., the required attenuation is high), the number of stopband zeros should be increased. Note that the iteration will not be able to move a zero from the passband to the stopband or vice versa; it will only find a local minimum of the error function with the initially specified distribution.

For right-half s-plane zeros, the magnitude of the real parts should be approximately twice the real part of the lowest-Q pole.

A1.1.2   Calculation and Iteration

After the first guess has been entered, the computer calculates and prints out the response and error. For a pure amplitude fit, the output is self-explanatory. For an amplitude and phase fit, it actually calculates the real and imaginary part of the response; the contribution at each frequency in the passband to the RMS error is then

$$RSE = [(real - want)^2 + (imag)^2]^{1/2}$$

"Number of Iterations (−1 for New Problem, 0 for Manual)." If the initial response is hopeless, some manual changes may be needed. Type 0, and the computer will prompt you as you change any number of factors. Some general guidelines for this can be given:

1. The number of ripples in the error in the passband should be approximately equal to the number of poles; if it is much less, increase the Q of the poles in the frequency region(s) where the ripples are missing.
2. If the amplitude tilts too much one way, change the frequencies of the poles accordingly.
3. If the error in the real part is reasonable but the imaginary part is mostly positive (or negative), increase (or decrease) the Q of the phase-equalizing zeros and decrease (or increase) the Q of the high-Q poles.

If the response is in the ballpark, try two iterations with a step size of 0.1; there should be a dramatic reduction in the error. If there is not, you may be near a local minimum. I have never seen this situation, however, and it is impossible to generalize about what to do.

If the response is almost acceptable, refine it for several iterations with a step size of about 0.02.

---

† "Equalize" is in quotes because there is no phase equalizer as such; all the singularities contribute to both the amplitude and phase.

During the iterations the computer prints out, for each parameter in turn, the error for the nominal value, and for the nominal value increased and decreased by the scaled step size. These errors are fitted by a parabola, and an improved value of the parameter is found by interpolating or extrapolating† to the minimum of the parabola.

Because the algorithm works on only one parameter at a time, it is slow; however, it is guaranteed to converge to something. Despite the worries of theoreticians about local minima and multiple "solutions," these have seldom been found; with an intelligent first guess (all the poles and zeros in the appropriate regions—passband or stopband, right or left half-plane, etc.), the iterations have nearly always converged to what is clearly the global minimum. The design of a LPF/BPF combination for a V.22 bis modem took about 45 minutes on an Apple IIe personal computer.

A1.2 COMPIR: A PROGRAM FOR CALCULATING A COMPLEX IMPULSE RESPONSE

This program accepts a definition of attenuation and delay distortion—either numerically, frequency by frequency, or according to a user-provided formula —and first calculates the sampling time for BECM as the slope of the chord between the phases at the two band edges. Then, with this or any other time offset, and a prescribed demodulating carrier phase, it calculates $h_p(t)$ and $h_q(t)$ at a number of equally spaced sampling times.

Then, if asked to, it will calculate the folded power spectrum and the decibel reduction of SNR that results from equalizing this spectrum for minimum MSE with a prescribed input SNR.

A1.3 SIMEQ: A PROGRAM FOR SOLVING SIMULTANEOUS EQUATIONS

Most investigations of adaptive equalizers begin with an offline (i.e., nonadaptive) design that requires the solution of a set of simultaneous equations. The solution is algebraically trivial and would not seem to be worth a section of explanation.

However, if one used a general-purpose program that did not take advantage of the repetitive nature of the coefficients [see equation (8.3)], the input of all the coefficients would be a very tedious chore. To design an N-tap equalizer, this program therefore accepts just the M terms of an impulse response and sets up the $N(N + M - 1)$ matrix. It also accepts a value for the noise variance and adds σ^2 to the main-diagonal terms of the auto-correlation matrix, before finding a least-squares solution.

For a $T/2$ FSE, only every other output must be forced to zero, and for a

---

† The range of the extrapolation is limited to just twice the scaled step.

DFE, only those outputs before the main one. Therefore, after the basic matrix has been constructed, the program allows any number of rows to be deleted. The right-hand sides of these equations (not very interesting for an FSE, but the feedback terms of a DFE) are then calculated after the reduced matrix has been inverted.

The program also allows you to iterate toward the eigenvalues in a very simple way. For any entered guess at an eigenvalue, it calculates the determinant of $(A - \lambda I)$; it is a simple matter to find all the approximate values of λ for which this becomes zero.

The listed program handles only real coefficients; extension to complex coefficients would be straightforward.

A1.4 PROGRAM LISTINGS

A1.4.1 FILPROX

```
C       FILPROX: THIS PROGRAM ITERATES THE COEFFICIENTS OF BIQUADRATIC
C       FACTORS OF REAL, COMPLEX OR MODULATED FILTERS, IN THE S OR Z
C       PLANE, TO FIT AMPLITUDE OR REAL AND IMAGINARY PARTS
        DIMENSION F(50),SF(50),CF(50),U(50),W(50),AMP(50),AMPF(50),PH(50),
        DIMENSION PHF(50),A(20,2),S(20),C(20),NEXP(20),MCOMP(20),FMOD(20)
10      WRITE (3,15)
15      FORMAT ('' NUMBER OF FREQS IN LSB,PB,USB (0,0,0 IF ALREADY INPUT ')
        READ (3,2000) N1,N2,N3
        IF (N2.EQ.0) GO TO 70
        NLSB=N1
        NPB=N2
        NUSB=N3
        N1PB=NLSB+1
        N2PB=NLSB+NPB
        N1USB=N2PB+1
        NF=N2PB+NUSB
        WRITE (2,20)
20      FORMAT ('' N     FREQ      IN        WANT')
        WRITE (3,21)
21      FORMAT ('' SIMPLE FORMAT OF INPUT? ')
        READ (3,2000) KIN
        IF (KIN.EQ.0) GO TO 29
        IF (NLSB.EQ.0) GO TO 23
        WRITE (3,22)
22      FORMAT ('' FSTART, FSTEP FOR LSB,PB,USB ')
        WRITE (3,1999)
        READ (3,2010) FLSB,DFLSB,FPB,DFPB,FUSB,DFUSB
        GO TO 25
23      WRITE (3,24)
24      FORMAT ('' FSTART, FSTEP FOR PB,USB ')
        WRITE (3,1999)
        READ (3,2010) FPB,DFPB,FUSB,DFUSB
25        DO 28 J=1,NF
          U(J)=1.
          W(J)=0
          IF (J.GT.NLSB) GO TO 26
          F(J)=FLSB+(J-1)*DFLSB
          GO TO 27
26        IF (J.GT.N2PB) GO TO 27
          F(J)=FPB+(J-N1PB)*DFPB
          W(J)=1.
          GO TO 28
27        F(J)=FUSB+(J-N1USB)*DFUSB
```

```
28        CONTINUE
          GO TO 35
29        WRITE (3,30)
30        FORMAT (' TYPE FREQ, INPUT AMP, WANTED AMP')
          WRITE (3,1999)
35        DO 40 J=1,NF
          IF (KIN.EQ.1) GO TO 37
          READ (3,2010) F(J),U(J),W(J)
37        WRITE (2,2020) J,F(J),U(J),W(J)
          IF ((J.NE.NLSB).AND.(J.NE.N2PB)) GO TO 40
          WRITE (2,1999)
40        CONTINUE
          WRITE (3,50)
50        FORMAT (' MISTAKES? (0 FOR NONE, N TO CHANGE LINE N) ')
60        READ (3,2000) NMIST
          IF (NMIST.EQ.0) GO TO 70
          READ (3,2010) F(NMIST),U(NMIST),W(NMIST)
          WRITE (2,2020) NMIST, F(NMIST),U(NMIST),W(NMIST)
          GO TO 60
70        WRITE (3,80)
80        FORMAT (' AMP FIT (1) OR AMP AND PHASE (2)?')
          READ (3,2000) MFIT
          IF (MFIT.EQ.1) GO TO 112
          WRITE (3,85)
85        FORMAT (' BAND-EDGE TIMING (1) OR LMS PHASE FIT (2)?')
          READ (3,2000) MPHAS
          IF (MPHAS.EQ.2) GO TO 112
          IF (NLSB.GT.0) GO TO 100
          WRITE (3,90)
90        FORMAT (' LINE NUMBER OF BAND-EDGE FREQ?')
          NF1=1
          READ (3,2000) NF2
          GO TO 110

100       WRITE (3,105)
105       FORMAT (' LINE NUMBERS OF BAND-EDGE FREQS?')
          READ (3,2000) NF1,NF2
110       BWABS=F(NF2)-F(NF1)
112       IF (F(1).GT.0) GO TO 130
          F(1)=.00001
          BWREL=1
          WRITE (3,115)
115       FORMAT (' STOP-BAND WEIGHT? ')
          READ (3,2010) WUSB
          WRITE (2,120)
120       FORMAT (' MFIT    WUSB')
          WRITE (2,2020) MFIT,WUSB
          GO TO 145

130       NMID=(N1PB+N2PB)/2
          BWREL=.5*(F(N2PB)-F(N1PB))/F(NMID)
          WRITE (3,135)
135       FORMAT (' STOP-BAND WEIGHTS? ')
          READ (3,2010) WLSB,WUSB
          WRITE (2,140)
140       FORMAT (' MFIT    WLSB    WUSB')
          WRITE (2,2020) MFIT,WLSB,WUSB

C     INITIAL INPUT ROUTINE:
145       WRITE (3,150)
150       FORMAT (' NUMBER OF FACTORS IN S PLANE, Z PLANE
     1(0,0 IF ALREADY STORED)? ')
          WRITE (3,1999)
          READ (3,2000) NS,NZ
          N=NS+NZ
```

```
          IF (N.EQ.0) GO TO 183
          NSFACT=NS
          NZFACT=NZ
          NSP1=NSFACT+1
          NFACT=N
          IF (NZFACT.EQ.0) GO TO 165
          WRITE (3,155)
155       FORMAT (' SAMPLING FREQUENCY? ')
          READ (3,2010) FSAMP
             DO 160 J=1,NF
             P=6.2832*F(J)/FSAMP
             CF(J)=COS(P)
160          SF(J)=SIN(P)
165       WRITE (3,170)
170       FORMAT (' TYPE FIRST GUESSES AT FACTORS')
          IF (NSFACT.EQ.0) GO TO 172
          WRITE (3,2100)
172       WRITE (2,1999)
          WRITE (3,1999)
             DO 180 I=1,NFACT
             FMOD(I)=0
             IF (I.NE.NSP1) GO TO 173
             WRITE (3,2110)
             WRITE (3,1999)
173       READ (3,2025) A(I,1),A(I,2),NEXP(I),MCOMP(I)
          IF ((MCOMP(I).GT.-2.5).AND.(MCOMP(I).LT.2.5)) GO TO 180
          WRITE (3,175)
175       FORMAT (' MODULATION FREQ? ')
          READ (3,2010) FMOD(I)
          IF (I.LE.NSFACT) GO TO 180
          FMOD(I)=6.2832*FMOD(I)/FSAMP
180          CONTINUE
183       KRET1=1
          GO TO 1500

189       GO TO (190,210,230),MFIT
190       WRITE (2,200)
          WRITE (3,200)
200       FORMAT ('    FREQ      AMP       DB')
          GO TO 250
210       WRITE (2,220)
          WRITE (3,220)
220       FORMAT ('    FREQ      REAL      IMAG     REAL ERR   RSE      DB')
          GO TO 250
230       WRITE (2,240)
          WRITE (3,240)
240       FORMAT (' FREQ  REAL  IMAG   DR    DI    DB')

C         INITIAL RESPONSE AND ERRORS
250          DO 260 J=1,NF
             AMP(J)=U(J)
             PH(J)=0
260          CONTINUE
          KRET=1
             DO 270 I=1,NFACT
             GO TO 1000
270          CONTINUE
          KPRINT=1
          KRET=2
          GO TO 1200

280       WRITE (3,300)
300       FORMAT (' NUMBER OF ITERATIONS (-1 FOR NEW PROBLEM, 0 FOR MANUAL
     1)')
          READ (3,2000) NITER
          IF (NITER.LT.0) GO TO 10
          IF (NITER.EQ.0) GO TO 1400
```

```
          WRITE (2,305)
          WRITE (3,305)
  305     FORMAT (' STEP SIZE?')
          READ (3,2010) STEP
          WRITE (2,2010) STEP
          WRITE (2,306)
  306     FORMAT (' NF      E-      E0      E+      PARAM')
          KPRINT=0

C         CYCLIC ITERATION OF PARAMETERS
          DO 600 ITER=1,NITER
            DO 500 I=1,NFACT
              DO 400 IRI=1,2
              IF ((A(I,IRI).EQ.0).OR.(MCOMP(I).LT.0)) GO TO 400
              IS=I
              E2=ETOT
              KRET=3
              GO TO 1000

  310         DA=A(I,IRI)*BWREL*STEP
              IF ((IRI.EQ.1).OR.(NEXP(I).GT.0)) GO TO 320
              DA=DA*2*A(I,1)
  320         A(I,IRI)=A(I,IRI)-DA
              KRET=4
              GO TO 900

  330         E1=ETOT
              A(I,IRI)=A(I,IRI)+2*DA
              KRET=5
              GO TO 900

  340         E3=ETOT
              EMIN=AMIN1(E1,E2,E3)
              CONC=E3+E1-2*E2
              IF (CONC.LE.0) GO TO 345
              ASTEP=.5*(E1-E3)/CONC
              IF (ABS(ASTEP).LT.2.0) GO TO 350
  345         ASTEP=SIGN(2.0,(E1-E3))
  350         A(I,IRI)=A(I,IRI)-DA+ASTEP*DA
  370         KRET=6
              GO TO 900

C       RECONSIDERATION
  375         IF (ETOT.LT.EMIN) GO TO 392
              BACK=DA*ASTEP
              ETOT=E2
              IF ((E1.GT.E2).OR.(E1.GT.E3)) GO TO 380
              BACK=BACK+DA
              ETOT=E1
              GO TO 390

  380         IF (E3.GT.E2) GO TO 390
              BACK=BACK-DA
              ETOT=E3
  390         A(I,IRI)=A(I,IRI)-BACK
  392         IF (DA.LT.0) GO TO 395
              WRITE (2,2020) I,E1,E2,E3,A(I,IRI)
              GO TO 400
  395         WRITE (2,2020) I,E3,E2,E1,A(I,IRI)
  400         CONTINUE
  500       CONTINUE
          WRITE (2,1999)
  600     CONTINUE
        KRET1=2
        GO TO 1500

  610   KPRINT =1
        I=IS
```

```
        KRET=7
        GO TO 900

C       PRE-CALCULATION ROUTINE
900        DO 910 J=1,NF
           AMP(J)=AMP(J)/AMPF(J)
           PH(J)=PH(J)-PHF(J)
910        CONTINUE

C       Z-PLANE ROUTINE TO CALCULATE SIN AND COS
1000    IF (I.LE.NSFACT) GO TO 1100
        C(I)=(A(I,1)+1)*COS(A(I,2))
        S(I)=(A(I,1)+1)*SIN(A(I,2))

C       CALCULATION OF RESPONSE OF FACTOR
1100    DPF=0
        DPFC=0
           DO 1190 J=1,NF
           IF (I.GT.NSFACT) GO TO 1120
C          S-PLANE
           X=A(I,1)
           Y=FMOD(I)+A(I,2)-F(J)
           Y1=FMOD(I)-A(I,2)-F(J)
           GO TO 1130

C          Z-PLANE
1120       X=C(I)-CF(J)
           Y=FMOD(I)+S(I)-SF(J)
           Y1=FMOD(I)-S(I)-SF(J)

C          CALCULATE AMPLITUDE AT ALL FREQUENCIES
1130       AF=X*X+Y*Y
           IF ((MCOMP(I).EQ.1).OR.(MCOMP(I).EQ.-1)) GO TO 1135
           AF=AF*(X*X+Y1*Y1)
1135       AMPF(J)=SQRT(AF)
           IF (NEXP(I).EQ.1) GO TO 1137
           AMPF(J)=1/AMPF(J)

C          CALCULATE PHASE ONLY IN PASS-BAND, AND ONLY IF MFIT=2
1137       IF ((J.LT.N1PB).OR.(J.GT.N2PB)) GO TO 1190
           IF (MFIT.EQ.1) GO TO 1190
           IF (X.NE.0) GO TO 1140
           X=.00001
1140       PF=ATAN(Y/X)
           IF (J.EQ.N1PB) GO TO 1142
           IF (X*XP.GT.0) GO TO 1142
           DPF=SIGN(3.14159,PF)
1142       PFC=0
           IF ((MCOMP(I).EQ.1).OR.(MCOMP(I).EQ.-1)) GO TO 1150
           PFC=ATAN(Y1/X)
           IF (J.EQ.N1PB) GO TO 1150
           IF (X*XP.GT.0) GO TO 1150
           DPFC=SIGN(3.14159,PFC)
1150       XP=X
           PHF(J)=NEXP(I)*(PF-DPF+PFC-DPFC)
1190       CONTINUE
        IF (KRET.EQ.3) GO TO 310

C       CALCULATION OF TOTAL RESPONSE
           DO 1195 J=1,NF
           AMP(J)=AMP(J)*AMPF(J)
1195       PH(J)=PH(J)+PHF(J)
        IF (KRET.EQ.1) GO TO 270

C       CALCULATION OF GAIN AND SAMPLING PHASE
1200    WASIG=0
```

```
         ASIG=0
         IF (MFIT.EQ.2) GO TO 1206
           DO 1205 J=N1PB,N2PB
           WASIG=WASIG+W(J)*AMP(J)
1205       ASIG=ASIG+AMP(J)**2
         GO TO 1215

1206     IF (MPHAS.EQ.2) GO TO 1207
         TAU=(PH(NF2)-PH(NF1))/BWABS
         PH0=(PH(NF1)*F(NF2)-PH(NF2)*F(NF1))/BWABS
         GO TO 1210
1207     S1=0
         S2=0
         S3=0
         S4=0
         S5=0
           DO 1208 J=N1PB,N2PB
           X=AMP(J)*W(J)
           S1=S1+X*W(J)
           S2=S2+X*F(J)
           S3=S3+X*F(J)*F(J)
           S4=S4+X*PH(J)
1208       S5=S5+X*PH(J)*F(J)
         IF (NLSB.GT.0) GO TO 1209
         PH0=0
         TAU=S5/S3
         GO TO 1210
1209     D=S1*S3-S2*S2
         PH0=(S2*S5-S3*S4)/D
         TAU=(S2*S4-S1*S5)/D
1210     DO 1212 J=N1PB,N2PB
           PHSAM=PH(J)-PH0-F(J)*TAU
           WASIG=WASIG+W(J)*AMP(J)*COS(PHSAM)
1212     ASIG=ASIG+AMP(J)**2
1215     G=WASIG/ASIG

C        CALCULATION OF ERROR
         ELSB=0
         EPB=0
         EUSB=0
         IF (NLSB.EQ.0) GO TO 1230
           DO 1220 J=1,NLSB
           X=.5
           IF ((J.EQ.1).OR.(J.EQ.NLSB)) GO TO 1216
           X=1.
1216       ELSB=ELSB+X*(WLSB*G*AMP(J))**2
           IF (KPRINT.NE.1) GO TO 1220
           DB=20*ALOG10(G*AMP(J)+.00001)
           IF (MFIT.EQ.2) GO TO 1217
           WRITE (2,2040) F(J),DB
           WRITE (3,2040) F(J),DB
           GO TO 1220
1217       WRITE (2,2050) F(J),DB
           WRITE (3,2050) F(J),DB
1220       CONTINUE
1230     IF (MFIT.EQ.2) GO TO 1250
           DO 1240 J=N1PB,N2PB
           GA=G*AMP(J)
           X=.5
           IF ((J.EQ.N1PB).OR.(J.EQ.N2PB)) GO TO 1235
           X=1.
1235       EPB=EPB+X*(W(J)-GA)**2
           IF (KPRINT.NE.1) GO TO 1240
           WRITE (2,2010) F(J),GA
           WRITE (3,2010) F(J),GA
1240       CONTINUE
         GO TO 1280
```

```
1250      DO 1260 J=N1PB,N2PB
          PHSAM=PH(J)-PHO-F(J)*TAU
          X=G*AMP(J)*COS(PHSAM)
          Y=G*AMP(J)*SIN(PHSAM)
          RERR=X-W(J)
          XM=.5
          ESQ=RERR**2+Y**2
          IF ((J.EQ.N1PB).OR.(J.EQ.N2PB)) GO TO 1255
          XM=1.
1255      RSE=SQRT(ESQ)
          EPB=EPB+XM*ESQ
          IF (KPRINT.NE.1) GO TO 1260
          WRITE (2,2010) F(J),X,Y,RERR,RSE
          WRITE (3,2010) F(J),X,Y,RERR,RSE
1260      CONTINUE

1280      DO 1290 J=N1USB,NF
          X=.5
          IF ((J.EQ.N1USB).OR.(J.EQ.NF)) GO TO 1282
          X=1.
1282      EUSB=EUSB+X*(WUSB*G*AMP(J))**2
          IF (KPRINT.NE.1) GO TO 1290
          DB=20*ALOG10(G*AMP(J)+.00001)
          IF (MFIT.EQ.2) GO TO 1285
          WRITE (2,2040) F(J),DB
          WRITE (3,2040) F(J),DB
          GO TO 1290
1285      WRITE (2,2050) F(J),DB
          WRITE (3,2050) F(J),DB
1290      CONTINUE
          IF (NLSB.EQ.0) GO TO 1295
          ELSB=ELSB/NLSB
1295      EPB=EPB/NPB
          EUSB=EUSB/NUSB
          ETOT=SQRT(ELSB+EPB+EUSB)
          ELSB=SQRT(ELSB)
          EPB=SQRT(EPB)
          EUSB=SQRT(EUSB)
          IF (KPRINT.NE.1) GO TO 1300
          WRITE (2,2120)
          WRITE (2,2010) ELSB,EPB,EUSB,ETOT
1300      IF (KRET.EQ.2) GO TO 280
          IF (KRET.EQ.4) GO TO 330
          IF (KRET.EQ.5) GO TO 340
          IF (KRET.EQ.6) GO TO 375
          GO TO 280

C     MANUAL INPUT ROUTINE:
1400  WRITE (3,1410)
1410  FORMAT (' FACTOR  NUMBER? (0 FOR NO MORE CHANGES) ')
          READ (3,2000) I
          IF (I.EQ.0) GO TO 183
          IF (I.GT.NSFACT) GO TO 1420
          WRITE (3,2100)
          GO TO 1430
1420  WRITE (3,2110)
1430  READ (3,2025) A(I,1),A(I,2),NEXP(I),MCOMP(I)
          GO TO 1400

C     ROUTINE TO PRINT FACTORS:
1500  IF (NSFACT.EQ.0) GO TO 1525
          WRITE (2,1510)
1510  FORMAT (' NF     REAL     IMAG   EXP C/CC/M')
          DO 1520 I=1,NSFACT
          WRITE (2,2030) I,A(I,1),A(I,2),NEXP(I),MCOMP(I)
1520      CONTINUE
          IF (N2FACT.EQ.0) GO TO 1560
```

```
1525   WRITE (2,1530)
1530   FORMAT (' FSAMP')
       WRITE (2,2010) FSAMP
       WRITE (2,1540)
1540   FORMAT (' NF    RAD-1    THETA  EXP C/CC/M')
          DO 1550 I=NSP1,NFACT
          WRITE (2,2030) I,A(I,1),A(I,2),NEXP(I),MCOMP(I)
1550   CONTINUE
1560   IF (KRET1.EQ.1) GO TO 189
       GO TO 610

1999   FORMAT (' ')
2000   FORMAT (4I4)
2010   FORMAT (6F9.4)
2020   FORMAT (I3,5F10.4)
2025   FORMAT (2F9.4,2I4)
2030   FORMAT (I3,2F9.4,3I4)
2040   FORMAT (F9.4,10X,F9.4)
2050   FORMAT (F9.4,40X,F9.4)
2100   FORMAT (' S-PLANE: REAL,IMAG,EXP,C/CC/M ')
2110   FORMAT (' Z-PLANE: (RAD-1),THETA,EXP,C/CC/M ')
2120   FORMAT (' LSB RMSE  PB RMSE  USB RMSE  TOT RMSE')
       END
```

A1.4.2 COMPIR

```
C      COMPIR: THIS PROGRAM CALCULATES THE REAL AND IMAGINARY PARTS OF
C      AN IMPULSE RESPONSE GIVEN THE SHAPING, AND THE PASS-BAND
C      ATTENUATION AND DELAY. IT ALSO CALCULATES THE FOLDED POWER
C      SPECTRUM AND THE NOISE ENHANCEMENT (REDUCTION OF INPUT SNR) WITH
C      INFINITE EQUALIZERS (TSE AND FSE)

       DIMENSION A(60),DB(60),DEL(60),PH(60),F(60),S(60)
10     WRITE (3,20)
20     FORMAT (' NLN,NC,NUN,NF,ALPHA  ')

C      FREQUENCIES NUMBERED FROM 1 TO NF. NLN, NC, AND NUN ARE ORDINALS
C      OF LOWER NYQUIST, CENTER, AND UPPER NYQUIST FREQUENCIES.
C      ALPHA IS EXCESS BANDWIDTH FACTOR.

       READ (3,510) NLN,NC,NUN,NF,ALPHA
       WRITE (3,30)
30     FORMAT (' START FREQ, FREQ SPACING  ')
       READ (3,520) F1,DF
       PH(1)=0
       DEL2=0
       WRITE (3,40)
40     FORMAT (' ATTEN AND PHASE FROM FORMULAS (0) OR TYPED IN (1)?  ')
       READ (3,510) KIN
       IF (KIN.EQ.1) GO TO 100
       WRITE (3,50)
50     FORMAT (' LINEAR (0) OR MULTIPATH (1)? ')

C      THESE ARE TWO USEFUL STANDARD DISTORTION SHAPES; USER CAN WRITE
C      ANY OTHERS TO SUIT A PARTICULAR MEDIUM.

       READ (3,510) KMP
       IF (KMP.EQ.0) GO TO 70
       WRITE (3,60)
60     FORMAT (' B, TAU, F0 ')
       READ (3,520) B,TAU,F0
       GO TO 100
70     WRITE (3,80)
80     FORMAT (' ATTEN BREAK FREQ, SLOPE    ')
       READ (3,520) FA,DDB
       WRITE (3,90)
```

```
90      FORMAT (' DELAY CENTER FREQ, SCALING  ')
        READ (3,520) FD,DD
100     NFB=NUN-NLN
        FB=DF*NFB
        WRITE (2,110)
110     FORMAT ('        FREQ    DB     DELAY    AMPL    PHASE')
        DANG=1.5708/(ALPHA*(NC-NLN))
          DO 190 I=1,NF
        F(I)=F1+(I-1)*DF
        S(I)=1.
        ANG=(1-NLN)*DANG
        IF (ANG.LE.1.5708) GO TO 115
        ANG=(NUN-I)*DANG
        IF (ANG.GT.1.5708) GO TO 130
115     IF (ANG.GE.-1.5708) GO TO 120
        ANG=-1.5708
120     S(I)=(1+SIN(ANG))/2
130     IF (KIN.EQ.1) GO TO 160
        IF (KMP.EQ.0) GO TO 140
        THETA=6.2832*TAU*(F(I)-F0)
        C=COS(THETA)
        AMPSQ=1+B*B-2*B*C
        DEL(I)=B*TAU*(C-B)/AMPSQ
        DB(I)=10*ALOG10(AMPSQ)
        A(I)=S(I)*SQRT(AMPSQ)
        PH(I)=ATAN(B*SIN(THETA)/(1-B*C))
        GO TO 180
140     DB(I)=0
        IF (F(I).LT.FA) GO TO 150
        DB(I)=-DDB*(F(I)-FA)
150     DEL(I)=DD*(F(I)-FD)**2
        GO TO 170
160     WRITE (3,550) F(I)
        READ (3,520) DB(I),DEL(I)
170     A(I)=S(I)*(10**(DB(I)/20))
        PH(I)=PH(I-1)-3.14159*DF*(DEL(I)+DEL2)
180     WRITE (2,540) I,F(I),DB(I),DEL(I),A(I),PH(I)
        DEL2=DEL(I)
190       CONTINUE

C       IN CASE YOU MADE A TYPING MISTAKE
200     WRITE (3,210)
210     FORMAT (' TYPE I TO CHANGE LINE I, 0 FOR ALL CORRECT ' )
        READ (3,510) I
        IF (I.EQ.0) GO TO 250
        WRITE (3,550) F(I)
        READ (3,520) DB(I),DEL(I)
        A(I)=S(I)*(10**(DB(I)/20))
        WRITE (2,540) I,F(I),DB(I),DEL(I)
        GO TO 200
220       DO 240 I=1,NF
          IF (I.EQ.1) GO TO 230
          PH(I)=PH(I-1)-3.14159*DF*(DEL(I)+DEL2)
230       WRITE (2,540) I,F(I),DB(I),DEL(I),A(I),PH(I)
          DEL2=DEL(I)
240       CONTINUE
250     TBE=.159155*(PH(NLN)-PH(NUN))
        WRITE (2,260)
260     FORMAT ('       TBE')
        WRITE (2,520) TBE
C       SAMPLING TIME FOR BECM
270     WRITE (2,500)
        WRITE (3,280)
280     FORMAT (' PERC,TSAMP,TDELT,NTERMS(ODD) ')
        READ (3,530) PERC,TSAMP,TDELT,NTERMS
C       FOR NTERMS IF IR CENTERED AROUND TSAMP
C       USE TDELT=1 FOR T-SPACED SAMPLES.
```

```
300    SIGAB=0
       EN=0
       NL=(NTERMS-1)/2
       T=TSAMP-NL*TDELT
       PHC=6.28318*PERC
       WRITE (2,500)
       WRITE (2,305)
305    FORMAT ('      T         HI      HQ ')
          DO 330 J=1,NTERMS
       T2PI=6.28318*T/FB
       HI=0
       HQ=0
          DO 310 I=1,NF
          P=T2PI*(F(I)-F(NC))+PH(I)-PHC
          HI=HI+A(I)*COS(P)
          HQ=HQ+A(I)*SIN(P)
310       CONTINUE
       HI=HI*DF/FB
       HQ=HQ*DF/FB
       SIGAB=SIGAB+ABS(HI)+ABS(HQ)
       EN=EN+HI*HI+HQ*HQ
       IF (J.NE.(NL+1)) GO TO 320
       HI0=HI
       HQ0=HQ
320    WRITE (2,520) T,HI,HQ
       T=T+TDELT
330    CONTINUE

       WRITE (2,340)
340    FORMAT ('   PERC    TSAMP    HI0    HQ0    EYE     EN     RMSE')
       D=SIGAB-ABS(HI0)
       EYE=SIGAB-2*D
       RMSE=SQRT(EN-HI0*HI0)
       WRITE (2,520) PERC,TSAMP,HI0,HQ0,EYE,EN,RMSE
       WRITE (2,500)
       WRITE (3,350)
350    FORMAT (' FOLDED SPECTRUM?  ')
       READ (3,510) KPF
       IF (KPF.EQ.0) GO TO 430
360    WRITE (3,370)
370    FORMAT (' SNR IN DB ')
       READ (3,520) SNR
       SIGSQ=EN*(10**(-SNR/10))
       POW=0
       POWM=0
       PINV0=0
       PINVN=0
       PMINV0=0
       PMINVN=0
       WRITE (2,375)
375    FORMAT ('     F      EN     MAX EN')
          DO 410 I=NLN,NUN
       A2=0
       P2=6.28318*TSAMP
       I2=I+NFB
       IF (I.LT.NC) GO TO 380
       I2=I-NFB
       P2=-P2
380    IF ((I2.LT.1).OR.(I2.GT.NF)) GO TO 390
       A2=A(I2)
       P2=P2+PH(I2)
390    ARG=PH(I)-P2
       POWF=A(I)**2+A2**2+2*A(I)*A2*COS(ARG)
       POWMF=(A(I)+A2)**2
       XM=.5*DF/FB
       IF ((I.EQ.NLN).OR.(I.EQ.NUN)) GO TO 400
       XM=DF/FB
```

```
400     POW=POW+XM*POWF
        POWM=POWM+XM*POWMF
        PINVN=PINVN+XM/(POWF+SIGSQ)
        PINVO=PINVO+XM/POWF
        PMINVN=PMINVN+XM/(POWMF+SIGSQ)
        PMINVO=PMINVO+XM/POWMF
        WRITE (2,520) F(I),POWF,POWMF
410     CONTINUE
        WRITE (2,500)
        WRITE (2,415)
415     FORMAT ('                        USING TSE       USING FSE')
        WRITE (2,420)
420     FORMAT ('   EN    MAX EN   IN SNR   DSNRO   DSNRN   DSNRO   DSNRN')

C       EN IS TOTAL IN SAMPLED PULSE, MAX EN IS MAXIMUM AVAILABLE.
C       DSNRO AND DSNRN ARE CHANGES OF SNR DUE TO INFINITE TSE AND FSE
C       WITHOUT AND WITH NOISE.

        DSNRO=-10*ALOG10(PINVO*POWM)
        DSNRN=-10*ALOG10(PINVN*POWM)
        DMSNRO=-10*ALOG10(PMINVO*POWM)
        DMSNRN=-10*ALOG10(PMINVN*POWM)
        WRITE (2,520) POW,POWM,SNR,DSNRO,DSNRN,DMSNRO,DMSNRN
430     WRITE (3,440)
440     FORMAT (' ANOTHER SNR (1), NEW PERC OR TSTART (2), OR NEW PROBLEM
       1(3) ')
        READ (3,510) KBACK
        GO TO (360,270,10), KBACK

500     FORMAT (' ')
510     FORMAT (4I3,F8.3)
520     FORMAT (7F8.3)
530     FORMAT (3F8.3,2I3)
540     FORMAT (I3,5F8.3)
550     FORMAT (F8.3,2X)
        END
```

A1.4.3 SIMEQ

```
C       SIMEQ: THIS PROGRAM SOLVES SIMULTANEOUS EQUATIONS EXACTLY
C       OR, IF THEY ARE OVER-DETERMINED, BY THE LS METHOD.
C       IT ALSO FINDS THE EIGENVALUES OF THE AUTO-CORRELATION MATRIX
        DIMENSION A(50,50),B(30,30),BW(30,30),R(50),X(40),U(30)
10      WRITE (3,20)
20      FORMAT (' NIR (0 FOR REGULAR INPUT)? ')
        READ (3,600) NIR
        IF (NIR.EQ.0) GO TO 180

C       SPECIALIZED INPUT OF IMPULSE RESPONSE
        WRITE (3,30)
30      FORMAT (' TYPE IMPULSE RESPONSE ')
        READ (3,610) (U(I), I=1,NIR)
        EN=0
        WRITE (3,40)
40      FORMAT (' NORMALIZE? ')
        READ (3,600) KNORM
        IF (KNORM.EQ.0) GO TO 70
        DO 50 I=1,NIR
        EN=EN+U(I)*U(I)
50      CONTINUE
        SNORM=1/SQRT(EN)
        DO 60 I=1,NIR
        U(I)=SNORM*U(I)
60      CONTINUE
70      WRITE (2,610) (U(I), I=1,NIR)
80      WRITE (3,90)
90      FORMAT (' NUMBER OF TAPS, CENTER TAP ')
```

```
         READ (3,600) NUO,NCT
         NEO=NUO+NIR-1
           DO 110 I=1,NEO
             DO 100 J=1,NUO
             A(I,J)=0
             IJ=I-J+1
             IF ((IJ.LT.1).OR.(IJ.GT.NIR)) GO TO 100
             A(I,J)=U(IJ)
100          CONTINUE
110        CONTINUE

C        OPTION TO DELETE ROWS FOR FSEs OR DFEs
         WRITE (3,120)
120      FORMAT (' NUMBER OF ROWS TO DELETE ')
         READ (3,600) NRD
         NE=NEO-NRD
         IF (NRD.EQ.0) GO TO 160
           DO 150 ID=1,NRD
           READ (3,600) MRD
           MRM=MRD-ID+2
             DO 140 J=1,NUO
             Z=A(MRM-1,J)
               DO 130 I=MRM,NEO
               A(I-1,J)=A(I,J)
130            CONTINUE
             A(NEO,J)=Z
140          CONTINUE
150        CONTINUE
160      N1=NU+1
         WRITE (2,170)
170      FORMAT ('   NIR   NT  NCT  NEQ')
         WRITE (2,600) NIR,NU,NCT,NE
         GO TO 240

C        GENERALIZED INPUT OF ALL COEFFICIENTS
180      WRITE (3,190)
190      FORMAT (' NUMBER OF UNKNOWNS, EQUATIONS?')
200      READ (3,600) NU,NE
         NEO=NE
         N1=NU+1
         WRITE (3,210)
210      FORMAT (' TYPE ROWS OF COEFFS. ')
220        DO 230 I=1,NE
           READ (3,610) (A(I,J),J=1,NU)
230        CONTINUE

240      WRITE (3,250)
250      FORMAT (' TYPE RHS ')
         READ (3,610) (A(I,N1), I=1,NE)
         WRITE (2,610) (A(I,N1), I=1,NE)
         WRITE (3,260)
         WRITE (2,260)
260      FORMAT (' NOISE SIGMA ')
         READ (3,610) SIGMA
         WRITE (2,610) SIGMA
         WRITE (3,270)
270      FORMAT (' PRINT EQUATIONS? ')
         READ (3,600) KP
         IF (KP.EQ.0) GO TO 300
           DO 280 I=1,NE
           WRITE (2,610) (A(I,J),J=1,N1)
280        CONTINUE
         IF (NRD.EQ.0) GO TO 300
         WRITE (2,650)
         ISP=NE+1
           DO 290 I=ISP,NEO
           WRITE (2,610) (A(I,J), J=1,NU)
```

```
290       CONTINUE
300       DO 330 I=1,NU
            DO 320 J=1,N1
            B(I,J)=A(I,J)
            IF ((NE.EQ.NU).AND.(SIGMA.EQ.0)) GO TO 320
            B(I,J)=0
              DO 310 K=1,NE
              B(I,J)=B(I,J)+A(K,I)*A(K,J)
310           CONTINUE
320         CONTINUE
            B(I,I)=B(I,I)+SIGMA*SIGMA
330       CONTINUE
          IF (KP.EQ.0) GO TO 350
            DO 340 I=1,NU
            WRITE (2,610) (B(I,J), J=1,N1)
340       CONTINUE
350       WRITE (3,360)
360       FORMAT (' MAT INV (0), OR EIGENVALUES (1)? ')
          READ (3,600) KGS
          IF (KGS.EQ.1) GO TO 470

C     LU FACTORIZATION
          DO 390 K=2,NU
            DO 380 I=K,NU
              DO 370 J=K,N1
              B(I,J)=B(I,J)-B(K-1,J)*B(I,K-1)/B(K-1,K-1)
370           CONTINUE
380         CONTINUE
390       CONTINUE

C     BACK SUBSTITUTION
          DO 420 IR=1,NU
          I=NU-IR+1
          Z=B(I,N1)
          IF (I.EQ.NU) GO TO 410
          IP1=I+1
            DO 400 J=IP1,NU
            Z=Z-B(I,J)*X(J)
400         CONTINUE
410       X(I)=Z/B(I,I)
420         CONTINUE

          WRITE (2,630)
          WRITE (2,610) (X(J), J=1,NU)
          E2=0
            DO 440 I=1,NEO
            R(I)=0
              DO 430 J=1,NU
              R(I)=R(I)+A(I,J)*X(J)
430           CONTINUE
            IF (I.GT.NE) GO TO 440
            E2=E2+(R(I)-A(I,N1))**2
440         CONTINUE
          E=SQRT(E2)
          X2=0
            DO 450 J=1,NU
            X2=X2+X(J)*X(J)
450         CONTINUE
          WRITE (2,620)
          WRITE (2,610) (R(I), I=1,NEO)
          WRITE (2,640)
          WRITE (2,610) E,X2
          WRITE (3,460)
460       FORMAT (' NEW PROBLEM (1) OR SAME IR (2) ? ')
          READ (3,600) KBACK
          GO TO (10,80), KBACK
```

```
C     FINDING EIGENVALUES
470      DO 490 I=1,NU
            DO 480 J=1,NU
            BW(I,J)=B(I,J)
480         CONTINUE
490      CONTINUE
      WRITE (3,500)
500   FORMAT (' GUESS AT EIGENVALUE--NEGATIVE FOR NEW PROBLEM ')
      READ (3,610) GEVM
      IF (GEV.LT.0) GO TO 10
         DO 510 I=1,NU
         BW(I,I)=BW(I,I)-GEV
510      CONTINUE
      WRITE (2,520)
520   FORMAT ('  GEV     DET ')
      DET=BW(1,1)
         DO 550 K=2,NU
            DO 540 I=K,NU
               DO 530 J=K,NU
               BW(I,J)=BW(I,J)-BW(K-1,J)*BW(I,K-1)/BW(K-1,K-1)
530            CONTINUE
540         CONTINUE
         DET=DET*BW(K,K)
550      CONTINUE
      WRITE (3,660) DET
      WRITE (2,660) ALAM,DET
      GO TO 470
600   FORMAT (8I5)
610   FORMAT (10F8.4)
620   FORMAT (' RHS')
630   FORMAT (' X')
640   FORMAT ('   RMSE    SIGXSQ ')
650   FORMAT (' ')
660   FORMAT (2E14.7)
      END
```

Appendix II

Simple Diagnostic Tests

Most of the tests described here are analog in nature and thus might seem to be applicable only to modems using analog signal processing. However, since a signal display on an oscilloscope is often more informative than a print-out of numbers, it is recommended that the development system for a modem that uses DSP be designed to provide an analog read-out of most of the critical signals. This can usually be done by attaching a data trap with a manually programmable address to the bus, and using this to drive a pair of DACs.

A2.1 EYE PATTERNS WITH SIMPLE REPEATED DATA

180° Phase Changes. This is the most useful data pattern for full-response signals. The two baseband signals are sine waves at $f_s/2$ that are nominally in phase or 180° out of phase. That is,

$$x_p(t) = X_p \sin(\pi f_s t)$$

and
$$x_q(t) = X_q \sin(\pi f_s t + k\pi + \phi'), \qquad k = 0 \text{ or } 1$$

An error θ_e in the phase of the demodulating carrier manifests itself as an amplitude difference between the two sinewaves so that

$$\frac{X_p}{X_q} = \frac{1 - \tan \theta_e}{1 + \tan \theta_e}$$

and amplitude distortion of the passband signal is seen as a phase difference between the two baseband sinewaves:

$$\sin \phi' = \frac{1 - (H_+/H_-)^2}{1 + (H_+/H_-)^2}$$

where H_+ and H_- are the amplitudes of the channel (including filters) response at $(f_c \pm f_s/2)$.

If the signals seem to want to stabilize with zero signal on one channel, the sign of the gain in the carrier loop is probably wrong—a very common mistake in the early stages of design of a loop.

90° Phase Changes. This is the most severe pattern; for any system with less than 50% excess bandwidth, it will result in a single tone in the passband; carrier and timing recovery become completely dependent.

A2.2 RECEIVE SIGNAL CONSTELLATIONS

The received constellation can be displayed by applying the sampled-and-held baseband signals to the X and Y inputs of a scope. With random data and only remanent ISI and noise, the signal points shown in, say, Fig. 4.7 for 16QAM, would be blurred equally in all directions. Two more informative degraded constellations are shown in Fig. A2.1; the first is symptomatic of phase jitter— either of the received signal or of the demodulating carrier—and the second of amplitude jitter—almost certainly from the AGC.

If the constellation is either continuously rotating or merely rotated by some constant arbitrary phase,† this means the carrier loop is defective.

A2.3 THWARTING A SCRAMBLER

In order to study the performance of a modem (transmit spectrum, eye patterns, etc.) with simple repeated data patterns, it is necessary to disable the scrambler; if it is buried in hardware or software, this may be difficult to do. The same effect, however, can sometimes be achieved by prescrambling the input data using feedback from the receiver. A transmitter T, which includes a scrambler, is connected directly to a receiver R, which includes a descrambler, by either analog loopback of one modem or the "back-to-back" connection of two; the set-up is shown in Fig. A2.2 for the special, but common, case of just two feedback taps (at k_1 and K).

---

† This can happen in the analog loop-back case, where both transmitter and receiver are driven by the same crystal clock.

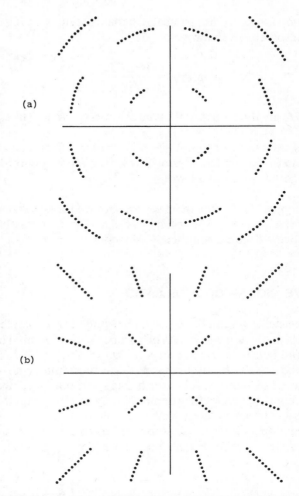

(a)

(b)

Figure A2.1 Sixteen-point constellation degraded by *(a)* phase jitter; *(b)* AGC jitter.

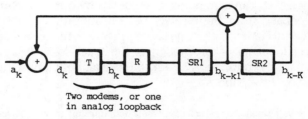

Figure A2.2 Thwarting a scrambler so as to directly control transmit data.

The critical factor is the total delay through transmitter and receiver; if this is less than or equal to k_1/f_b, the delay to the first tap of the scrambler, then the first shift register (SR_1) can build out the delay so that the transmitter, receiver, and SR_1 together emulate the first k_1 stages of a scrambler. Then the prescrambled data is

$$d_k = a_k \oplus b_{k-k_1} \oplus b_{k-K}$$

and the output of the modem's scrambler (the transmitted data) becomes

$$b_k = d_k \oplus b_{k-k_1} \oplus b_{k-K}$$

which is equal to the original input data a_k.

The delay through a modem depends on the symbol period rather than just the bit period, and the biggest contributor to the delay is usually the adaptive equalizer. Therefore this stratagem is only possible for receivers with no adaptive equalizer—or at most, a very short one.

A2.4   CHECKING ASCs AND SACs

The operation of both converters can be checked by displaying random, character-formatted receive data on a scope, triggering on a negative edge, and adjusting the time scale so as to synchronize on alternate start bits. (The complete traverse time of the beam is approximately equal to two character periods.) Slow and fast input data should produce patterns as shown in Figs. A2.3a and A2.3b—with occasional double full-length stop bits with slow data, and single $\frac{7}{8}$ (or $\frac{3}{4}$, if the ASC is in over-speed mode) length stop bits with fast data.

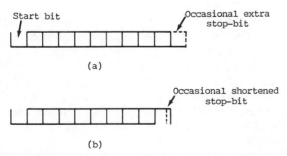

Start bit

Occasional extra
stop–bit

(a)

Occasional shortened
stop–bit

(b)

Figure A2.3 Received character with occasional "aberrations": *(a)* slow data. *(b)* fast data.

References

[A1] H. von Alven, "Designing Telephone and Data Equipment for the New Competitive Environment," *IEEE Intl. Conf. Commun. Record,* vol. 1, pp. 405–409, 1983.

[An] I. N. Andersen, "Sample-Whitened Matched Filters," *IEEE Trans. Inform. Theory,* vol. **IT-19**, pp. 653–660, Sept. 1973.

[Au] M. E. Austin, "Decision-Feedback Equalization for Digital Communication over Dispersive Channels," Tech. Rep. No. 437, MIT Lincoln Lab., Cambridge, MA, Aug. 1967.

[AA&S] J. Anderson, T. Aulin, and C.-E. Sundberg, *Digital Phase Modulation,* Plenum, New York, 1986.

[ACHM] M. C. Austin, M. U. Chang, D. F. Horwood, and R. A. Maslov, "QPSK, Staggered QPSK, and MSK—A Comparative Evaluation," *IEEE Trans. Commun.,* vol. **COM-31,** pp. 171–182, Feb. 1983.

[AC&N] J.-O. Andersson, B. Carlqvist, and G. Nilsson, "A Field Trial with Three Methods for Digital Two-Wire Transmission," *Proc. Intl. Symp. Subscriber Loops and Services,* Sept. 1982.

[A&F] R. R. Anderson and G. J. Foschini, "The Minimum Distance for MLSE Digital Data Systems of Limited Complexity," *IEEE Trans. Inform. Theory,* vol. **IT-25**, pp. 544–551, Sept. 1975.

[AG&N] A. A. Alexander, R. M. Gryb, and D. W. Nast, "Capabilities of the Telephone Network for Data Transmission," *Bell Syst. Tech. J.,* vol. 39, pp. 431–476, May 1960.

[AM&H] O. Agazzi, D. G. Messerschmitt, and D. A. Hodges, "Nonlinear Echo Cancellation of Data Signals," *IEEE Trans. Commun.,* vol. **COM-30,** pp. 2421–2433, Nov. 1982.

[AM&S]   B. Aschrafi, P. Meschkat, and K. Szechenyi, "Field Trial of a Comparison of Time Separation, Echo Cancellation, and Four-wire Digital Subscriber Loops," *Proc. Intl. Symp. Subscriber Loops and Services,* Sept. 1982.

[AR&S]   T. Aulin, N. Rydbeck, and C.-E. W. Sundberg, "Continuous Phase Modulation — Part II: Partial Response Signaling," *IEEE Trans. Commun.,* vol. **COM-29,** pp. 210–225, March 1981.

[A&S1]   T. Aulin and C.-E. W. Sunderberg, "Continuous Phase Modulation — Part I: Full Response Signaling," *IEEE Trans. Commun.,* vol. **COM-29,** pp. 196–209, March 1981.

[A&S2]   ———, "Partially Coherent Detection of Digital Full-Response Continuous Phase Modulated Signals," *IEEE Trans. Commun.,* vol. **COM-30,** pp. 1096–1117, May 1982.

[ATMH]   O. Agazzi, C.-P. J. Tzeng, D. G. Messerschmitt, and D. A. Hodges, "Timing Recovery in Digital Subscriber Loops," *IEEE Trans. Commun.,* vol. **COM-33,** pp. 558–569, June 1985.

[AT&T]   A.T.&T. Information Systems, "A Trellis-Coded Modulation Scheme that Includes Differential Encoding for 9600 bit/s, Full-Duplex, Two-Wire Modems," CCITT SGXVII, Contribution No. D159, Aug. 1983.

[Ba]   P. Baran, "Packetized Ensemble Modem," U.S. Patent No. 4,438,511, March 1984.

[Be1]   R. E. Bellman, *Dynamic Programming,* Princeton Univ. Press, Princeton, NJ, 1957.

[Be2]   T. F. Benewicz, "Measuring Line Level on Telephone Systems," *Bell Labs Record,* vol. 38, p. 96, March 1960.

[Be3]   C. T. Beare, "The Choice of the Desired Impulse Response in Combined Linear-Viterbi Algorithm Equalizers," *IEEE Trans. Commun.,* vol. **COM-26,** pp. 1301–1307, Aug. 1978.

[Be4]   J. W. M. Bergmans, "Partial Response Feedback Equalization," *IEEE Intl. Conf. Commun. Record* pp. 7.2.1–7.2.4, June 1986.

[Bi1]   J. A. C. Bingham, "Simultaneous and Rapid Regeneration of Carrier and Data Timing from Polyphase Signals," *IEEE Intl. Conf. Commun. Record,* pp. 44.23–26, June, 1976.

[Bi2]   ———, "Design of Carrier Loop Filters to Track Phase Jitter," Asilomar Conference on Circuits, Systems and Computers, Monterey, CA, Aug. 1982.

[Bi3]   ———, "Switched-Capacitor Modulator," U.S. Patent No. 4,295,105, Oct. 1981.

[Bi4]   ———, "Switched-Capacitor Modulators," IEEE Intl. Symp. Circuits and Systems, Newport Beach, CA, May 1983.

[Bi5]   ———, "Switched-Capacitor Modulator for Quadrature Modulation," U.S. Patent No. 4,458,216, July 1984.

[Bi6]   ———, "Improved Methods of Accelerating the Convergence of Adaptive Equalizers for Partial-Response Signals," *IEEE Trans. Commun.,* vol. **COM-35,** pp. 257–260, March 1987.

[B&B]   D. V. Batorsky and M. E. Burke, "1980 Bell System Noise Survey of the Loop Plant," *AT&T Bell Labs. Tech. J.,* vol. 63, pp. 775–818, May–June 1984.

[B&D]   W. R. Bennett and J. R. Davey, *Data Transmission,* McGraw-Hill, New York, 1965.

[BLR]   "Notice of Limited Trials of Dataphone Service in Three Bell System Areas," *Bell Labs. Record,* vol. 36, p. 148, April 1958.

[BM&T]   A. Brosio, C. Mogavero, and A. Tofanelli, "Echo Canceler Burst Mode; a New Technique for Digital Transmission on Subscriber Lines," *Proc. IEEE Intl. Conf. Commun.,* June 1985.

[BS1]   Bell System Technical Reference, "Data Communications Using Voice-Band Private Line Channels," Pub. 41004, Oct. 1973.

[BS2]   ———, "Attenuation and Envelope Delay Characteristics from the 1969–70 Switched Telecommunications Network Survey," Pub. 41006, June 1973.

[BS3]   ———, "Analog Parameters Affecting Voiceband Data Transmission—Description of Parameters," Pub. 41008, Oct. 1971.

[BS4]   ———, "Transmission Parameters Affecting Voiceband Data Transmission—Measuring Techniques," Pub. 41009, May 1975.

[B . . 1] G. Brand, E. A. Lea, N. S. Lin, D. A. Hodges, V. Madisetti, and D. G. Messerschmitt, "Comparison of Line Codes and Proposal for Modified Duobinary," ANSI Telecomm. Commit., Contrib. T1D1 3-85-237, Nov. 1985.

[B . . 2] ———, "Comparison of Line Codes with Optimal DFE Design," ibid., Contrib. T1D1.3-86-018.

[Ch1]   R. W. Chang, "High-Speed Multichannel Data Transmission with Band-limited Orthogonal Signals," *Bell Syst. Tech. J.,* vol. 45, pp. 1775–1796, Dec. 1966.

[Ch2]   ———, "A New Equalizer Structure for Fast Start-Up Digital Communication," *Bell Syst. Tech. J.,* vol. 50, pp. 1969–2014, July–Aug. 1971.

[Ch3]   K.-S. Chung, "Generalized Tamed Frequency Modulation and Its Application for Mobile Radio Communications," *IEEE J. SAC.,* vol. SAC-2, pp. 487–497, July 1984.

[Cl1]   A. P. Clark, *Advanced Data Transmission Systems,* Pentech Press, London, 1977.

[Cl2]   ———, *Equalizers for Digital Modems,* Wiley, New York, 1985.

[C&C]   G. C. Clark and J. B. Cain, *Error-Correction Coding for Digital Communications,* Plenum, New York, 1981.

[CDSI]   Concord Data Systems, Inc., "Preliminary Specification of a 9600/300b/s Asymmetric Modem," U.S. Modem Working Party, 86–63, July 1986.

[CFLP]   J. L. Crouch, J. R. Faulkner, O. Loosme, and L. R. Putham, "Electrocardiagrams by Telephone," *Bell Labs. Record,* vol. 44, pp. 43–47, Feb. 1966.

[CFR]   "Connections of Terminal Equipment to the Telephone Network," Code of Federal Regulations, Title 47, Part 68.

[C&H]   R. W. Chang and E. Y. Ho, "On Fast Start-Up Data Communication Systems Using Pseudo-Random Training Sequences," *Bell Syst. Tech. J.,* vol. 51, pp. 2013–2027, Nov. 1972.

[CH&D]   A. P. Clark, J. D. Harvey, and J. P. Driscoll, "Near Maximum-Likelihood

Detection Processes for Distorted Digital Signals," *Radio and Electronic Engineer,* vol. 48, pp. 301–309, June 1978.

[C&K] J. M. Cioffi and T. Kailath, "Fast, Recursive-Least-Squares Transversal Filters for Adaptive Filtering," *IEEE Trans. ASSP,* vol. **ASSP-32,** pp. 304–337, April 1984.

[CK&H] A. P. Clark, C. P. Kwong, and J. D. Harvey, "Detection Processes for Severely Distorted Digital Signals," *Electronics Circuits & Systems,* vol. 3, pp. 27–37, Jan. 1979.

[C&L] J. R. Cessna and D. M. Levy, "Phase Noise and Transient Times for a Binary Quantized Digital Phase-Locked Loop in White Gaussian Noise," *IEEE Trans. Commun.,* vol. **COM-20,** pp. 94–104, April 1972.

[CL&M] A. P. Clark, L. H. Lee, and R. S. Marshall, "Developments of the Conventional Nonlinear Equalizer," *Proc. IEE,* vol. 129, pt. F, pp. 85–94, April 1982.

[CLPW] C. Coleman, D. Lindholm, D. Petersen, and R. Wood, "High Data Rate Magnetic Recording in a Single Channel," *Proc. IERE Intl. Conf. Video and Data Recording,* pp. 151–157, Southampton, England, April 1984.

[CM&K] G. Carayannis, D. Manolakis, and N. Kalouptsidis, "A Fast Sequential Algorithm for Least Squares Filtering and Prediction," *IEEE Trans. ASSP,* vol. ASSP-31, pp. 227–239, Dec. 1983.

[C&N] M. F. Choquet and H. J. Nussbaumer, "Generation of Synchronous Data Transmission Signals by Digital Echo Modulation," *IBM J. Res. & Develop.,* pp. 364–377, Sept. 1971.

[C&P] A. Croisier and J. M. Pierret, "High Efficiency Data Transmission Through Digital Echo Modulation," *IEEE Intl. Conf. Commun. Record,* pp. 29.9–14, June 1969.

[C . . 1] S. Crozier, M. Wilson, K. W. Moreland, J. Camelon, and P. J. McLane, "Microprocessor-Based Implementation and Testing of a Simple Viterbi Detector," *Can. Elec. Eng. J.,* vol. 6, no. 3, 1981.

[C . . 2] M. B. Carey, H.-T. Chen, A. Descloux, J. F. Ingle, and K. I. Park, "1982/83 End Office Connection Study: Analog Voice and Voiceband Data Transmission Performance Characterization of the Public Switched Network," *AT&T Bell Labs. Tech. J.,* vol. 63, pp. 2059–2119, Nov. 1984.

[Dr] S. E. Dreyfus, *Dynamic Programming and the Calculus of Variations,* Academic, New York, 1965.

[Dw] H. B. Dwight, *Tables of Integrals and Other Mathematical Data,* Macmillan, New York, 1961.

[D&H] M. L. Doelz and E. H. Heald, "Minimum Shift Data Communication System," U.S. Patent No. 2,977,417, March 1961.

[DH&M] M. L. Doelz, E. T. Heald, and D. L. Martin, "Binary Data Transmission Techniques for Linear Systems," *Proc. IRE,* vol. 45, pp. 656–661, May 1957.

[DI&K] J. Dodo, S. Iwai, and K. Kawai, "Computer Simulation of Carrier Recovery Circuit in Microwave PCM-PSK System," *Fujitsu Sci. & Tech. J.,* pp. 59–82, Sept. 1972.

[D&L]   J. A. Drager and T. R. Lawrence, "Linking Data to Telephones Acoustically," *Bell Labs. Record,* vol. 47, pp. 266–270, Sept. 1969.

[D&M]   N. A. D'Andrea and U. Mengali, "A Simulation Study of Clock Recovery in QPSK and 9QPRS Systems," *IEEE Trans. Commun.,* vol. **COM-23,** pp. 1139–1142, Oct. 1985.

[D&T]   F. P. Duffy and T. W. Thatcher, "Analog Transmission Performance on the Switched Telephone Network," *Bell Syst. Tech. J.,* vol. 50, pp. 1311–1347, April 1971.

[Fa1]   D. D. Falconer, "Jointly Adaptive Equalization and Carrier Recovery in Two-Dimensional Communications Systems," *Bell Syst. Tech. J.,* vol. 55, pp. 317–334, March 1976.

[Fa2]   ——— , "Adaptive Reference Echo Cancellation," *IEEE Trans. Commun.,* vol. **COM-30,** pp. 2083–2094, Sept. 1982.

[Fe1]   J. H. Fennick, "Amplitude Distributions of Telephone Channel Noise and a Model for Impulse Noise," *Bell Syst. Tech. J.,* vol. 47, pp. 1001 *et seq.,* July 1968.

[Fe2]   M. J. Ferguson, "Optimal Reception for Binary Partial Response Channels," *Bell Syst. Tech. J.,* vol. 51, pp. 493–505, Feb. 1972.

[Fe3]   K. Feher, *Digital Communications: Satellite/Earth Station Enginering,* Prentice-Hall, Englewood Cliffs, NJ, 1981.

[Fo1]   G. D. Forney Jr., "Maximum Likelihood Sequence Estimation of Digital Sequences in the Presence of Inter-Symbol Interference," *IEEE Trans Inform. Theory,* vol. **IT-18,** pp. 363–378, May 1972.

[Fo2]   ——— , "The Viterbi Algorithm", Proc. IEEE, vol. 61, pp. 268–278, March 1973.

[Fo3]   ——— , "Training Adaptive Linear Filters," U.S. Patent No. 3,723,91, March 1973.

[Fo4]   H. C. Foults (Ed.), *A Compilation of Data Communication Standards,* McGraw-Hill, New York, 1986.

[Fr1]   S. A. Fredricsson, "A Reduced State Viterbi Detector for Multi-Level Partial Response Channels," Tech. Rep. No. 86, Royal Institute of Techn., Stockholm, Sweden, Sept. 1974.

[Fr2]   ——— , "Joint Optimization of Transmitter and Receiver Filters in Digital PAM Systems with a Viterbi Detector," *IEEE Trans. Inform. Theory,* vol. **IT-22,** pp. 200–209, March, 1976.

[Fr3]   L. E. Franks, "Carrier and Bit Synchronization in Data Communication—a Tutorial Review," *IEEE Trans. Commun,* vol. **COM-28,** pp. 1107–1121, Aug. 1980.

[Fr4]   R. L. Freeman, *Telecommunication Transmission Handbook,* Wiley, New York, 1981.

[F&B]   L. E. Franks and J. P. Bubrouski, "Statistical Properties of Timing Jitter in a PAM Timing Recovery Scheme," *IEEE Trans. Commun.,* vol. **COM-22,** pp. 913–920, July 1974.

[F&G]   G. D. Forney and R. G. Gallager, "Signal Structures for Double-Sideband Quadrature Carrier Modulation," U.S. Patent No. 3,887,768, 1973.

[F&L1] G. A. Franco and G. Lachs, "An Orthogonal Coding Technique for Communications," *IRE Intl. Conv. Rec.,* vol. 9, pt. 8, pp. 126–133, 1961.

[F&L2] D. D. Falconer and L. Ljung, "Application of Fast Kalman Estimation to Adaptive Equalization," *IEEE Trans. Commun.,* vol. **COM-26,** pp. 1439–1446, Oct. 1978.

[F&M1] D. D. Falconer and F. R. Magee, Jr., "Adaptive Channel Memory Truncation for Maximum Likelihood Sequence Estimation," *Bell Syst. Tech. J.,* vol. 52, pp. 1541–1562, Nov. 1973.

[F&M2] ———, "Evaluation of Decision Feedback Equalization and Viterbi Algorithm Detection for Voiceband Data Transmission: Parts I and II," *IEEE Trans. Commun.,* vol. **COM-24,** pp. 1130–1139, 1238–1245, Oct. and Nov. 1976.

[F&M3] D. D. Falconer and K. H. Mueller, "Adaptive Echo Cancellation/AGC Structures for Two-Wire FDX Data Transmission," *Bell Syst. Tech. J.,* pp. 1593–1616, Sept. 1979.

[FMPC] N. J. P. Frenette, P. J. McLane, L. E. Peppard, and F. Cotter, "Implementation of a Viterbi Processor for a Digital Communications System with a Time-Dispersive Channel," *IEEE J. SAC,* vol. **SAC-4,** pp. 160–167, Jan. 1986.

[F&T] K. Feher and G. S. Takhar, "A New Symbol Timing Recovery Technique for Burst Modem Applications," *IEEE Trans. Commun.,* vol. **COM-26,** pp. 100–108, Jan. 1978.

[FT&M] M. Fukuda, T. Tsuda, and K. Murano, "Digital Subscriber Loop Transmission Using Echo Canceler and Balancing Networks," *IEEE Intl. Conf. Commun.,* June 1985.

[F&W] P. E. Fleischer and L. Wu, "The DI43 Line Code and its Spectral Properties," Contribution No. T1D1.3/85-184, Aug. 1985.

[F . . .] G. D. Forney, R. G. Gallager, G. R. Lang, F. M. Longstaff, and S. U. Qureshi, "Efficient Modulation for Band-Limited Channels," *IEEE J. SAC,* vol. **SAC-2,** pp. 659–672, Sept. 1984.

[Ga1] R. G. Gallager, *Information Theory and Reliable Communication,* Wiley, New York, 1968.

[Ga2] F. M. Gardner, "Carrier and Clock Synchronization for TDMA Digital Communications," European Space Agency Preliminary Report, June 1974.

[Ga3] ———, "Hang-up in Phase-Lock Loops," *IEEE Trans. Commun.,* vol. **COM-25,** pp. 1210–1214, Oct. 1977.

[Ga4] ———*"Phaselock Techniques,"* 2nd ed., Wiley, New York, 1979.

[Ga5] ———*"Self-Noise in Synchronizers,"* *IEEE Trans. Commun.,* vol. **COM-28,** pp. 1159–1163, Aug. 1980.

[Ga6] ———, "Equivocation as a Cause of PLL Hangup," *IEEE Trans. Commun.,* vol. **COM-30,** pp. 2242–2243, Oct. 1982.

[Ge] A. Gersho, "Adaptive Equalization of Highly Dispersive Channels," *Bell Syst. Tech. J.,* vol. 48, pp. 55–70, Jan. 1969.

[Go1] D. Godard, "Channel Equalization Using a Kalman Filter for Fast Data Transmission," *IBM J. Res. & Develop.,* vol. 18, pp. 267–273, May 1974.

[Go2]   ———, "Pass-Band Timing Recovery in an All-Digital Modem Receiver," *IEEE Trans. Commun.,* vol. **COM-26**, pp. 517–523, May 1978.

[Gr]   P. A. Gresh, "Physical and Transmission Characteristics of Customer Loop Plant," *Bell Syst. Tech. J.,* vol. 48, pp. 3337–3385, Dec. 1969.

[Gu]   L. Guidoux, "Egaliseur autoadaptiv a double échantillonage," *L'Onde Electrique,* vol. 55, pp. 9–13, Jan. 1975.

[G&C]   L. J. Greenstein and B. A. Czekaj-Augun, "Performance Comparisons among Digital Radio Techniques Subjected to Multipath Fading," *IEEE Trans. Commun.,* vol. **COM-30**, pp. 1185–1197, May 1982.

[G&L]   J. F. Gunn and J. A. Lombardi, "Error Detection for Partial Response Systems," *IEEE Trans. Commun. Tech.,* vol. **COM-17**, pp. 734–737, Dec. 1969.

[G&M1]   S. A. Gronemeyer and A. L. McBride, "MSK and Offset QPSK modulation," *IEEE Trans. Commun.,* vol. **COM-24**, pp. 809–820, Aug. 1976.

[GM&T]   R. D. Gitlin, J. E. Mazo, and M. G. Taylor, "On the Design of Gradient Algorithms for Digitally Implemented Adaptive Filters," *IEEE Trans. Circuit Theory,* vol. **CT-20**, pp. 125–136, March 1973.

[GM&W]   R. D. Gitlin, H. C. Meadows, and S. B. Weinstein, "The Tap-Leakage Algorithm: An Algorithm for the Stable Operation of a Digitally Implemented Fractionally-Spaced Adaptive Equalizer," *Bell Syst. Tech. J.,* vol. 61, pp. 1817–1839, Oct. 1982.

[G&S]   P. G. Gulak and E. Shwedyk, "VLSI Structures for Viterbi Receivers," *IEEE J. SAC,* vol. **SAC-4**, pp. 142–159, Jan. 1986.

[GS&B]   J. A. Guinea, C. Stacey, and R. Batruni, "Digital Transmission in the Subscriber Loop," *IEEE Circuits and Devices Mag.,* pp. 14–27, Sept. 1986.

[G&W1]   P. J. van Gerwen and P. van der Wurf, "Data Modems with Integrated Digital Filters and Modulators," *IEEE Trans. Commun. Tech.,* vol. **COM-18**, pp. 214–222, June 1970.

[G&W2]   R. D. Gitlin and S. B. Weinstein, "On the Required Tap-Weight Precision for Digitally-Implemented Adaptive Mean-Square Equalizers," *Bell Syst. Tech. J.,* vol. 58, pp. 301–321, Feb. 1979.

[G&W3]   ———, "Fractionally-Spaced Equalization: An Improved Digital Transversal Equalizer," *Bell Syst. Tech. J.,* vol. 60, pp. 275–296, Feb. 1981.

[Ha1]   J. V. Hartley, U.S. Patent No. 1,666,206, 1928.

[Ha2]   J. F. Hayes, "The Viterbi Algorithm Applied to Digital Data Transmission," *IEEE Commun. Soc. Mag.,* vol. 13, pp. 15–20, Mar. 1975.

[Hi1]   B. Hirosaki, "An Analysis of Automatic Equalizers for Orthogonally Multiplexed QAM Systems," *IEEE Trans. Commun.,* vol. **COM-28**, pp. 73–83, Jan. 1980.

[Hi2]   ———, "An Orthogonally Multiplexed QAM System Using the Discrete Fourier Transform," *IEEE Trans. Commun.,* vol. **COM-29**, pp. 982–989, July 1981.

[Ho1]   J. L. Holsinger, "Digital Communication over Fixed Time-Continuous Channels with Memory—with Special Application to Telephone Channels," Lincoln Laboratory Technical Report No. 366, MIT, Cambridge, MA Oct. 1964.

[HC&R]   J. F. Hayes, T. M. Cover, and J. B. Riera, "Optimal Sequence Detection and Optimal Symbol-by-Symbol Detection: Similar Algorithms," *IEEE Trans. Commun.,* vol. **COM-30,** pp. 152–157, pt. 2, Jan. 1982.

[HH&S]   B. Hirosaki, S. Hasegawa, and A. Sabato, "Advanced Group-Band Modem Using Orthogonally Multiplexed QAM Technique," *IEEE Trans. Commun.,* vol. **COM-34,** pp. 587–592, June 1986.

[H&M]   M. L. Honig and D. G. Messerschmitt, *"Adaptive Filters: Structures, Algorithms and Applications,"* Kluwer Academic Publishers, Boston, 1984.

[H&V]   A. W. Horton and H. E. Vaughan, "The Transmission of Digital Information over Telephone Circuits," *Bell Syst. Tech. J.,* vol. 34, pp. 511–528, May 1955.

[Hi . . .]   B. Hirosaki, A. Yoshida, S. Hasegawa, O. Tanaka, K. Inoue, and K. Watanabe, "A 19.2Kb/s Voice-Band Data Modem Based on Orthogonally Multiplexed QAM Techniques," *IEEE GLOBECOM 1985 Record,* pp. 661–665.

[IBM1]   IBM, Europe, "Trellis-Coded Modulation Schemes for Use in Data Modems Transmitting 3–7 Bits per Modulation Interval," CCITT SGXVII, Contribution No. D114, April 1983.

[IBM2]   ——, "Trellis-Coded Modulation Schemes with 8-State Systematic Encoder and 90° Symmetry for Use in Data Modems Transmitting 3–7 Bits per Modulation Interval," CCITT SGXVII Contribution No. D180, Oct. 1983.

[I&S]   F. Itakura and S. Saito, "Digital Filtering Techniques for Speech Analysis and Synthesis," *Proc. 7th Intl. Cong. Acoust.,* Paper 25-C-1, pp. 261–264, Budapest, 1971.

[Ja]   L. B. Jackson, "On the Interaction of Roundoff Noise and Dynamic Range in Digital Filters," *Bell Syst. Tech. J.,* vol. 49, pp. 159 *et seq.,* Feb. 1970.

[J&D]   F. de Jager and C. B. Dekker, "Tamed Frequency Modulation—a Novel Method to Achieve Spectrum Economy in Digital Transmission," *IEEE Trans. Commun.,* vol. **COM-26,** pp. 534–542, May 1978.

[Ka]   M. Kalb, "ADPCM Performance Impact on Voice-Band Data," *IEEE GLOBECOM Record,* pp. 1133–1137, 1985.

[Ko1]   H. Kobayashi, "Simultaneous Adaptive Estimation and Decision Algorithm for Carrier Modulated Data Transmission Systems," *IEEE Trans. Commun.,* vol. **COM-19,** pp. 268–280, June 1971.

[Ko2]   ——, "Correlative Level Coding and Maximum-Likelihood Decoding," *IEEE Trans. Inf. Theory,* vol. **IT-17,** pp. 586–594, Sept. 1971.

[Ko3]   ——, "A Survey of Coding Schemes for Transmission or Recording of Digital Data," *IEEE Trans. Commun.,* vol. **COM-19,** pp. 1087–1100, Dec. 1971.

[Kr1]   E. R. Kretzmer, "An Efficient Binary Data Transmission System," *IEEE Trans. Commun. Systs.,* vol. **CS-12,** pp. 250–251, June 1964.

[Kr2)   ——, "Binary Data Transmission by Partial Response Transmission," *IEEE Intl. Conf. Commun.,* G-1E.5, pp. 451–455, June 1965.

[K&B]   J. Kurzweil and J. A. C. Bingham, "Adaptive Equalizer Design Considerations," *IEEE Intl. Conf. Commun.,* pp. 56.5.1–56.5.7, 1981.

[K&L]   J. L. Kelly and C. C. Lochbaum, "Speech Synthesis," *Proc. 4th Intl. Congr. Acoustics,* pp. 1–4, 1962.

[K&P]   P. Kabal and S. Pasupathy, "Partial-Response Signaling," *IEEE Trans. Commun.,* vol. **COM-23,** pp. 921–934, Sept. 1975.

[K&W]   V. G. Koll and S. B. Weinstein, "Simultaneous Two-Way Data Transmission over a Two-Wire Circuit," *IEEE Trans. Commun.,* vol. **COM-21,** pp. 143–147, Feb. 1973.

[K . . .]   H. Kurihara et al., "Carrier Recovery with Low Cycle Skipping Rate for CPSK/TDMA Systems," *Proc. 5th Intl. Conf. Digital Satellite Commun.,* pp. 319–324, Genoa, Italy, March 1981.

[Le1]   A. Lender, "The Duo-Binary Technique for High-Speed Data Transmission," *IEEE Trans. Commun. Electron.,* vol. 82, pp. 214–218, May 1963.

[Le2]   ———, "Correlative Digital Communication Techniques," *IEEE Trans. Commun. Tech.,* vol. **COM-12,** pp. 128–135, Dec. 1964.

[Le3]   ———, "Correlative Level Coding for Binary-Data Transmission," *IEEE Spectrum,* vol. 3, pp. 104–115, Feb. 1966.

[Le4]   ———, "Error Detection for Modified Duobinary systems," U.S. Patent No. 3,461,426, Aug. 1969.

[Le5]   C. C. Lee, "A New Switched-Capacitor Realization for Cyclic Analog-to-Digital Converter," *Proc. IEEE Intl. Symp. Circuits & Systems,* pp. 1261–1265, June 1983.

[Li1]   W. C. Lindsey, *Synchronization Systems in Communications and Control,* Prentice-Hall, Englewood Cliffs, NJ, 1972.

[Li2]   N.-S. Lin, "Low Sample-Rate Timing Recovery using Switched-Capacitor Techniques," Memo No. UCB/ERL M83/72, Univ. Calif., Dec. 1983.

[Lu1]   R. W. Lucky, "Automatic Equalization for Digital Communication," *Bell Syst. Tech. J.,* vol. 44, pp. 547–588, April 1965.

[Lu2]   ———, "Techniques for Adaptive Equalization of Digital Communication Systems," *Bell Syst. Tech. J.,* vol. 45, pp. 255–286, Feb. 1966.

[Ly1]   D. L. Lyon, "Timing Recovery in Synchronous Equalized Data Communications," *IEEE Trans. Commun.,* vol. **COM-23,** pp. 269–274, Feb. 1975.

[Ly2]   ———, "Envelope Derived Timing Recovery in QAM and SQAM Systems," *IEEE Trans. Commun.,* vol. **COM-23,** pp. 1327–1331, Nov. 1975.

[L&B]   G. R. Lang and P. O. Brackett, "Complex Analogue Filters," *Proc. Eur. Conf. Circuit Theory and Design,"* pp. 412–419, 1981.

[L&C1]   W. C. Lindsey and C. M. Chie, "A Survey of Digital Phase-Locked Loops," *Proc. IEEE,* vol. 69, pp. 410–431, April 1981.

[L&C2]   S. Lin and D. J. Costello, *Error Control Coding: Fundamentals and Applications,* Prentice Hall, Englewood Cliffs, NJ, 1983.

[L&F]   T. Le-Ngoc and K. Feher, "A Digital Approach to Symbol Timing Recovery Systems," *IEEE Trans. Commun.,* vol. **COM-28,** pp. 1993–1999, Dec. 1980.

[L&H]   W. U. Lee and F. S. Hill, "A Maximum Likelihood Estimator with Decision Feedback Equalization," *IEEE Trans. Commun.,* vol. **COM-25,** pp. 971–979, Sept. 1971.

[L&K]   S. Laufer and I. Kalet, "Optimization of Generalized Tamed Frequency Modulation," *IEEE GLOBECOM 1986 Record,* Session 29.4.

[LL&B]   K. K. Lee, T. Le-Ngoc, and V. K. Bhargava, "A New Feedforward Tracking System Bandpass Filter for Carrier Recovery Systems," *IEEE Intl. Conf. Commun. Record,* pp. 32.5.1–5, June 1985.

[L&P]   F. Ling and J. G. Proakis, "A Generalized Multi-Channel Least Squares Lattice Algorithm Based on Sequential Processing Stages," *IEEE Trans. ASSP,* vol. **ASSP-32,** pp. 381–389, April 1984.

[L&S1]   W. C. Lindsey and M. K. Simon, "Data-Aided Carrier Tracking Loops," *IEEE Trans. Commun.,* vol. **COM-19,** pp. 157–168, April 1971.

[L&S2]   ———, "Carrier Synchronization and Detection of Polyphase Signals," *IEEE Trans. Commun.,* vol. **COM-20,** pp. 441–454, June 1972.

[L&V]   A. Leclert and P. Vandamme, "Universal Carrier Recovery Loop for QASK and PSK Signal Sets," *IEEE Trans. Commun.,* vol. **COM-31,** pp. 130–136, Jan. 1983.

[LS&W]   R. W. Lucky, J. Salz, and E. J. Weldon, *"Principles of Data Communication,"* McGraw-Hill, New York, 1966.

[Ma1]   J. E. Mazo, "Optimum Timing Phase for an Infinite Equalizer," *Bell Syst. Tech. J.,* vol. 54, pp. 624–636, Jan. 1975.

[Ma2]   L. M. Manhire, "Physical and Transmission Characteristics of Customer Loop Plant," *Bell System Tech. J.,* vol. 57, pp. 35–59, Jan. 1978.

[Ma3]   J. Makhoul, "A Class of All-Zero Lattice Digital Filters: Properties and Applications," *IEEE Trans. ASSP,* vol. **ASSP-26,** pp. 304–314, Aug. 1978.

[Ma4]   ———, "Analysis of Decision-Directed Equalizer Convergence," *Bell Syst. Tech. J.,* vol. 59, pp. 1857–1876, Dec. 1980.

[Ma5]   J. L. Massey, "The How and Why of Channel Coding," *Intl. Seminar Data Commun.,* Swiss Fed. Inst. Technol., Zurich, March 1984.

[Me1]   U. Mengali, "Joint Phase and Timing Acquisition in Data Transmission," *IEEE Trans. Commun.,* vol. **COM-25,** pp. 1174–1185, Oct. 1977.

[Me2]   D. G. Messerschmitt, "Design of a Finite Impulse Response for the Viterbi Algorithm and Decision Feedback Equalizer," *IEEE Intl. Conf. Commun. Record,* pp. 37D.1–37D.5, June 1974.

[Me3]   ———, "Echo Cancellation in Speech and Data Transmission," *IEEE J. SAC,* vol. **SAC-2,** pp. 283–296, March 1984.

[Me4]   ———, "Asynchronous and Timing-Jitter-Insensitive Data Echo Cancellation," *IEEE Trans. Commun.,* vol. **COM-34,** pp. 1209–1217, Dec. 1986.

[Me5]   ———, "Echo Cancellation in Speech and Data Transmission," Chapter 4 in *Advanced Digital Communications and Signal Processing,"* K. Feher (Ed.), Prentice Hall, New York, 1987.

[Mo1]   P. Monsen, "Feedback Equalization for Fading Dispersive Channels," *IEEE Trans. Inform. Theory,* vol. **IT-17,** pp. 56–64, Jan. 1971.

[Mu1]   K. H. Mueller, "A New, Fast-Converging Mean-Square Algorithm for Adaptive Equalizers with Partial-Response Signaling," *Bell Syst. Tech. J.,* vol. 54, pp. 143–153, Jan. 1975.

[Mu2]   B. R. N. Murthy, "Crosstalk Requirements for PCM Transmission," *IEEE Trans. Commun.,* vol. **COM-24,** pp. 88–97, Jan. 1976.

[Mu3)   K. H. Mueller, "A New Digital Echo Canceler for Two-Wire Full-Duplex

Data Transmission," *IEEE Trans. Commun.,* vol. **COM-24,** pp. 956–962, Sept. 1976.

[M&C] C. M. Melas and J. M. Cioffi, "Evaluating the Performance of Maximum Likelihood Sequence Detection in a Magnetic Recording Channel," *IEEE GLOBECOM 1986 Record,* Session 30.4.

[M&G] R. S. Medaugh and L. J. Griffiths, "Further Results of a Least-Squares and Gradient Adaptive Lattice Algorithm Comparison," *Proc. IEEE Intl. Conf. ASSP,* pp. 1412–1415, May 1982.

[M&G2] D. G. Messerschmitt and P. R. Gray, "Design Issues in the ISDN U-Interface Transceiver," Internal Report, EECS Dept., Univ. Calif., Berkeley (privately received).

[M&M] K. H. Mueller and M. Muller, "Timing Recovery in Digital Synchronous Data Receivers," *IEEE Trans. Commun.,* vol. **COM-24,** pp. 516–521, May 1976

[M&P] H. Meyr and L. Popken, "Phase Acquisition Statistics for Phase-Locked Loops," *IEEE Trans. Commun.,* vol. **COM-28,** pp. 1365–1372, Aug. 1980.

[M&S1] K. H. Mueller and D. A. Spaulding, "Cyclic Equalization — A New Rapidly Converging Equalization Technique for Synchronous Data Communication," *Bell Syst. Tech. J.,* vol. 54, pp. 369–406, Feb. 1975.

[M&S2] K. Martin and A. S. Sedra, "Strays-Insensitive Switched-Capacitor Filters Based on Bilinear *z*-Transform," *Electronics Lett.,* vol. 15, no. 13, pp. 365–366, June 21, 1979.

[MS&T] G. R. McMillen, M. Shafi, and D. P. Taylor, "Simultaneous Adaptive Estimation of Carrier Phase, Symbol Timing and Data for a 49-QPRS DFE Radio Receiver," *IEEE Trans. Commun.,* vol. **COM-32,** pp. 429–443, April 1984.

[Ny] H. Nyquist, "Certain Topics in Telegraph Transmission Theory," *AIEE Trans.,* vol. 47, pp. 617–644, April 1928

[N&P] E. A. Newcombe and S. Pasupathy, "Effects of Filtering Allocation on the Performance of a Modified Duobinary System," *IEEE Trans. Commun.,* vol. **COM-28,** pp. 749–752, May 1980.

[NM&M] R. A. Northrup, D. M. Motley, and G. K. McAuliffe, "Implementation and Performance of ADEM," *IEEE Intl. Conf. Commun. Record,* Sess. 12/1, pp. 1–9, 1968.

[Oe] M. Oerder, "Rotationally Invariant Trellis Codes for mPSK Modulation," *IEEE Intl. Conf. Commun. Record,* pp. 552–556, June 1985.

[Pa] S. Pasupathy, "Minimum Shift Keying: A Spectrally Efficient Modulation," *IEEE Commun. Mag.,* pp. 14–22, July 1979.

[Pe] R. J. Peck, "Digital Modulator," U.S. Patent No. 4,049,909, 1977.

[Pr1] R. Price, "Nonlinearly Feedback-Equalized PAM vs Capacity for Noisy Filter Channels," *IEEE Intl. Conf. Commun. Record,* pp. 22.12–16, June 1970.

[Pr2] J. G. Proakis, *"Digital Communications,"* McGraw-Hill, New York, 1983.

[P&M] J. G. Proakis and J. H. Miller, "Adaptive Receiver for Digital Signaling through Channels with Inter-Symbol Interference," *IEEE Trans. Inform. Theory,* vol. **IT-15,** pp. 484–497, July 1969.

[P&W1]   W. W. Peterson and E. J. Weldon, *Error-Correcting Codes,* MIT Press, Cambridge, MA, 1972.

[P&W2]   S. V. Pizzi and S. G. Wilson, "Convolutional Coding Combined with Continuous Phase Modulation," *IEEE Trans. Commun.,* vol. **COM-33,** pp. 20–29, Jan. 1985.

[Qu1]   S. U. H. Qureshi, "An Adaptive Decision-Feedback Receiver Using Maximum-Likelihood Sequence Estimation," *IEEE Intl. Conf. Commun. Record,* June 1973.

[Qu2]   ———, "Adjustment of the Position of the Reference Tap of an Adaptive Equalizer," *IEEE Trans. Commun.,* vol. **COM-21,** pp. 1046–1052, Sept. 1973.

[Qu3]   ———, "Timing Recovery for Equalized Partial Response Systems," *IEEE Trans. Commun.,* vol. **COM-24,** pp. 1326–1331, Dec. 1976.

[Qu4]   ———, "Fast Start-up Equalization with Periodic Training Sequences," *IEEE Trans. Inform. Theory,* vol. **IT-23,** pp. 553–563, Sept. 1977.

[Qu5]   ———, "Digital Modem Transmitter," U.S. Patent No. 4,358,853, Nov. 1982.

[Qu6]   ———, "Adaptive Equalization," *Proc. IEEE,* vol. 73, pp. 1349–1387, Sept. 1985.

[Q&F]   S. U. H. Qureshi and D. Forney, "Performance and Properties of a $T/2$ Equalizer," *Proc., Nat. Telecomm. Conf.,* pp. 11:1,1–9, Dec. 1977.

[Q&N]   S. U. H. Qureshi and E. E. Newhall, "An Adaptive Receiver for Data Transmission over Time-Dispersive Channels," *IEEE Trans. Inf. Theory,* vol. **IT-19,** pp. 448–457, July 1973.

[Ri1]   P. N. Ridout, "A New Method of Timing Recovery for $(1, 0, -1)$ Multiple-Response Channels," *Electronics Lett.,* vol. 14, no. 12, pp. 353–355, June 1978.

[Ri2]   D. B. Richards, "Simulation of an Adaptive Channel Truncation Scheme for the Viterbi Algorithm," Internal Report, EECS Dept., Univ. Calif., Berkeley, Dec. 1983.

[Ru1]   W. D. Rummler, "A New Selective Fading Model: Application to Propagation Data," *Bell Syst. Tech. J.,* vol. 58, pp. 1032–1071, May–June 1979.

[Ru2]   ———, "More on the Multipath Fading Channel Model," *IEEE Trans. Commun.,* vol. **COM-29,** pp. 346–352, March 1981.

[R&S]   L. R. Rabiner and R. W. Schafer, *Digital Processing of Speech Signals,* Prentice-Hall, Englewood Cliffs, NJ, 1978.

[R . . .]   B. Ross et al., "Microprocessor Realization of the Adaptive Viterbi Detector," *Proc. Natl. Telecommun. Conf.,* Dec. 1980.

[SA1]   B. R. Saltzberg, "Timing Recovery for Synchronous Binary Data Transmission," *Bell Syst. Tech. J.,* vol. 46, pp. 593–622, March 1967.

[Sa2]   ———, "Performance of an Efficient Parallel Data Transmission System," *IEEE Trans. Commun. Tech.,* vol. **COM-15,** pp. 805–811, Dec. 1967.

[Sa3]   J. Salz, "Optimum Mean-Square Decision Feedback Equalization," *Bell Syst. Tech. J.,* vol. 52, pp. 1341–1373, Oct. 1973.

[Sa4]   H. Sari, "Baseband Equalizer Performance in the Presence of Selective Fading," *IEEE GLOBECOM 1983 Record,* pp. 1–7.

[Sc]   G. Schollmeier, "A Transmission Line Code for the Network Side of the NT—Basic Access," Contribution No. T1D1.3/85-069, May 1985.

[Sh1]   C. E. Shannon, "The Mathematical Theory of Communication," *Bell Syst. Tech. J.,* vol. 27, pp. 379–423, July 1948.

[Si1]   M. K. Simon, "An MSK Approach to Offset QASK," *IEEE Trans. Commun.,* vol. **COM-24**, pp. 921–923, Aug. 1976.

[Si2]   ———, "Tracking Performance of Costas Loops with Hard-Limited In-Phase Channel," *IEEE Trans. Commun.,* vol. **COM-26,** pp. 420–432, April 1978.

[Si3]   ———, "Optimum Receiver Structures of Phase-Multiplexed Modulations," *IEEE Trans. Commun.,* vol. **COM-26,** pp. 865–872, June 1978.

[Si4]   ———, "On the Optimality of the MAP Estimation Loop for Carrier Phase Tracking BPSK and QPSK Signals," *IEEE Trans. Commun.,* vol. **COM-27,** pp. 158–165, Jan. 1979.

[Si5]   ———, "False Lock Performance of Quadri-Phase Receivers," *IEEE Trans. Commun.,* vol. **COM-27,** pp. 1660–1670, Nov. 1979.

[Sp]   J. J. Spilker, *Digital Communications by Satellite,* Prentice Hall, Englewood Cliffs, NJ, 1979.

[S&A]   E. H. Satorius and S. T. Alexander, "Channel Equalization using Adaptive Lattice Algorithms," *IEEE Trans. Commun.,* vol. **COM-27,** pp. 899–905, June 1979.

[S&M]   M. Shafi and D. J. Moore, "Further Results on Adaptive Equalizer Improvement for 16QAM and 64QAM Digital Radio," *IEEE Trans. Commun.,* vol. **COM-34,** pp. 59–66, Jan. 1986.

[SD&M]   H. Sari, L. Desperben, and S. Moridi, "Minimum MSE Timing Recovery Schemes for Digital Equalizers," *IEEE Trans. Commun.,* vol. **COM-34,** pp. 694–702, July 1986.

[S&P]   E. H. Satorius and J. D. Pack, "Application of Least-Squares Lattice Algorithms to Adaptive Equalization," *IEEE Trans. Commun.,* vol. **COM-29,** pp. 136–142, Feb. 1981.

[S&S1]   M. K. Simon and J. G. Smith, "Carrier Synchronization and Detection of QASK Signal Sets," *IEEE Trans. Commun.,* vol. **COM-22,** pp. 98–106, Feb. 1974.

[S&S2]   ———, "Offset Quadrature Communication with Decision Feedback Synchronization," *IEEE Trans. Commun.,* vol. **COM-22,** pp. 1576–1584, Oct. 1974.

[S&S3]   W. M. Snelgrove and A. S. Sedra, "State-Space Synthesis of Complex Analog Filters," *Proc. Eur. Conf. Circuit Theory and Design,* pp. 420–424, 1981.

[SSLB]   W. M. Snelgrove, A. S. Sedra, G. R. Lang, and P. O. Brackett, "Complex Analog Filters," Tech. Report in Electrical Engineering Dept., Univ. Toronto, 1981.

[STOT]   T. Suzuki, H. Takatori, M. Ogawa, and K. Tomooka, "Line Equalizer for a Digital Subscriber Loop Employing Switched Capacitor Technology," *IEEE Trans. Commun.,* vol. **COM-30,** pp. 2074–2082, Sept. 1982.

[Th]   H. Thapar, "Real-Time Application of Trellis Coding to High-Speed Voice

Band Data Transmission," *IEEE J. SAC,* vol. **SAC-2,** pp. 648–658, Sept. 1984.

[Tu]   D. G. Tucker, "The History of the Homodyne and Synchrodyne," *J. Br. IRE,* pp. 143–154, April 1954.

[TC]   Telebit Corporation, "Multicarrier Modulation for the High-Speed Asymmetrical Modem Recommendation," CCITT WP XVII/1 Doc. No. 87-02-013, Dec. 1986.

[T&H]   D. P. Taylor and P. R. Hartmann, "Telecommunications by Microwave Digital Radio," *IEEE Commun. Mag.,* vol. 24, pp. 11–16, Aug. 1986.

[TH&M] C.-P. Tzeng, D. A. Hodges, and D. G. Messerschmitt, "Timing Recovery in Digital Subscriber Loops using Baud-Rate Sampling," *IEEE J. SAC,* vol. **SAC-4,** pp. 1302–1311, Nov. 1986.

[T&P]   H. Thapar and A. M. Patel, "A Class of P-R Systems for Increasing Storage Density in Magnetic Recording," *IEEE Trans. Mag. Record,* Sept. 1987.

[T&S]   D. P. Taylor and M. Shafi, "Decision Feedback Equalization for Multipath Induced Interference in Digital Microwave LOS Links," *IEEE Trans. Commun.,* vol. **COM-32,** pp. 267–279, March 1984.

[TT&M] D. P. Taylor, S. K. Tang, and S. Mariuz, "The Limit-Switched Loop: A Phase-Locked Loop for Burst Mode Operation," *IEEE Trans. Commun.,* vol. **COM-30,** pp. 396–407, Feb. 1982.

[TW&D] C. M. Thomas, M. Y. Weidner, and S. H. Durrani, "Digital Amplitude Phase Shift Keying M-ary Alphabets," *IEEE Trans. Commun.,* vol. **COM-22,** pp. 168–180, Feb. 1974.

[Un1]   G. Ungerboeck, "Theory on the Speed of Convergence in Adaptive Equalizers for Data Communication," *IBM J. Res. Devel.,* vol. 16, pp. 546–555, Nov. 1972.

[Un2]   ———, "Adaptive Maximum Likelihood Receiver for Carrier Modulated Data Transmission Systems," *IEEE Trans. Commun.,* vol. **COM-22,** pp. 624–636, May 1974.

[Un3]   ———, "Fractional-Tap Spacing Equalizer and Consequences for Clock Recovery in Data Modems," *IEEE Trans. Commun.,* vol. **COM-24,** pp. 856–864, Aug. 1976.

[Un4]   ———, "Channel Coding with Multilevel/Phase Signals," *IEEE Trans. Inform. Theory,* vol. **IT-28,** pp. 55–67, Jan. 1982.

[Un5]   ———, "Trellis-Coded Modulation with Redundant Signal Sets—Part I: Introduction, Part II: State of the Art," *IEEE Commun. Mag.,* vol. 25, pp. 5–21, Feb. 1987.

[UH&A] G. Ungerboeck, J. Hagenauer, and T. Abdel Nabi, "Coded 8-PSK Experimental Modem for the INTELSAT SCPC System," *Proc. 7th Intl. Conf. Digital Satellite Commun.,* pp. 299–304, May 1986.

[U&S]   T. Uyematsu and K. Sakaniwa, "A New Tap-Adjustment Algorithm for the Fractionally-Spaced Equalizer," *IEEE Intl. Conf. Commun. Record,* pp. 46.3.1–4, June 1985.

[Vi1]   A. J. Viterbi, "Error Bounds for Convolutional Codes and an Asymptotically Optimum Decoding Algorithm," *IEEE Trans. Inform. Theory,* vol. **IT-13,** pp. 260–269, April 1967.

[Vi2]   ———, "Convolutional Codes and their Performance in Communications Systems," *IEEE Trans. Commun. Technol.,* vol. **COM-19,** pp. 751–772, Oct. 1971.

[Vo]   J. Voelcker, "Helping Computers Communicate," *IEEE Spectrum,* pp. 61–70, March 1986.

[V&O]   A. J. Viterbi and J. K. Omura, *Principles of Digital Communications and Coding,* McGraw-Hill, New York, 1979.

[We1]   D. K. Weaver Jr., "A Third Method of Generation and Detection of Single-Sideband Signals," *Proc. IRE,* pp. 1703–1705, Dec. 1956.

[We2]   S. B. Weinstein, "A Baseband Data-Driven Echo Canceler for Full-Duplex Transmission on Two-Wire Circuits," *IEEE Trans. Commun.,* vol. **COM-25,** pp. 654–666, July 1977.

[We3]   L.-F. Wei, "Rotationally Invariant Convolutional Channel Coding with Expanded Signal Space—Part II: Non-Linear Codes," *IEEE J. SAC.,* vol. **SAC-2,** pp. 672–686, Sept. 1984.

[Wh]   H. A. Wheeler, "The Interpretation of Amplitude and Phase Distortion in Terms of Paired Echoes," *Proc. IRE,* pp. 359–385, June 1939.

[W&A]   C. L. Weber and W. K. Alem, "Demod-Remod Tracking Receiver for QPSK and SQPSK," *IEEE Trans. Commun.,* vol. **COM-28,** pp. 1945–1968, Dec. 1980.

[W&E]   S. B. Weinstein and P. M. Ebert, "Data Transmission by Frequency-Division Multiplexing Using the Discrete Fourier Transform," *IEEE Trans. Commun. Tech.,* vol. **COM-19,** pp. 628–634, Oct. 1971.

[W&U]   J. K. Wolf and G. Ungerboeck, "Trellis Coding for Partial-Response Channels," *IEEE Trans. Commun.,* vol. **COM-34,** pp. 765–773, Aug. 1986.

[YH&W]   H. Yamamoto, K. Hirade, and Y. Watanabe, "Carrier Synchronizer for High-Speed Four-Phase-Shift-Keyed Signals," *IEEE Trans. Commun.,* vol. **COM-20,** pp. 803–808, Aug. 1972.

[Y&T]   K. Yoshida and N. Tamaki, "Subscriber Loop Noise Considerations and the Estimated TCM Application Range," *IEEE Intl. Conf. Commun. Record,* June 1985.

[Z&R]   R. E. Ziemer and C. R. Ryan, "Minimum-Shift Keyed Modem Implementations for High Data Rates," *IEEE Commun. Mag.,* pp. 28–37, Oct. 1983.

INDEX

Page numbers appearing in **boldface** indicate main source.